Advanced Structured Materials

Volume 235

Series Editors

Common engineering materials are reaching their limits in many applications, and new developments are required to meet the increasing demands on engineering materials. The performance of materials can be improved by combining different materials to achieve better properties than with a single constituent, or by shaping the material or constituents into a specific structure. The interaction between material and structure can occur at different length scales, such as the micro, meso, or macro scale, and offers potential applications in very different fields.

This book series addresses the fundamental relationships between materials and their structure on overall properties (e.g., mechanical, thermal, chemical, electrical, or magnetic properties, etc.). Experimental data and procedures are presented, as well as methods for modeling structures and materials using numerical and analytical approaches. In addition, the series shows how these materials engineering and design processes are implemented and how new technologies can be used to optimize materials and processes.

Advanced Structured Materials is indexed in Google Scholar and Scopus.

P. M. Visakh
Editor

Engineering Applications of Polymer based Nano Blends

Springer

Editor
P. M. Visakh
Natural Bioactive Materials Laboratory
Department of Bioengineering
Ege University
Bornova, Türkiye

ISSN 1869-8433 ISSN 1869-8441 (electronic)
Advanced Structured Materials
ISBN 978-3-031-90978-8 ISBN 978-3-031-90979-5 (eBook)
https://doi.org/10.1007/978-3-031-90979-5

This Springer imprint is published by the registered company Springer Nature Switzerland AG
The registered company address is: Gewerbestrasse 11, 6330 Cham, Switzerland

Contents

About the Editor

Dr. P. M. Visakh (M.Sc., M.Phil, Ph.D.) is a prolific editor with more than **50** books already published. Currently he is working in the Natural Bioactive Materials Laboratory Department of Bioengineering, Ege University, Bornova/Izmir, Turkey. He was working as Associate Professor in TUSUR University, Tomsk, Russia from 2017 till 2022. He did his Post doc. Research in Tomsk Polytechnic University, Tomsk, Russia (2014–2017). He obtained his Ph.D., M.Phil and M.Sc. degrees from School of Chemical Sciences, Mahatma Gandhi University, Kottayam, kerala, India. He edited **50 books** from Scrivener (Wiley), Springer, Royal Society of Chemistry, Elsevier and more than **25 books** in press, (from Wiley, Springer, Royal Society of Chemistry and Elsevier). He has been invited as a visiting researcher in Russia (2014–2022), Portugal (2013, 2014), Czech Republic (2012, 2013) Italy (2009, 2012), Argentina (2010) Sweden (2010–2012), Switzerland (2010), Spain (2011, 2012), Slovenia (2011), France (2011), Belgium (2012) and Austria (2012) for his research work. He visited 13 countries; He visited 15 universities in Europe. He published 23 publications, four reviews and more than **55** book chapters. He has attended and presented more than 28 conferences, he has **3193 citations** and his **h-index is 27**. He acts as guest editor for four International Journals.

Chapter 1
Engineering Applications of Polymer-Based Nanoblends Challenges and Opportunities

Shiv Kumari Panda and P. M. Visakh

Abstract In this chapter providing and short version of all chapters, here I am writing about the different chapter topics such as poly(Lactic Acid) blends in biomedical applications, bioactive glasses/polylactic acid for bone tissue engineering, rubber-based nanoblends: medical applications, polymer/rubber-based nanoblends, eco-friendly fabrication of nanoparticles (NP) particles and fibrils using polymer blends as templates, application of nano znO-blended polymeric membrane, rubber-based nanoblends, polylactic acid and their blends for biomedical application, therapeutic applications of sustainable chitin and chitosan derivatives and Its nanocomposites, thermoplastic starch-based LLDPE films for active packaging.

1.1 Poly(Lactic Acid) Blends in Biomedical Applications

Biodegradable polymers like poly(lactic acid) (PLA), poly(glycolic acid) (PGA), and their copolymer PLGA are particularly valuable in applications where temporary functionality is needed (Benatti et al. 2019). These materials degrade into non-toxic byproducts, such as water or carbon dioxide, which are safely eliminated by the body (Prajapati et al. 2019). Polymers can be processed into various forms, such as films, fibers, hydrogels, and 3D structures, using scalable and cost-effective methods (Gong et al. 2020). This versatility makes them suitable for an extensive range of medical applications (Kirillova et al. 2021). The mechanical, thermal, and chemical properties of polymers can be finely tuned by adjusting their molecular structure or by blending them with other materials (Yu et al. 2006). This adaptability ensures their

S. K. Panda (✉)
Department of Chemistry, Udayanath Autonomous College of Science & Technology, Adaspur, Cuttack, India
e-mail: shivkumaripanda@gmail.com

P. M. Visakh
Natural Bioactive Materials Laboratory Department of Bioengineering, Ege University, Izmir, Turkey

P. M. Visakh (ed.), *Engineering Applications of Polymer based Nano Blends*, Advanced Structured Materials 235, https://doi.org/10.1007/978-3-031-90979-5_1

functionality in specific biomedical scenarios (Kumar et al. 2018). Common examples include polyethylene and polypropylene for surgical sutures and prosthetics, polylactic acid (PLA), and polyglycolic acid (PGA) for biodegradable implants, and silicone for medical tubing and implants (He and Benson 2017). Synthetic polymers are also crucial in drug delivery systems, where materials like polyethylene glycol (PEG) are used to improve drug solubility and stability (Srivastava et al. 2015). Their ability to be sterilized and manufactured at scale ensures reliability in clinical settings. However, while they provide significant advantages, synthetic polymers must be carefully designed to ensure biocompatibility and avoid long-term adverse effects in the body (Arif et al. 2019). Another advantage of synthetic polymers is their controlled biodegradability (Gunatillake et al. 2006). Polymers such as polylactic acid (PLA), polyglycolic acid (PGA), and their copolymer poly(lactic-co-glycolic acid) (PLGA) can be engineered to degrade at a specific rate, making them ideal for temporary medical devices like sutures, drug delivery systems, and tissue scaffolds. These materials break down into non-toxic byproducts, such as carbon dioxide and water, which are naturally eliminated by the body (Anju et al. 2020). This eliminates the need for secondary surgical procedures to remove implants, reducing patient risk and healthcare costs.

PLA's optical properties can be adjusted by controlling its degree of crystallization, where higher crystallinity often reduces transparency and imparts a milky appearance. This tunability allows PLA to be used in both clear and opaque products. Additionally, PLA can exhibit optical activity due to its chiral structure, depending on the proportion of L- and D-isomers in the polymer chain. These optical properties make PLA ideal for packaging, films, and other applications where aesthetics and transparency are critical (Farah et al. 2016). PLA is widely used for making biodegradable containers, cups, bottles, and trays for food and beverages due to its transparency, rigidity, and safety for food contact (Mangaraj et al. 2019). Thin PLA films are used for packaging fresh produce, baked goods, and other perishable items, providing an eco-friendly alternative to petroleum-based plastics (Lunt and Shafer 2000). It is used in producing biodegradable fibers for clothing, upholstery, and non-woven materials. Its properties, such as low flammability and resistance to UV light, make it suitable for technical textiles (Khouri et al. 2024). It is used in sutures, stents, screws, and pins due to its biocompatibility and ability to degrade naturally in the body. It is employed in controlled drug release applications, as it can encapsulate and deliver active compounds over time (Martín-Closas et al. 2017). While polylactic acid (PLA) is a promising biodegradable and renewable polymer, it has several disadvantages that limit its wider adoption in certain industries. One of the primary drawbacks is its brittleness and low-impact resistance, which make it unsuitable for applications requiring toughness and flexibility, such as durable goods or high-stress components (Marano et al. 2022). PLA also suffers from a low glass transition temperature (55–65 °C) and poor thermal stability, meaning it cannot withstand high-temperature environments, such as hot food containers or automotive parts, without modifications. Additionally, PLA has poor barrier properties, limiting its effectiveness in packaging applications where resistance to moisture and gas permeation is crucial. Its slow crystallization rate during processing can lead to dimensional instability, longer manufacturing times, and difficulties in achieving high heat resistance in products (Clarizio and Tatara 2012).

Polylactic acid (PLA) blends are a class of materials created by combining PLA with other polymers or additives to enhance its properties, overcome its limitations, and broaden its application range. While PLA is valued for its biodegradability, renewable nature, and eco-friendliness, it has several drawbacks, such as brittleness, low thermal resistance, slow crystallization rates, and poor barrier properties. By blending PLA with other materials, these shortcomings can be addressed, making PLA more versatile for various industrial applications (Thomas et al. 2014). PLA blends are typically designed to improve mechanical performance, enhance thermal stability, and facilitate processing. Blending PLA with cellulose creates a biocomposite that leverages the renewable and biodegradable properties of both materials while enhancing PLA's mechanical performance and thermal stability. Cellulose, derived from plant fibers, serves as a natural reinforcing agent, improving the strength and stiffness of PLA. Additionally, cellulose enhances the material's biodegradability and reduces its environmental impact (Makhijani et al. 2015). PLA blends are at the forefront of advancements in tissue engineering, offering customizable materials that address the specific mechanical, structural, and biological requirements of different tissues. Poly(lactic acid)/Polyethylene glycol blends enhance hydrophilicity and facilitate the encapsulation and sustained release of hydrophilic drugs (Sharma et al. 2021), while poly(lactic acid)/polycaprolactone blends improve flexibility and adjust degradation rates for controlled release over extended periods (Aijaz et al. 2022). PLA blends are used for screws, pins, plates, and scaffolds that provide temporary mechanical support while the body regenerates natural tissue (Perinelli et al. 2019). The ability to control the degradation rate of PLA blends ensures that the implant maintains structural integrity during the healing process and degrades completely once its function is fulfilled, eliminating the need for secondary surgeries to remove the implant (DeStefano et al. 2020).

1.2 Bioactive Glasses/Polylactic Acid for Bone Tissue Engineering

Bones and the connective tissue around cartilage are vital components of the human body serving a variety of purposes, such as nourishing and protecting the human body, storing essential nutrition, and producing blood cells (Xue et al. 2022). The mechanical characteristics of the scaffold should closely resemble those of the tissue it aims to replace, ensuring adequate support throughout the regeneration process. *Interconnected porous structure*: BTE scaffolds must possess a high porosity of over 80% and a pore size greater than 100 μm. These conditions are optimal for promoting vascularization, allowing cell infiltration, and facilitating tissue growth (Ylänen 2011). BG and PLA composites have noteworthy synergistic qualities in biomedical applications. BGs, known for their osteoconductive properties, and PLA, known for their biodegradability, have drawn a considerable amount of attention in the field of biomaterials because they have the potential to improve both mechanical and biological efficiency when combined (Tinco et al. 2020). Bioactive glasses are highly appealing materials for creating scaffolds aimed at bone regeneration because of their adaptable properties and flexible composition. A key characteristic that makes BGs ideal for BTE is their capacity to promote revascularization, facilitate osteoblast adhesion, increase enzyme activity, and support the differentiation of mesenchymal stem cells and osteoprogenitor cells (Kaur 2017). Pant et al. (2022) have used pneumatic extrusion-based advanced 3D bioprinting to fabricate PLA and PLA/BG composite scaffolds given in Fig. 11.3. The preparation of PLA bio-ink involves the dissolution of PLA pellets in an appropriate amount of solvent chloroform and evaporating slowly to obtain 3D bio-ink based on neat PLA. A melt-blending method was used by Lee et al. using a kneader at 190C for 10 min, resulting in BG/PLLA composite pellets. The volume ratios of BG powders in the composites were set to 10 and 30 vol%. The volumes of PLLA and BG powders were calculated from their density. The densities were measured by Archimedes' method using acetone and water as immersion fluid for BG and PLLA, respectively, at 25C ($n = 3$) (Lee et al. 2019).

1.3 Rubber-Based Nanoblends: Medical Applications

Rubber, well-known for its excellent elasticity, resilience, and water resistance, has a long history and a substantial economic impact (Sisanth et al. 2017; Rodgers and Waddell 2005; Song 2017). Its discovery and advancement changed the automotive and medical sectors, and it is still essential in various uses today, from tires to medical equipment. Rubber comes in two main varieties: natural and synthetic. Each has unique qualities and uses. NR's exceptional elasticity, which enables it to expand considerably without breaking and then regain its original shape, and its resistance to abrasion, which makes it perfect for high-wear items like tires, are two of its key characteristics. A further factor in its endurance is its high tensile strength, which allows it to tolerate significant strain. NR's comparatively low heat conductivity also makes it an excellent thermal insulator (Mooibroek and Cornish 2000; Venkatachalam et al. 2013; Bokobza 2018; Cruz-Morales et al. 2023). Composite materials known as rubber blends are made by mixing two or more varieties of rubber to produce a unique combination of qualities. Manufacturers can customize materials to maximize performance, satisfy budgetary constraints, and even enhance sustainability by combining rubbers (Kaliyathan et al. 2020). NR has superior tensile strength and suppleness, but it is not resistant to chemicals, oils, or extremely high or low temperatures. By combining NR with synthetic rubber, such as SBR or nitrile rubber (NBR), these shortcomings are made up for, resulting in a composite that is more elastic and has better chemical resistance (Vayyaprontavida Kaliyathan et al. 2020; Sidkey et al. 2002). Better Mechanical Properties: Rubber blends give producers the perfect mix of durability, strength, flexibility, and resistance to wear. SBR can improve

NR's abrasion resistance, for instance, making it a more durable material for high-wear applications. The blending procedure greatly influences the material's ultimate performance properties, such as its mechanical strength, elasticity, thermal stability, and chemical resistance. This section examines the most widely used methods for blending rubber, emphasizing their benefits, uses, and guiding principles. Achieving consistent dispersion of nanoparticles is essential for optimizing the performance of polymer-based nanoblends (Albdiry 2024; Imoisili et al. 2020a). Specialized blending processes guarantee uniform dispersion of nanoparticles such as silica, graphene, or carbon nanotubes. Applications include thermal insulation, sophisticated coatings, and high-performance tires (Bheemappa and Gurumurthy 2022). Rubber blends are extensively utilized in many industries because of their unique qualities, which combine the benefits of several rubbers to satisfy specific application needs. By combining natural and synthetic rubbers, producers can produce materials with the ideal ratio of flexibility, toughness, and resistance to weathering, chemicals, and temperature fluctuations (Rodgers and Halasa 2011; Popescu et al. 2017).

A previous study claims that producers can enhance material qualities for specific industrial applications by combining different types of rubber. In the automotive sector, rubber blends combine the advantages of synthetic and NR to satisfy the high-performance requirements of contemporary automobiles. These blends contribute to the overall efficiency and durability of the vehicle while providing optimal performance in various applications, guaranteeing the safety and comfort of the occupants (Ramarad et al. 2015; Zanchet et al. 2012). For engine seals, fuel hoses, and transmission gaskets, NBR is perfect since it can be combined with other rubbers to increase their flexibility, heat resistance, and compression set qualities (Dahi Taleghani and Santos 2023; Periyasamy et al. 2024). To create seals that can resist water and antifreeze fluids, as well as harsh weather, UV rays, and ozone depletion, EPDM is frequently combined with NBR. Vehicles are guaranteed to retain their integrity throughout a variety of operating situations, including high temperatures, the presence of chemicals, and severe environmental stressors, thanks to blended rubbers in gaskets and seals.

1.4 Polymer/Rubber-Based Nanoblends

Natural and synthetic polymers are the two primary groups into which polymers are often divided. While synthetic polymers like polystyrene, polyvinyl chloride (PVC), and polypropylene are produced by industrial chemical processes, natural polymers like cellulose, proteins, and rubber come from biological origins. Developments in polymer science have made it possible to develop ecologically friendly and functional materials. Bioplastics and biodegradable polymers, e.g., provide substitutes for conventional plastics while promoting sustainability and lessening their negative effects on the environment (Sionkowska 2011; Bhatia and Bhatia 2016; Wojtowicz et al. 2009). The latex of the Hevea brasiliensis tree, which is mostly found in tropical areas, is the source of natural rubber (NR). It can stretch and bounce because of its molecular structure, which comprises long, coiled polymer chains that are weakly cross-linked. NR has long been an important resource because of its durability; it was first employed in crude ways and became a mainstay in the developing industrial sector in the nineteenth and twentieth centuries. NR has become the preferred material in many industries due to its tensile strength, abrasion resistance, and capacity to retain elasticity across a broad temperature range (Bokobza 2018; Cruz-Morales et al. 2023; Ansari et al. 2021). Developing polymer/rubber blends, especially those at the nanoscale, or nanoblends, has been a focus of significant research and innovation as the need for high-performance materials increases. One benefit of blending rubber and polymers is that the desired qualities of both materials can be combined to create new materials with better mechanical, thermal, and chemical qualities. For example, combining rubber and a rigid polymer can create a flexible and durable material, making it perfect for applications requiring resilience and impact resistance (Shah et al. 2013; Dhanorkar et al. 2021; Jourabchi and Ostad Movahed 2024). Polymer/rubber mixes are increasingly used in the medical industry for applications, including flexible medical tubing, implants, and wearable technology that need to be durable and biocompatible (Archibong et al. 2023).

Another critical area of research for rubber and polymers is sustainability. Rubber and synthetic polymers, mostly made from non-renewable resources, harm the environment, prompting attempts to develop sustainable substitutes. The creation of recyclable polymer/rubber composites and advancements in biopolymers have the potential to save resources and cut waste. For instance, the environmental issues related to traditional rubber products may be lessened by employing NR or biodegradable polymers in particular applications (Singh et al. 2023a). Some common synthetic polymers include polyethylene, polypropylene, polystyrene, and polyvinyl chloride (PVC), each offering unique properties suited to different applications (Caldona et al. 2017; Abbas-Abadi et al. 2021; Costa et al. 2010; Ciliz et al. 2004). On the other hand, rubber is a specific category of polymer known for its high elasticity and resilience. Its molecular structure of coiled chains with cross-linked points allows it to stretch and return to its original shape. The types of polymers and rubbers utilized, the mix ratio, and the processing techniques can all substantially impact the properties of polymer/rubber blends. Blends of polymers and rubber are made to combine the best qualities of both materials, frequently leading to improved processing, mechanical, thermal, and chemical properties (Abdul Sattar 2024; Paul et al. 2009; Rane et al. 2017). Incorporating nanotechnology has revolutionized the blending of polymers and rubbers by enabling scientists to work with these materials at the nanoscale. To improve the blend's mechanical, thermal, and electrical qualities, nano blends use nanoscale fillers like graphene, silica nanoparticles, carbon nanotubes, or nanoclay (Imoisili et al. 2020a, b; Okafor et al. 2021). Blends of polymers and rubber are recognized for their unique qualities, such as flexibility, durability, chemical resistance, electrical insulation, and thermal stability, making them indispensable across a variety of industries. Their adaptability and resilience allow for customization to meet specific requirements in sectors like automotive, aerospace, health care, consumer goods, and electronics. Advances in material science have further enhanced the versatility of these blends, enabling them to address the evolving demands of modern applications (Singha et al. 2019).

1.5 Eco-friendly Fabrication of Nanoparticles (NP) Particles and Fibrils Using Polymer Blends as Templates

Plastic trash is widely present in the environment due to the quick development and wide-ranging use of plastic-based materials in recent decades. Because of their longevity and resistance to deterioration, traditional plastics lead to the buildup of persistent debris, which presents serious health and environmental risks (Walker and Fequet 2023). This has spurred a global drive to manage plastic garbage sustainably and, more importantly, to develop eco-friendly substitutes through materials science innovation (Imoisili and Jen 2023). The utilization of polymer blends as templates in the manufacturing of NP presents a method in which bio-based polymers, polylactic acid (PLA), polyhydroxyalkanoates (PHAs), and other eco-friendly polymers act as matrices that allow the formation of nano-plastic fibrils and particles (Li et al. 2023; Tan et al. 2021). Various methods are used to fabricate NP fibrils and particles to attain the appropriate morphologies, sizes, and environmental compatibility. The following strategies, which highlight the usage of polymer blends and low-impact processing technologies, are examples of sustainable practices (Imoisili et al. 2020a; Rane et al. 2018; Imoisili and Jen 2021). Environmental risks are reduced when surfactants and initiators with low toxicity and biodegradability are used. Without adding long-lasting chemicals, the emulsion can be stabilized by using natural surfactants like lecithin or chitosan (Zhang et al. 2023). Electrospinning is a potent method that allows for fine control over morphology, scalability, and compatibility with biodegradable materials when creating environmentally friendly NP fibers and particles. Using biodegradable polymer blends, green solvents, and energy efficiency optimization, this method fits nicely with sustainable objectives and is a viable way to develop environmentally friendly nanomaterials. The versatility of electrospinning makes it appropriate for a wide range of industries where controlled release, tenable degradation, and large surface area are desired (Bavatharani et al. 2021; Li et al. 2021). The development of fibers or particles is greatly influenced by the polymer solution's concentration, viscosity, and conductivity. For example, a higher viscosity promotes continuous fiber creation, while a lower viscosity promotes particle formation. Template-assisted synthesis is an effective and adaptable technique for creating environmentally acceptable NP fibrils and particles. This method offers fine control over nanostructure morphology and is in line with sustainability goals by using renewable polymers, low-toxicity solvents, and biodegradable or reusable templates. Applications for template-assisted synthesis can be found in a variety of fields, such as environmental remediation and biomedicine, where environmentally friendly nanostructures help create a more sustainable future (Kim et al. 2023; Shafqat et al. 2023).

The template is selected by selecting an appropriate material, such as biodegradable templates (e.g., cellulose generated from plants), porous membranes, or nanoporous alumina. The template defines the nanostructures' dimensions, which acts as a mold. Waste can be reduced for environmentally responsible synthesis by using reusable or biodegradable templates (Su et al. 2021). For applications that care about the environment, biodegradable or renewable polymers such as polylactic acid (PLA), polyhydroxyalkanoates (PHA), or naturally occurring polymers like cellulose, are preferred. The mechanical strength, rate of deterioration, and environmental suitability of the resultant nanostructures are all influenced by the characteristics of the polymer (Pavlenko et al. 2022; Nwanna et al. 2020). Polymer-Type Versatility: This technique works with various polymers, including biodegradable ones. Because bio-based polymers are made to break down without remaining in the environment, using them is a good way to achieve sustainability goals (Kaur et al. 2022). Polymer mixtures can be used to create nanostructures for biosensing, tissue engineering, and drug delivery. For instance, nanofibrous scaffolds that promote cell development and break down over time in tandem with tissue healing processes can be produced using biodegradable templates (Yu et al. 2022). High surface area and adjustable porosity are advantageous in applications such as battery electrodes or catalytic materials and are provided by nanoporous or fibrous structures made from polymer blend templates (Liu et al. 2019).

1.6 Application of Nano ZnO-Blended Polymeric Membrane

The Greek numerical prefix nano, which denotes a billionth, and the word technology are combined to form the term nanotechnology. As a result, nanotechnology is usually thought of as being smaller than 100 nm (Ratner and Ratner 2003). Renowned physicist Richard Feynman predicted the theoretical potential of nanotechnology in his well-known and memorable speech in 1959. "There's plenty of room at the bottom," through which he described the potential for very small-scale manipulation and controlling of objects (Feynman 1959). With the advent of nanotechnology, material science and engineering have advanced quickly, especially in the design and production of hybrid materials. Among these developments, polymeric membranes combined with nanoparticles are a noteworthy advance. By combining nanoparticles into polymer matrices, these membranes are produced, creating hybrid systems that take advantage of both the inherent benefits of polymers and the special properties of nanomaterials (Zhu et al. 2018; Goh et al. 2015; Zhang et al. 2016). This combination has made it possible to create materials with specific qualities, tackling difficult problems in a range of scientific and industrial fields. A multidisciplinary approach is used in the synthesis and development of nanomaterials-blended polymeric membranes, using ideas from materials engineering, nanotechnology, and polymer science. Achieving a uniform dispersion of nanoparticles within the polymer matrix is an essential step in this process since agglomeration can seriously impair the performance of the membrane and decrease the efficiency of the nanomaterials (Ng et al. 2013; Hiremath et al. 2021).

Membrane technology is essential to many industries, such as chemical processing, medical filtration, and water purification (Baker 2023; Zhang et al. 2024). Because of their improved physical, chemical, and antibacterial qualities, blended membranes that incorporate nanoparticles have recently drawn interest. The incorporation of nano zinc oxide (ZnO) into polymeric membranes is one such development. Nano ZnO-polymeric-blended membranes combine the flexibility and adaptability of polymeric materials with the beneficial characteristics of ZnO nanoparticles, including enhanced mechanical strength, UV resistance, and antibacterial activity (Tian et al. 2023; Sahu et al. 2021). One class of metal oxide nanoparticles that is well-known for its distinct chemical and physical characteristics is zinc oxide (ZnO) nanoparticles (NPs). ZnO nanoparticles are being researched extensively for use in electronics, optics, medicine, and environmental remediation because of its large surface area, size-dependent properties, and multifunctional nature (Sharma et al. 2022). ZnO nanoparticles are inherently antimicrobial due to their ability to generate ROS, release Zn^{2+} ions, and disrupt microbial membranes. When embedded in polymeric membranes, they provide long-term protection against bacterial and fungal growth, which are major causes of biofouling in water treatment systems. It is mainly used in the prevention of biofilm formation in membrane bioreactors (MBRs) and ultrafiltration systems (Warsinger et al. 2018). Polymer nanocomposites membranes offer superior protective characteristics against moisture, oxygen, and UV rays, which are significant causes of food spoiling, in addition to antimicrobial protection (Guerraf et al. 2024; Zhao et al. 2023; Shen et al. 2023).

By decreasing the permeability of the polymer matrix, ZnO nanoparticle integration stops oxygen and water vapor from penetrating, which might cause oxidation or microbial growth. At the same time, its capacity to block UV rays keeps delicate food items like dairy and oils safe from photodegradation, maintaining their nutritional value and flavor (Irede et al. 2024). Nano ZnO-polymeric-blended membranes have emerged as valuable materials for protective coatings and textiles, owing to their multifunctional properties such as antimicrobial activity, UV-blocking ability, and mechanical durability. In protective coatings, these membranes are applied to surfaces to enhance resistance to environmental factors, including microbial contamination, UV radiation, and chemical exposure (Rivero et al. 2015; Singh et al. 2023b). The incorporation of ZnO nanoparticles into the polymer matrix imparts significant UV-shielding properties, protecting underlying materials from photodegradation. This feature is particularly useful for coatings applied to outdoor structures, vehicles, and electronic devices, where prolonged UV exposure can lead to material deterioration (Shah et al. 2022; Alam et al. 2021; Zhu et al. 2022).

1.7 Rubber-Based Nanoblends

Natural rubber is an elastomer obtained by the polymerization of latex obtained from the rubber tree, Hevea brasiliensis, mainly as polyisoprene $(C_5H_8)_n$. Its polymer structure was formed by long-chain hydrocarbons, which are responsible for the material's key properties, such as elasticity and flexibility. The linking of the isoprene units, 2-methyl-1,3-butadiene, in a cis manner within natural rubber makes the material more flexible. The synthetic rubbers have similar polymer structures but can be prepared by the polymerization of different monomers, which makes possible variations in their chemical, thermal, and mechanical properties (Vayyaprontavida Kaliyathan et al. 2020). In the past two years, a number of nanomaterials have been proposed as effective fillers in rubber. These include nanoclay, carbon nanotubes, graphene, and cellulose nanofibers/nanocrystals, CNF/CNC, which are derived from renewable resources. CNF/CNC has outstanding properties of stress transfer and reinforcement owing to its highly extended surface area, high aspect ratio, and ability to form strong interfacial bonds with various polymer matrices. However, CNF/CNC is inherently hydrophilic, which makes it a big challenge to incorporate it into the hydrophobic polymer matrices and to induce good fiber-polymer bonding. Thus, surface functionalization becomes an important task. A number of publications have demonstrated the potential of functionalized CNF/CNC as reinforcement fillers in rubber. For example, Trovatti et al. added BC nanofibers and polystyrene-decorated BC into natural rubber latex, improving the tensile modulus and strength of the resulting nanocomposites. This esterification with polystyrene enhanced the compatibility of BC with the hydrophobic NR, improving the interface bonding and dispersion (Sinclair et al. 2019). Rubber blends use additives such as fillers, plasticizers, and stabilizers, among other chemical compounds. These are added into rubber to improve its processability and performance. As such, carbon black is used as the filler in improving the abrasion resistance of rubber, while various chemical stabilizers are added into extending the life of the rubber through improving its resistance to environmental elements (Urtekin et al. 2023).

Rubber compounds are essential to modern-day materials research and in several industrial applications. Fundamentally, these materials comprise carefully matched mixtures of various rubber forms or rubber and other polymers to enhance specific performance attributes (Sisanth et al. 2017). As a single type of rubber frequently does not meet all application demands, blends are an essential solution to address these constraints by delivering advantages such as cost efficiency, superior mechanical qualities, increased processing attributes, and tailored performance (E. Mark James EBREF 2019). The performance characteristics of rubber compounds depend on factors like the ratio of rubber types, the mixing procedure, and additives (Vayyaprontavida Kaliyathan et al. 2020). Mixing procedures are crucial for ensuring uniform ingredient distribution and consistency in the mixture. Common methods involve high heat and specialized mixers for this purpose. Rubber compounds have specialized usages in multiple sectors, including construction, automotive, footwear, and medical (Rodgers and Waddell 2005). Car tires are the most recognized product, balancing road grip, fuel efficiency, and wear resistance. Specially blended rubber compounds are also widely used in vehicle components like hoses, seals, and belts (Bandyopadhyay 2024). The applications of rubber compounds in the building and construction industries include sealants, vibration dampers, and roofing membranes. Some other uses of rubber compounds in the footwear industry are soles. Medical uses include surgical gloves, catheters, and drug delivery devices. Nanotechnology components with nano-size dispersion in the SBR matrix overcame these disadvantages (Šupová et al. 2011). These include carbon nanotubes (Soliman et al. 2012), graphene (Khan et al. 2022), nanoclays (Rajasekar et al. 2009), silica (Wei et al. 2015), and various metal oxides (El-Sabbagh et al. 2017).

When formulating SBR-based nanoblends, the main objective is to sharply improve the mechanical and thermal properties by adding minimal nanoparticle quantities. The elevated surface area-to-volume ratio and distinctive characteristics of nanoparticles result in robust interactions with the SBR matrix and enhanced performance. Using carbon nanotubes in SBR can improve tensile strength, hardness, and wear resistance (He et al. 2017). Natural rubber-based nanocomposites improve the performance of natural rubber, which is used in various industries due to its remarkable flexibility, increased tensile strength, and biodegradable nature. Using nanoparticles could help to lower restrictions, including limited abrasion resistance, poor heat resistance, and weathering sensitivity. These nanocomposites integrate the benefits of natural rubber with the specific properties of nanoparticles, resulting in a synergistic impact (Bokobza 2018). Apart from traditional nanomaterials, including carbon nanotubes, graphene, nanoclays, and silica, other nanomaterials can also be formed into the natural rubber matrix to offer different purposes to the material (Feldman 2017). The main nanomaterial used in nitrile rubber latex composites is silica nanoparticles. These particles improve the abrasion resistance and tear strength of the material, enabling the manufacturing of durable goods with great longevity (Ghamarpoor and Jamshidi 2022). Moreover, silica nanoparticles improve the surface properties of the material, thereby improving adhesion and coating effectiveness. As was already indicated, these features increase the use of nitrile rubber latex mixtures, especially in the automotive and construction sectors (Sala et al. 2014).

1.8 Polylactic Acid and Their Blends for Biomedical Application

PLA was discovered in the early 1800s by Pelouze who produced low molecular weight PLA by lactic acid condensation through a distillation process of water (Sin et al. 2013). In 1932, almost a century later, Carothers (DuPont) produced a low molecular weight product by heating lactic acid under vacuum. Due to the high cost of manufacture, mainly coming from the purification step, low molecular weight and low availability, the initial uses were limited to medical and pharmaceutical applications. Biodegradability is a key feature of PLA and is an important factor for both the environment and biomedical field, particularly in non-removable or resorbable implants. The degradation of this biopolymer involves different mechanisms, such as hydrolytic, oxidative, thermal, microbial, enzymatic, chemical, and photodegradative processes, each contributing to its breakdown. Among them, hydrolysis plays a fundamental role in PLA degradation, producing the formation of oligomers and monomers of lactic acid by the cleavage of ester bonds (Khouri et al. 2024). LA is a naturally occurring organic acid because it is a metabolic by-product, then in the biomedical applications, it can be excreted from the body by the normal cellular activity through kidney filtration and urine or as carbon dioxide (Silva et al. 2018). PLA also exhibits characteristically autocatalytic degradation. The carboxylic groups produced during cleavage of ester bond acts as a catalyst for hydrolysis during the degradation of the PLA (Shasteen and Bin Choy 2011). Since the generated carboxyl groups are confined within the bulk of the PLA and cannot escape into the surrounding environment, their autocatalytic effect often leads to accelerated degradation within the interior of a PLA component. This causes PLA to typically degrade from the inside out, resulting in hollow and porous structures before it is fully degraded. Useful to blend PLA with some natural polymers in order to avoid degradation and inter-chain chemical reactions that can occur during the melt blending of PLA, particularly for biomedical applications (Hamad et al. 2018).

Different solvents are commonly used during the preparation of PLA blends via the solution blending method, including chloroform, dichloromethane (Vogel et al. 2009), dimethylformamide (Zhao et al. 2020), tetrahydrofuran (Peesan et al. 2006), and 1,4-dioxan (Meereboer et al. 2020). The type of solvent can play a crucial role in determining the final morphology of the PLA blend prepared via this method, as it was demonstrated for PLA/chitosan blends. Additionally, other strategies were explored in literature to enhance the compatibility in PLA blends (Zeng et al. 2015). Adding micro and nanoparticles to an immiscible blend offers an alternative approach to enhancing compatibility (Fredi and Dorigato 2024). Micro-nanoparticles typically position themselves at the interface between the blend components, serving as interfacial modifiers that strengthen adhesion and improve the overall performance of the blend. They help prevent the coalescence of finely dispersed phase particles during melt processing, thereby stabilizing the morphology and preserving the blend's properties. Unlike polymer compatibilizers, nanoparticles are not specific to the nature of the immiscible components and can be easily incorporated through blending. Thus, Hoidy et al. (2010) compatibilized PLA/PCL blends using organoclay (OMMT) as a filler. Their findings revealed that OMMT not only improved the dispersion and interfacial adhesion within the polymer matrix but also enhanced the mechanical properties and thermal stability of the PLA/PCL blends. In the same way, organoclay (Cloisite® 30B) was used to compatibilize PLA blend on poly(butylene succinate) (PBS) (Risse et al. 2014) and poly[(butylene adipate)-co-terephthalate] (PBAT) (Nofar et al. 2016). As mentioned above, PLA blends could be miscible or immiscible, for the first ones both polymers are compatible with interfacial tension close to zero, while for the second ones an interphase and interfacial interactions are presents (Nofar et al. 2019). For content lower than 10 wt%, it has been reported a quite independent dynamic melt viscoelastic performance to the blending composition, morphology and interfacial interactions. This feature has been observed for immiscible blends of PLA with PBAT, PBSA, PCL, PHBV or starch. PLA/starch blends displayed droplet-like, rod-like, or co-continuous morphologies with highly similar dynamic viscoelastic performance. Blends of compatible high and low molecular weight PLA displayed storage modulus almost independent of the composition into the analyzed frequency range, while the melt viscosity decreased at low frequencies (Yang and Du 2024).

1.9 Therapeutic Applications of Sustainable Chitin and Chitosan Derivatives and Its Nanocomposites

Chitin a natural biopolymer is the second most abundant organic material on Earth, derived primarily from the exoskeletons of arthropods (e.g., shrimp, crabs, and insects), fungal cell walls, and mollusk shells (Pal and Kahawas 2024). Chitin excels in wound healing, as a biodegradable matrix and in filtration technologies, while chitosan finds use as an antimicrobial agent, drug delivery vehicle, and bioplastic (Rai et al. 2021). The biocompatibility, renewability and functional versatility of these biomaterials have positioned them as key resources in sustainable technology and advanced scientific research. One of the most prominent therapeutic applications of chitin, chitosan, and its derivatives is in the domain of drug delivery systems. Their inherent mucoadhesive properties and the ability to open tight junctions in epithelial cells facilitate the transport of therapeutic molecules across biological membranes (Mura et al. 2022). In the realm of antimicrobial treatments, chitin offers broad-spectrum activity against pathogenic bacteria, fungi, and even viruses. Their antimicrobial mechanisms, combined with their ability to enhance the action of encapsulated bio-actives, make them indispensable in creating coatings for surgical implants, devices, and hospital environments (Croitoru et al. 2024). Chitin and its derivatives and its nanoparticles have been widely explored for mucosal vaccine delivery, particularly for nasal and oral immunization. Their mucoadhesive properties facilitate prolonged retention at mucosal sites, enhancing antigen absorption and immune activation. Chitin-based adjuvants have shown promise in improving the efficacy of vaccines against infectious diseases, including influenza, tuberculosis and Covid-19 (Wike et al. 2023).

Chitin can neutralize chitinase-like proteins (CLPs), reducing immunosuppression and enhancing the efficacy of immune checkpoint blockade therapies. The use of chitin in conjunction with other immunotherapies has shown promise in overcoming resistance mechanisms, particularly in triple-negative breast cancer models (Salembier et al. 2024). With continued research, chitin-based adjuvants could revolutionize vaccine formulations, providing safer, more effective, and sustainable immunization strategies. The advancement in biopolymer technology has positioned chitosan derivatives and their nanocomposites as pivotal elements in therapeutic applications, leveraging their sustainability, biocompatibility, and adaptability. Derived from the deacetylation of chitin, chitosan possesses unique properties such as biodegradability, mucoadhesiveness and antimicrobial activity, making it suitable for diverse medical and pharmaceutical innovations. Chitosan is structurally composed of glucosamine and N-acetylglucosamine units linked by β-(1 → 4) glycosidic bonds. The ability to modify its structure through chemical and enzymatic modifications enhances its application potential (Gao et al. 2023). Chitosan derivatives have gained significant traction in drug delivery systems owing to their ability to form nanoparticles and microspheres, which provide controlled and sustained drug release. These properties are particularly advantageous in cancer therapy, where chitosan nanocarriers can encapsulate chemotherapeutic agents, ensuring their stability and targeted release to tumor cells while minimizing systemic toxicity and chitosan nanoparticles serve as effective drug delivery agents in brain delivery, improving bioavailability and minimizing systemic side effects. They facilitate drug transport across the blood–brain barrier, making them promising for treating neurodegenerative diseases and other medical conditions (Omidian et al. 2024).

1.10 Thermoplastic Starch-Based LLDPE Films for Active Packaging

LLDPE exhibits remarkable flexibility and adaptability, enabling it to function effectively under a variety of conditions. Its excellent impact resistance is essential for protecting sensitive materials during packaging processes. In addition, LLDPE's strong chemical resistance enhances its application in the chemical sector. Thermal stability is another key strength (Khumkomgool et al. 2020). LLDPE is widely used in the packaging industry to produce base films, plastic sheets, and also required bags. Despite all the unique characteristics of LLDPE, this polymer exhibits weak barrier properties against oxygen and ultraviolet radiation, which exacerbates concerns about plastic pollution and its environmental impact. Given these challenges, various studies have sought to alternative petroleum-based plastics with biopolymers derived from Rechargeable resources such as chitosan, starch, and cellulose. Blending TPS with LLDPE enhances processing capabilities and improves the stability of biomaterials for potential commercialization. Also, pyrogallol (PG) is a natural phenolic compound found in hardwoods such as oak and eucalyptus (Mohammed-Ziegler and Billes 2002). Its molecular structure consists of a benzene ring with three hydroxyl groups, which allows the formation of non-covalent hydrogen bonds and hydrophobic interactions. The auto-oxidation of hydroxyl groups can lead to the consumption of dissolved oxygen under alkaline conditions, thereby removing free oxygen from the surrounding environment (Shin et al. 2019). Research has shown that the auto-oxidation of PG in a liquid medium leads to the formation of various byproducts, including hydrogen peroxide and quinone derivatives (Ramasarma et al. 2015). Chatkitanan and Harnkarnsujarit created a thermoplastic film based on TPS and LLDPE that incorporated 5% sodium nitrite. The film effectively stabilized oxidation and microbial growth in vacuum-packed pork stored at 4 °C for six days, resulting in a reduction in thiobarbituric acid reactive substances (TBARS), methemoglobin content, total viable count (TVC), and other microbial indices (Jacinto-Valderrama et al. 2023). Notably, nitrite levels in encapsulated pork were measured at approximately 1.05 ppm, significantly below the FDA limit of 200 ppm and Chinese regulations of 150 ppm for meat products (Chatkitanan and Harnkarnsujarit 2020).

References

Abbas-Abadi MS, Van Geem KM, Fathi M, Bazgir H, Ghadiri M (2021) The pyrolysis of oak with polyethylene, polypropylene and polystyrene using fixed bed and stirred reactors and TGA instrument. Energy 232:121085

Abdul Sattar M (2024) Surface activated pyrolytic carbon black: a dual functional sustainable filler for natural rubber composites. Chemsuschem 17(2):e202301001

Aijaz MO, Yang SB, Karim MR, Othman MHD, Alnaser IA (2022) Preparation and characterization of poly (lactic acid)/poly (ethylene glycol)-poly (propyl glycol)-poly (ethylene glycol) blended nanofiber membranes for fog collection. Membranes 13(1):32

Alam MA, Samad UA, Anis A, Alam M, Ubaidullah M, Al-Zahrani SM (2021) Effects of SiO_2 and ZnO nanoparticles on epoxy coatings and its performance investigation using thermal and nanoindentation technique. Polymers 13(9):1490

Albdiry M (2024) Effect of melt blending processing on mechanical properties of polymer nanocomposites: a review. Polym Bull 81(7):5793–5821

Anju S, Prajitha N, Sukanya VS, Mohanan PV (2020) Complicity of degradable polymers in health-care applications. Mater Today Chem 16:100236

Ansari AH, Jakarni FM, Muniandy R, Hassim S, Elahi Z (2021) Natural rubber as a renewable and sustainable bio-modifier for pavement applications: a review. J Clean Prod 289:125727

Archibong FN, Orakwe LC, Ogah OA, Mbam SO, Ajah SA, Okechukwu ME, Igberi CO, Okafor KJ, Chima MO, Ikelle II (2023) Emerging progress in montmorillonite rubber/polymer nanocomposites: a review. J Mater Sci 58(6):2396–2429

Arif U, Haider S, Haider A, Khan N, Alghyamah AA, Jamila N, Khan MI, Almasry WA, Kang IK (2019) Biocompatible polymers and their potential biomedical applications: a review. Curr Pharm des 25(34):3608–3619

Baker RW (2023) Membrane technology and applications. Wiley

Bandyopadhyay S (2024) Fundamentals of rubber compounding. Rubber Products: Technology and Cost Optimisation 1–52

Bavatharani C, Muthusankar E, Wabaidur SM, Alothman ZA, Alsheetan KM, mana AL-Anazy M, Ragupathy D (2021) Electrospinning technique for production of polyaniline nanocomposites/nanofibres for multi-functional applications: a review. Synth Metals 271:116609

Benatti ACB, Pattaro AF, Rodrigues AA, Xavier MV, Kaasi A, Barbosa MIR, Jardini AL, Maciel Filho R, Kharmandayan P (2019) Bioreabsorbable polymers for tissue engineering: PLA, PGA, and their copolymers. In: Materials for biomedical engineering. Elsevier, pp 83–116

Bhatia S, Bhatia S (2016) Natural polymers vs synthetic polymer. Natural polymer drug delivery systems: nanoparticles, plants, and algae, pp 95–118

Bheemappa S, Gurumurthy H (2022) Recent advances in fabrication and characterization of nanofiller filled epoxy nanocomposites. Trends Fabr Polym Polym Compos 2022:1–40

Bokobza L (2018) Natural rubber nanocomposites: a review. Nanomaterials 9(1):12

Bokobza L (2018) Natural rubber nanocomposites: a review. Nanomaterials 9(1):12

Caldona EB, De Leon ACC, Pajarito BB, Advincula RC (2017) A review on rubber-enhanced polymeric materials. Polym Rev 57(2):311–338

Chatkitanan T, Harnkarnsujarit N (2020) Development of nitrite compounded starch-based films to improve color and quality of vacuum-packaged pork. Food Packag Shelf Life 25:100521

Ciliz NK, Ekinci E, Snape CE (2004) Pyrolysis of virgin and waste polypropylene and its mixtures with waste polyethylene and polystyrene. Waste Manage 24(2):173–181

Clarizio SC, Tatara RA (2012) Tensile strength, elongation, hardness, and tensile and flexural moduli of PLA filled with glycerol-plasticized DDGS. J Polym Environ 20:638–646

Costa P, Pinto F, Ramos AM, Gulyurtlu I, Cabrita I, Bernardo MS (2010) Study of the pyrolysis kinetics of a mixture of polyethylene, polypropylene, and polystyrene. Energy Fuels 24(12):6239–6247

Croitoru GA, Pîrvulescu DC, Niculescu AG, Rădulescu M, Grumezescu AM, Nicolae CL (2024) Advancements in aerogel technology for antimicrobial therapy: a review. Nanomaterials 14:1110. https://doi.org/10.3390/nano14131110

Cruz-Morales JA, Gutiérrez-Flores C, Zárate-Saldaña D, Burelo M, García-Ortega H, Gutiérrez S (2023) Synthetic polyisoprene rubber as a mimic of natural rubber: recent advances on synthesis, nanocomposites, and applications. Polymers 15(20):4074

da Silva D, Kaduri M, Poley M, Adir O, Krinsky N, Shainsky-Roitman J, Schroeder A (2018) Biocompatibility, biodegradation and excretion of polylactic acid (PLA) in medical implants and theranostic systems. Chem Eng J 340:9–14. https://doi.org/10.1016/j.cej.2018.01.010

Dahi Taleghani A, Santos L (2023) Elastomeric sealing assemblies. Wellbore integrity: from theory to practice, pp 165–193

DeStefano V, Khan S, Tabada A (2020) Applications of PLA in modern medicine. Eng Regeneration 1:76–87

Dhanorkar RJ, Mohanty S, Gupta VK (2021) Synthesis of functionalized styrene butadiene rubber and its applications in SBR–silica composites for high performance tire applications. Ind Eng Chem Res 60(12):4517–4535

E. Mark James EBREF (2019) Science and Technology of RUBBER Third Edition

El Guerraf A, Ziani I, Ben Jadi S, El Bachiri A, Bazzaoui M, Bazzaoui EA, Sher F (2024) Smart conducting polymer innovations for sustainable and safe food packaging technologies. Compr Rev Food Sci Food Saf 23(6):e70045

El-Sabbagh SH, Ahmed NM, Turky GM, Selim MM (2017) Rubber nanocomposites with new core-shell metal oxides as nanofillers. Prog Rubber Nanocomposites 249–283

Farah S, Anderson DG, Langer R (2016) Physical and mechanical properties of PLA, and their functions in widespread applications—a comprehensive review. Adv Drug Deliv Rev 107:367–392

Feldman D (2017) Natural rubber nanocomposites. J Macromol Sci Part A 54:629–634

Feynman R (1959) There is plenty of room at the bottom: An invitation to enter a new field of physics. In: Lecture at American physical society meeting, vol 29

Fredi G, Dorigato A (2024) Compatibilization of biopolymer blends: a review. Adv Ind Eng Polym Res 7:373–404. https://doi.org/10.1016/j.aiepr.2023.11.002

Gao K, Qin Y, Liu S, Wang L, Xing R, Yu H, Chen X, Li P (2023) A review of the preparation, derivatization and functions of glucosamine and N-acetyl-glucosamine from chitin. Carbohydr Polym Technol Appl 5:100296

Ghamarpoor R, Jamshidi M (2022) Synthesis of vinyl-based silica nanoparticles by sol–gel method and their influences on network microstructure and dynamic mechanical properties of nitrile rubber nanocomposites. Sci Rep 2022 12:1 12:1–15

Goh PS, Ng BC, Lau WJ, Ismail AF (2015) Inorganic nanomaterials in polymeric ultrafiltration membranes for water treatment. Sep Purif Rev 44(3):216–249

Gong M, Zhang L, Wan P (2020) Polymer nanocomposite meshes for flexible electronic devices. Prog Polym Sci 107:101279

Gunatillake P, Mayadunne R, Adhikari R (2006) Recent developments in biodegradable synthetic polymers. Biotechnol Annu Rev 12:301–347

Hamad K, Kaseem M, Ayyoob M, Joo J, Deri F (2018) Polylactic acid blends: the future of green, light and tough. Prog Polym Sci 85:83–127. https://doi.org/10.1016/j.progpolymsci.2018.07.001

He Y, Gao J, Gong X, Xu J (2017) The role of carbon nanotubes in promoting the properties of carbon black-filled natural rubber/butadiene rubber composites. Results Phys 7:4352–4358

He W, Benson R (2017) Polymeric biomaterials. In: Applied plastics engineering handbook. William Andrew Publishing, pp 145–164

Hiremath A, Murthy AA, Thipperudrappa S, KN, B. (2021) Nanoparticles filled polymer nanocomposites: a technological review. Cogent Eng 8(1):1991229

Hoidy WH, Al-Mulla EAJ, Al-Janabi KW (2010) Mechanical and thermal properties of PLLA/PCL modified clay nanocomposites. J Polym Environ 18:608–616. https://doi.org/10.1007/S10924-010-0240-X/METRICS

Imoisili PE, Jen T-C (2021) Facile extraction and characterization of silica nanoparticles from corn stalk by sol-gel hydrothermal methods. IOP Conf Series Mater Sci Eng 1107(1):012030

Imoisili PE, Jen TC (2023) Synthesis and characterization of bioplastic films from potato peel starch; effect of glycerol as plasticizer. Mater Today: Proceed

Imoisili PE, Ukoba KO, Jen T-C (2020a) Green technology extraction and characterisation of silica nanoparticles from palm kernel shell ash via sol–gel. J Mater Res Technol 9(1):307–313

Imoisili PE, Ukoba KO, Jen T-C (2020b) Synthesis and characterization of amorphous mesoporous silica from palm kernel shell ash. Boletín De La Sociedad Española De Cerámica y Vidrio. 59:159–164

Irede EL, Awoyemi RF, Owolabi B, Aworinde OR, Kajola RO, Hazeez A, Raji AA, Ganiyu LO, Onukwuli CO, Onivefu AP, Ifijen IH (2024) Cutting-edge developments in zinc oxide nanoparticles: synthesis and applications for enhanced antimicrobial and UV protection in healthcare solutions. RSC Adv 14(29):20992–21034

Jacinto-Valderrama RA et al (2023) Recent trends in active packaging using nanotechnology to inhibit oxidation and microbiological growth in muscle foods. Foods 12(19):3662

Jourabchi S, Ostad Movahed S (2024) The effect of microwave irradiation on the molecular diffusion of the ethyl acetate (EA) into the blend of styrene-butadiene rubber (SBR)/ethylene-propylene-diene monomer rubber (EPDM). Progr Rubber Plast Recycl Technol 14777606241262875

Kaliyathan AV, Rane AV, Huskic M, Kunaver M, Kalarikkal N, Rouxel D, Thomas S (2020) Carbon black distribution in natural rubber/butadiene rubber blend composites: distribution driven by morphology. Compos Sci Technol 200:108484

Kaur G (2017) Bioactive glasses: potential biomaterials for future therapy. Bioactive glasses: potential biomaterials for future therapy 1–332

Kaur A, Bajaj B, Kaushik A, Saini A, Sud D (2022) A review on template assisted synthesis of multi-functional metal oxide nanostructures: status and pro

Khan A, Kian LK, Jawaid M, Khan AAP, Alotaibi MM, Asiri AM, Marwani HM (2022) Preparation of styrene-butadiene rubber (SBR) composite incorporated with collagen-functionalized graphene oxide for green tire application. Gels 8:161

Khouri NG, Bahú JO, Blanco-Llamero C, Severino P, Concha VOC, Souto EB (2024) Polylactic acid (PLA): properties, synthesis, and biomedical applications—a review of the literature. J Mol Struct 1309:138243. https://doi.org/10.1016/j.molstruc.2024.138243

Khumkomgool A, Saneluksana T, Harnkarnsujarit N (2020) Active meat packaging from thermoplastic cassava starch containing sappan and cinnamon herbal extracts via LLDPE blown-film extrusion. Food Packag Shelf Life 26:100557

Kim KW, Park B, Kim J, Jo C, Kim JK (2023) Recent progress in block copo lymer s oft-template-assisted synthesis of versatile mesoporous materials for energy storage systems. J Mater Chem A 11(14):7358–7386

Kirillova A, Yeazel TR, Asheghali D, Petersen SR, Dort S, Gall K, Becker ML (2021) Fabrication of biomedical scaffolds using biodegradable polymers. Chem Rev 121(18):11238–11304

Kumar S, Nehra M, Dilbaghi N, Tankeshwar K, Kim KH (2018) Recent advances and remaining challenges for polymeric nanocomposites in healthcare applications. Prog Polym Sci 80:1–38

Lee S, Matsugaki A, Kasuga T, Nakano T (2019) Development of bifunctional oriented bioactive glass/poly(lactic acid) composite scaffolds to control osteoblast alignment and proliferation. J Biomed Mater Res Part A 107(5):1031–1041
Li M, Yu H, Xie Y, Guo Y, Cheng Y, Qian H, Yao W (2021) Fabrication of eugenol loaded gelatin nanofibers by electrospinning technique as active packaging material. Lwt 139:110800
Li X, Lin Y, Liu M, Meng L, Li C (2023) A review of research and application of polylactic acid composites. J Appl Polym Sci 140(7):e53477
Liu W, Ullah B, Kuo CC, Cai X (2019) Two-dimensional nanomaterials-based polymer composites: fabrication and energy storage applications. Adv Polym Technol 2019(1):4294306
Lunt J, Shafer AL (2000) Polylactic acid polymers from com. Applications in the textiles industry. J Ind Text 29(3):191–205
Makhijani K, Kumar R, Sharma SK (2015) Biodegradability of blended polymers: a comparison of various properties. Crit Rev Environ Sci Technol 45(16):1801–1825
Mangaraj S, Mohanty S, Swain S, Yadav A (2019) Development and characterization of commercial biodegradable film from PLA and corn starch for fresh produce packaging. J Packag Technol Res 3:127–140
Marano S, Laudadio E, Minnelli C, Stipa P (2022) Tailoring the barrier properties of PLA: a state-of-the-art review for food packaging applications. Polymers 14(8):1626
Martín-Closas L, Costa J, Pelacho AM (2017) Agronomic effects of biodegradable films on crop and field environment. Soil degradable bioplastics for a sustainable modern agriculture, pp 67–104
Meereboer KW, Misra M, Mohanty AK (2020) Review of recent advances in the biodegradability of polyhydroxyalkanoate (PHA) bioplastics and their composites. Green Chem 22:5519–5558. https://doi.org/10.1039/d0gc01647k
Mohammed-Ziegler I, Billes F (2002) Vibrational spectroscopic calculations on pyrogallol and gallic acid. J Mol Struct (Thoechem) 618(3):259–265
Mooibroek H, Cornish K (2000) Alternative sources of natural rubber. Appl Microbiol Biotechnol 53:355–365
Mura P, Maestrelli F, Cirri M, Mennini N (2022) Multiple roles of chitosan in mucosal drug delivery: an updated review. Mar Drugs 20:335
Ng LY, Mohammad AW, Leo CP, Hilal N (2013) Polymeric membranes incorporated with metal/metal oxide nanoparticles: a comprehensive review. Desalination 308:15–33
Nofar M, Heuzey MC, Carreau PJ, Kamal MR (2016) Effects of nanoclay and its localization on the morphology stabilization of PLA/PBAT blends under shear flow. Polymer (Guildf). 98:353–364. https://doi.org/10.1016/j.polymer.2016.06.044
Nofar M, Salehiyan R, Sinha Ray S (2019) Rheology of poly (lactic acid)-based systems. Polym Rev 59:465–509. https://doi.org/10.1080/15583724.2019.1572185
Nwanna EC, Imoisili PE, Jen T-C (2020) Fabrication and synthesis of SnOX thin films: a review. Int J Adv Manuf Technol 111(9–10):2809–2831
Omidian H, Gill EJ, Chowdhury SR, Cubeddu LX (2024) Chitosan nanoparticles for intranasal drug delivery. Pharmaceuticals 16:746
Okafor CE, Onovo AC, Imoisili PE, Kulkarni KM, Ihueze CC (2021) Optimal route to robust hybridization of banana-coir fibre particulate in polymer matrix for automotive applications. Materialia 101098
Pal D, Kahawas K (2024) Potential sources and uses of chitin and its polymers. J Discoveries 2
Pant S, Thomas S, Loganathan S, Valapa RB (2022) 3D bioprinted poly(lactic acid)/mesoporous bioactive glass based biomimetic scaffold with rapid apatite crystallization and in-vitro cytocompatability for bone tissue engineering. Int J Biol Macromol 217:979–997
Paul KT, Pabi SK, Chakraborty KK, Nando GB (2009) Nanostructured fly ash–styrene butadiene rubber hybrid nanocomposites. Polym Compos 30(11):1647–1656
Pavlenko V, Żółtowska S, Haruna AB, Zahid M, Mansurov Z, Supiyeva Z et al (2022) A comprehensive review of template-assisted porous carbons: modern preparation methods and advanced applications. Mater Sci Eng R Rep 149:100682

Peesan M, Rujiravanit R, Supaphol P (2006) Electrospinning of hexanoyl chitosan/polylactide blends. J Biomater Sci Polym Ed 17:547–565. https://doi.org/10.1163/156856206776986251

Perinelli DR, Cespi M, Bonacucina G, Palmieri GF (2019) PEGylated polylactide (PLA) and poly (lactic-co-glycolic acid)(PLGA) copolymers for the design of drug delivery systems. J Pharm Investig 49:443–458

Periyasamy M, Mubasshir AA, Kodra S, Chandramouli S, Campbell R, Kazmer DO, Mead JL (2024) From formulation to application: effects of plasticizer on the printability of fluoro elastomer compounds and additive manufacturing of specialized seals. Micromachines 15(5):622

Popescu RC, Popescu D, Grumezescu AM (2017) Applications of rubber-based blends. In: Recent developments in polymer macro, micro and nano blends. Woodhead Publishing, pp 75–109

Prajapati SK, Jain A, Jain A, Jain S (2019) Biodegradable polymers and constructs: a novel approach in drug delivery. Eur Polymer J 120:109191

Rai M, Dos Santos CA (eds) (2021) Biopolymer-based nano films: applications in food packaging and wound healing. Elsevier

Rajasekar R, Heinrich G, Das A, Das CK (2009) Development of SBR-nanoclay composites with epoxidized natural rubber as compatibilizer. J Nanotechnol 2009:405153

Ramarad S, Khalid M, Ratnam CT, Chuah AL, Rashmi W (2015) Waste tire rubber in polymer blends: a review on the evolution, properties and future. Prog Mater Sci 72:100–140

Ramasarma T et al (2015) New insights of superoxide dismutase inhibition of pyrogallol autoxidation. Mol Cell Biochem 400(1):277–285

Rane AV, Kanny K, Abitha VK, Patil SS, Thomas S (2017) Clay–polymer composites: design of clay polymer nanocomposite by mixing. In: Clay-polymer nanocomposites. Elsevier, pp 113–144

Rane AV, Kanny K, Abitha VK, Thomas S (2018) Methods for synthesis of nanoparticles and fabrication of nanocomposites. In: Synthesis of inorganic nanomaterials. Woodhead publishing, pp 121–139

Ratner MA, Ratner D (2003) Nanotechnology: a gentle introduction to the next big idea. Prentice Hall Professional

Risse S, Tighzert L, Berzin F, Vergnes B (2014) Microstructure, rheological behavior, and properties of poly(lactic acid)/poly(butylene succinate)/organoclay nanocomposites. J Appl Polym Sci 131:40364. https://doi.org/10.1002/APP.40364

Rivero PJ, Urrutia A, Goicoechea J, Arregui FJ (2015) Nanomaterials for functional textiles and fibers. Nanoscale Res Lett 10:1–22

Rodgers B, Halasa A (2011) Compounding and processing of rubber/rubber blends. Encycl Polym Blends 163–206

Rodgers B, Waddell W (2005) The science of rubber compounding. In: Science and technology of rubber. Academic Press, pp 401–454

Sahu DK, Ghosh G, Rath G (2021) Nanofibers in drug delivery. Nanopharmaceutical advanced delivery systems, pp 99–123

Sala RL, Arantes TM, Longo E, Leite ER, Paranhos CM, Camargo ER (2014) Evaluation of modified silica nanoparticles in carboxylated nitrile rubber nanocomposites. Colloids Surf A Physicochem Eng Asp 462:45–51

Salembier R, De Haes C, Bellemans J, Demeyere K, Van Den Broeck W, Sanders NN, Van Laere S, Lyons TR, Meyer E, Steenbrugge J (2024) Chitin-mediated blockade of chitinase-like proteins reduces tumor immunosuppression, inhibits lymphatic metastasis and enhances anti-PD-1 efficacy in complementary TNBC models. Breast Cancer Res 26:63

Shafqat SR, Bhawani SA, Bakhtiar S, Ibrahim MNM, Shafqat SS (2023) Template-assisted synthesis of molecularly imprinted polymers for the removal of methyl red from aqueous media. BMC Chem 17(1):46

Shah AA, Hasan F, Shah Z, Kanwal N, Zeb S (2013) Biodegradation of natural and synthetic rubbers: a review. Int Biodeterior Biodegradation 83:145–157

Shah V, Bhaliya J, Patel GM, Deshmukh K (2022) Advances in polymeric nanocomposites for automotive applications:a review. Polym Adv Technol 33(10):3023–3048

Sharma S, Sudhakara P, Singh J, Ilyas RA, Asyraf MRM, Razman MR (2021) Critical review of biodegradable and bioactive polymer composites for bone tissue engineering and drug delivery applications. Polymers 13(16):2623
Sharma DK, Shukla S, Sharma KK, Kumar V (2022) A review on ZnO: fundamental properties and applications. Mater Today Proceed 49:3028–3035
Shasteen C, Bin Choy Y (2011) Controlling degradation rate of poly(lactic acid) for its biomedical applications. Biomed Eng Lett 1:163–167. https://doi.org/10.1007/s13534-011-0025-8
Shen Y, Seidi F, Ahmad M, Liu Y, Saeb MR, Akbari A, Xiao H (2023) Recent advances in functional cellulose-based films with antimicrobial and antioxidant properties for food packaging. J Agric Food Chem 71(44):16469–16487
Shin M, Park E, Lee H (2019) Plant-inspired pyrogallol-containing functional materials. Adv Func Mater 29(43):1903022
Sidkey MA, Yehia AA, Abd El Malak NA, Gaafar MS (2002) Compatibility studies on some rubber blend systems by ultrasonic techniques. Mater Chem Phys 74(1):23–32
Sin LT, Rahmat AR, Rahman WAWA (2013) Overview of Poly(lactic Acid) (2013). https://doi.org/10.1016/b978-1-4377-4459-0.00001-9
Sinclair A, Zhou X, Tangpong S, Bajwa DS, Quadir M, Jiang L (2019) High-performance styrene-butadiene rubber nanocomposites reinforced by surface-modified cellulose nanofibers. ACS Omega 4:13189–13199
Singh P, Goli VSNS, Singh DN (2023a) Current practices and the way forward in waste rubber utilization for infrastructural and geoenvironmental engineering applications. Environ Technol Rev 12(1):2252164
Singh P, Ali SW, Kale RD (2023b) Antimicrobial nanomaterials as advanced coatings for self-sanitizing of textile clothing and personal protective equipment. ACS Omega 8(9):8159–8171
Singha NR, Mahapatra M, Karmakar M, Chattopadhyay PK (2019) Processing, characterization, and application of natural rubber-based environmentally friendly polymer composites. Sustainable polymer composites and nanocomposites, 855–897
Sionkowska A (2011) Current research on the blends of natural and synthetic polymers as new biomaterials. Prog Polym Sci 36(9):1254–1276
Sisanth KS, Thomas MG, Abraham J, Thomas S (2017) General introduction to rubber compounding. In: Progress in rubber nanocomposites. Woodhead Publishing, pp 1–39
Soliman EM, Kandil UF, Taha MMR (2012) The significance of carbon nanotubes on styrene butadiene rubber (SBR) and SBR modified mortar. Mater Struct/Materiaux Et Constr 45:803–816
Song K (2017) Micro-and nano-fillers used in the rubber industry. In: Progress in rubber nanocomposites. Woodhead Publishing, pp 41–80
Srivastava A, Yadav T, Sharma S, Nayak A, Kumari AA, Mishra N (2015) Polymers in drug delivery. J Biosci Med 4(1):69–84
Su Y, Wang F, Wu S, Fan Y, Bai W, Wang S, Sun H, Zhu Z, Liang W, Li A (2021) Template-assisted preparation of conjugated microporous polymers membranes for selective separation. Sep Purif Technol 259:118203
Šupová M, Martynková GS, Barabaszová K (2011) Effect of nanofillers dispersion in polymer matrices: a review. Sci Adv Mater 3:1–25
Tan D, Wang Y, Tong Y, Chen GQ (2021) Grand challenges for industrializing polyhydroxyalkanoates (PHAs). Trends Biotechnol 39(9):953–963
Thomas S, Grohens Y, Jyotishkumar P (eds) (2014) Characterization of polymer blends: miscibility, morphology and interfaces. Wiley
Tian J, Paterson TE, Zhang J, Li Y, Ouyang H, Asencio IO, Hatton PV, Zhao Y, Li Z (2023) Enhanced antibacterial ability of electrospun PCL scaffolds incorporating ZNO nanowires. Int J Mol Sci 24(19):14420
Di Tinco R, Sergi R, Bertani G, Pisciotta A, Bellucci D, Carnevale G, Cannillo V, Bertoni L (2020) Effects of a novel bioactive glass composition on biological properties of human dental pulp stem cells. Materials 13(18)

Urtekin G, Ullah MS, Yildirim R, Ozkoc G, Kodal M (2023) A comprehensive review of the recent developments in thermoplastics and rubber blends-based composites and nanocomposites. Polym Compos 44:8303–8329

Vayyaprontavida Kaliyathan A, Varghese KM, Nair AS, Thomas S (2020a) Rubber–rubber blends: a critical review. Prog Rubber Plast Recycl Technol 36(3):196–242

Venkatachalam P, Geetha N, Sangeetha P, Thulaseedharan A (2013) Natural rubber producing plants: an overview. Afr J Biotechnol 12(12)

Vogel C, Hoffmann GG, Siesler HW (2009) Rheo-optical FT-IR spectroscopy of poly(3-hydroxybutyrate)/poly(lactic acid) blend films. Vib Spectrosc 49:284–287. https://doi.org/10.1016/J.VIBSPEC.2008.10.014

Walker TR, Fequet L (2023) Current trends of unsustainable plastic production and micro (nano) plastic pollution. TrAC Trends Anal Chem 160:116984

Warsinger DM, Chakraborty S, Tow EW, Plumlee MH, Bellona C, Loutatidou S, Karimi L, Mikelonis AM, Achilli A, Ghassemi A, Padhye LP (2018) A review of polymeric membranes and processes for potable water reuse. Prog Polym Sci 81:209–237

Wei H, Guo L, Zheng J, Huang G, Li G (2015) Effect of nanosilica-based immobile antioxidant on thermal oxidative degradation of SBR. RSC Adv 5:62788–62796

Wike NY, Akinbo O, Olaniyan OT, Adetunji CO, Adetunji JB (2023) The role of nano chitosan for effective delivery of nutrients and drugs including hormones and vaccines in cattle. In: Next generation nanochitosan. Academic Press, pp 171–202

Wojtowicz A, Janssen LP, Moscicki L (2009) Blends of natural and synthetic polymers. Thermoplastic starch: a green material for various industries, 35–53

Xue N, Ding X, Huang R, Jiang R, Huang H, Pan X, Min W, Chen J, Duan JA, Liu P, Wang Y (2022) Bone tissue engineering in the treatment of bone defects. Pharmaceuticals (Basel) 15(7)

Yang H, Du J (2024) Crystallinity, rheology, and mechanical properties of low-/high-molecular-weight PLA blended systems, molecules. 29. https://doi.org/10.3390/molecules29010169

Ylänen HO (2011) Bioactive glasses materials, properties and applications introduction. Bioactive glasses: materials, properties and applications, Xiii–Xiv

Yu L, Dean K, Li L (2006) Polymer blends and composites from renewable resources. Prog Polym Sci 31(6):576–602

Yu CH, Betrehem UM, Ali N, Ali F, Nawaz S, Sajid M, Yang Y, Chen T, Bilal M (2022) Design strategies, surface functionalization, and environmental remediation potentialities of polymer-functionalized nanocomposites. Chemosphere 306:135656

Zanchet A, Carli LN, Giovanela M, Brandalise RN, Crespo JS (2012) Use of styrene butadiene rubber industrial waste devulcanized by microwave in rubber composites for automotive application. Mater des 39:437–443

Zeng JB, Li KA, Du AK (2015) Compatibilization strategies in poly(lactic acid)-based blends. RSC Adv 5:32546–32565. https://doi.org/10.1039/c5ra01655j

Zhang Y, Wu B, Xu H, Liu H, Wang M, He Y, Pan B (2016) Nanomaterials-enabled water and wastewater treatment. Nanoimpact 3:22–39

Zhang T, Xu G, Blum FD (2023) Eco-friendly room-temperature polymerization in emulsions and beyond. Polym Rev 63(4):852–865

Zhang Y, Zhang J, Li J, Shen J, Zhang X (2024) Thermosensitive ZnO-PMMA/PEG microgels: a smart sunscreen with adapting SPF to ambient temperature fluctuations. Mater Today Chem 37:102039

Zhao X, Hu H, Wang X, Yu X, Zhou W, Peng S (2020) Super tough poly(lactic acid) blends: a comprehensive review. RSC Adv 10:13316–13368. https://doi.org/10.1039/d0ra01801e

Zhao Y, Li B, Zhang W, Zhang L, Zhao H, Wang S, Huang C (2023) Recent advances in sustainable antimicrobial food packaging: insights into release mechanisms, design strategies, and applications in the food industry. J Agric Food Chem 71(31):11806–11833

Zhu J, Hou J, Zhang Y, Tian M, He T, Liu J, Chen V (2018) Polymeric antimicrobial membranes enabled by nanomaterials for water treatment. J Membr Sci 550:173–197

Zhu Q, Chua MH, Ong PJ, Lee JJC, Chin KLO, Wang S, Kai D, Ji R, Kong J, Dong Z, Xu J (2022) Recent advances in nanotechnology-based functional coatings for the built environment. Mater Today Adv 15:100270

Chapter 2
Poly(Lactic Acid) Blends in Biomedical Applications

R. Biju Bennie, C. Joel, S. Daniel Abraham, and A. Jerold Antony

Abstract This chapter titled "Poly(lactic acid) blends in biomedical applications" provides an in-depth study of the use of poly(lactic acid) (PLA) and its blends in various biomedical contexts. PLA is recognized for its biodegradability, biocompatibility, and non-toxicity, making it a preferred material in applications such as drug delivery, tissue engineering and the development of medical implants. However, pure PLA exhibits limitations, including hydrophobicity and insufficient mechanical toughness, which can hinder its performance in demanding biomedical applications. To address these challenges, the blending of PLA with natural or synthetic polymers is necessary. This blending process enhances PLA's properties and creates novel characteristics suitable for specific biomedical applications. This chapter also covers a variety of current applications of PLA blends, including their roles in drug delivery systems, sutures and scaffolds for tissue engineering. By examining recent advancements and ongoing research, this chapter aims to provide a comprehensive overview of the potential of PLA blends in the biomedical field, emphasizing their adaptability and the innovative approaches being developed to optimize their functionality in medical applications.

Keywords Poly(lactic acid) · Biomedical applications · Tissue engineering · Biocompatible · Drug Delivery · Medical implants

R. Biju Bennie (✉) · C. Joel
Department of Chemistry, St. John's College, Tirunelveli, Tamil Nadu, India
e-mail: biubennie1986@yahoo.com

S. Daniel Abraham
Department of Chemistry, Madras Christian College (Autonomous), Chennai, Tamil Nadu, India

A. Jerold Antony
Department of Chemistry, St. Xavier's College (Autonomous), Tirunelveli, Tamil Nadu, India

P. M. Visakh (ed.), *Engineering Applications of Polymer based Nano Blends*, Advanced Structured Materials 235, https://doi.org/10.1007/978-3-031-90979-5_2

2.1 Introduction

The biomedical field has witnessed tremendous advancements in the recent years, driven by the need for innovative materials that meet both functional and environmental demands (Alshangiti et al. 2023). Polymers have emerged as critical components in this domain due to their versatility, adaptability, and unique ability to fulfil a wide range of requirements (Saleh Alghamdi et al. 2021). These long-chain macromolecules, composed of repeating monomer units, offer unmatched flexibility in tailoring properties to suit specific applications (Causin and Causin 2015). Polymers can range from natural, bio-derived materials to synthetically engineered compounds, making them indispensable in medical devices, drug delivery systems (Mondal et al. 2023), tissue engineering (Chouhan et al. 2021), and regenerative medicine (Homaeigohar and Boccaccini 2022). The biomedical use of polymers has expanded significantly over the past decades, driven by advancements in polymer chemistry, processing techniques, and the growing demand for biocompatible and biodegradable materials (Song et al. 2018). Their applications are varied, spanning from everyday medical tools, such as surgical sutures and catheters, to sophisticated systems like bioresorbable implants, smart drug carriers, and three-dimensional (3D) scaffolds for tissue regeneration (Kenry and Liu, B. 2018). This introduction explores the critical role of polymers in biomedical science, highlighting their significance, categories and diverse applications.

Many polymers are designed to interact harmoniously with biological systems, causing minimal immune response or adverse reactions (Zhang et al. 2024). This property is crucial for materials used in direct contact with human tissues, such as implants and wound dressings (Teo et al. 2016). Biodegradable polymers like poly(lactic acid) (PLA), poly(glycolic acid) (PGA), and their copolymer PLGA are particularly valuable in applications where temporary functionality is needed (Benatti et al. 2019). These materials degrade into non-toxic by-products, such as water or carbon dioxide, which are safely eliminated by the body (Prajapati et al. 2019).

Polymers can be processed into various forms, such as films, fibres, hydrogels and 3D structures, using scalable and cost-effective methods (Gong et al. 2020). This versatility makes them suitable for an extensive range of medical applications (Kirillova et al. 2021). The mechanical, thermal and chemical properties of polymers can be finely tuned by adjusting their molecular structure or by blending them with other materials (Yu et al. 2006). This adaptability ensures their functionality in specific biomedical scenarios (Kumar et al. 2018).

Biomedical polymers can be broadly classified into two categories: natural polymers and synthetic polymers. Each type offers distinct advantages and challenges, which influence their suitability for various applications (Laurencin et al. 2014).

2.1.1 Natural Polymers

Natural polymers, often proteins or polysaccharides, provide structural and functional support in various biological systems, making them ideal candidates for medical and therapeutic purposes (Mohammed et al. 2021). The innate ability of natural polymers to interact with cells and promote biological processes such as healing and regeneration has positioned them as essential materials in tissue engineering, drug delivery, and wound care (Malafaya et al. 2007). Collagen, for instance, serves as a scaffold for regenerating skin, cartilage, and bone, while chitosan, derived from crustacean shells, is used for antimicrobial wound dressings and drug carriers (Elkhenany et al. 2024). Hyaluronic acid is essential in orthopaedic and cosmetic applications, promoting joint lubrication and skin hydration (Chen et al. 2018). Alginate and cellulose derivatives are applied in controlled drug release and as hydrogels for wound care (Abasalizadeh et al. 2020; Alven and Aderibigbe 2020). The natural origin and functionality of these polymers makes them ideal for creating medical materials that safely integrate with the human body, fostering healing and reducing environmental impact.

While natural polymers offer significant advantages in medical applications, they also have notable disadvantages that can limit their utility. One major drawback is their variability in composition and properties, as these depend on the source and extraction process, making standardization challenging (Nayak 1999). Natural polymers often have lower mechanical strength compared to synthetic alternatives, which can restrict their use in load-bearing applications (Piggott 2002). Their susceptibility to microbial degradation and environmental conditions, such as moisture and temperature, may compromise their stability and shelf life (Iñiguez-Moreno et al. 2021). The high cost of purification and limited scalability of production also pose challenges for widespread adoption in medical fields (Akinsemolu et al. 2024). These limitations necessitate careful processing, modification or combination with synthetic polymers to enhance their performance and reliability (Fig. 2.1).

2.1.2 Synthetic Polymers

Synthetic polymers created through chemical synthesis have revolutionized biomedical science with their versatility, durability and ability to meet highly specific requirements (Tian et al. 2012). Unlike natural polymers, synthetic polymers offer consistent quality, tunable properties and scalability, making them ideal for applications where precise performance is critical (Terzopoulou et al. 2022). These materials play a pivotal role in areas such as drug delivery, tissue engineering and medical devices, often addressing limitations posed by their natural counterparts (Bolívar-Monsalve et al. 2021). Common examples include polyethylene and polypropylene for surgical sutures and prosthetics, polylactic acid (PLA) and polyglycolic acid (PGA) for biodegradable implants, and silicone for medical tubing and implants (He

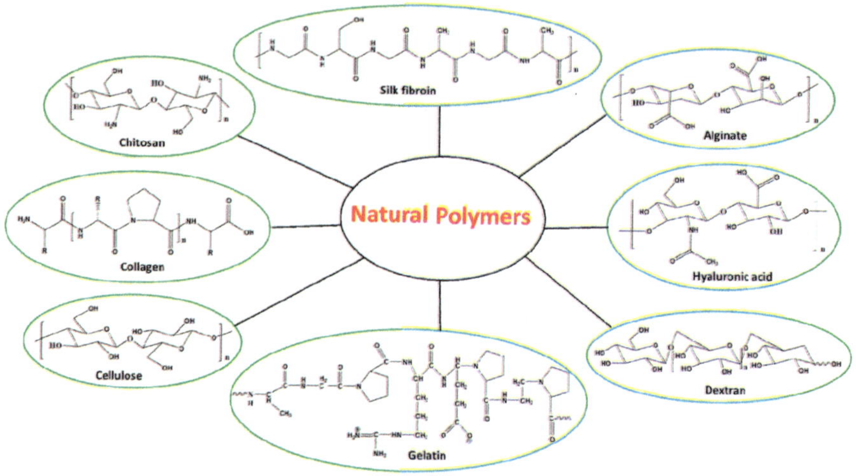

Fig. 2.1 Natural polymers used in biomedical applications

and Benson 2017). Synthetic polymers are also crucial in drug delivery systems, where materials like polyethylene glycol (PEG) are used to improve drug solubility and stability (Srivastava et al. 2015). Their ability to be sterilized and manufactured at scale ensures reliability in clinical settings. However, while they provide significant advantages, synthetic polymers must be carefully designed to ensure biocompatibility and avoid long-term adverse effects in the body (Arif et al. 2019) (Fig. 2.2).

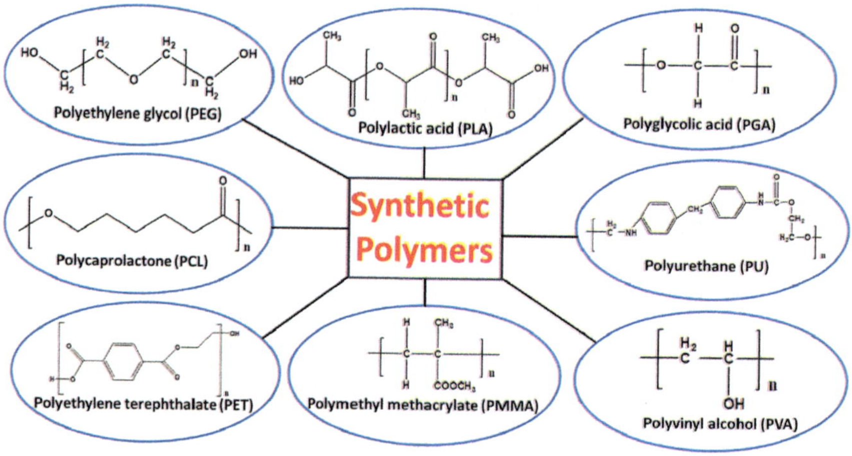

Fig. 2.2 Synthetic polymers used in biomedical applications

2.1.3 Advantages of Synthetic Polymers

Synthetic polymers are highly valued in medical applications due to their remarkable versatility and ability to be tailored for specific purposes (Maitz 2015). One of the most significant advantages is their tunable physical and mechanical properties, such as flexibility, strength, elasticity, and durability. These properties allow synthetic polymers to be designed for a variety of applications, from soft and flexible materials for catheters and tubing to rigid and strong materials for bone implants and prosthetics (Terzopoulou et al. 2022). This adaptability ensures that they can meet the demands of various medical environments, including those requiring high strength or intricate structural designs.

Another advantage of synthetic polymers is their controlled biodegradability (Gunatillake et al. 2006). Polymers such as polylactic acid (PLA), polyglycolic acid (PGA), and their copolymer poly(lactic-co-glycolic acid) (PLGA) can be engineered to degrade at a specific rate, making them ideal for temporary medical devices like sutures, drug delivery systems, and tissue scaffolds (Benatti et al. 2019). These materials break down into non-toxic by-products, such as carbon dioxide and water, which are naturally eliminated by the body (Anju et al. 2020). This eliminates the need for secondary surgical procedures to remove implants, reducing patient risk and healthcare costs.

Synthetic polymers are also advantageous due to their ability to be mass-produced with consistent quality and uniform properties (Bhatia and Bhatia 2016). Unlike natural polymers, which may exhibit variability depending on their source, synthetic polymers are manufactured in controlled environments, ensuring minimal batch-to-batch variation (Satchanska et al. 2024). This consistency is critical in medical applications, where reliability and precision are paramount. For example, polyethylene glycol (PEG) is commonly used to enhance drug delivery systems, and its predictable properties ensure reproducibility in clinical treatments (Wang et al. 2023).

In addition to consistency, synthetic polymers often exhibit excellent resistance to environmental factors, such as moisture, heat, and microbial growth (Luckachan and Pillai 2011). This stability increases the shelf life of medical products and devices, making them suitable for long-term storage and use in diverse healthcare settings. Polymers like silicone are widely used in medical implants, including heart valves and breast implants, due to their ability to maintain functionality and integrity over extended periods. Their non-reactive nature further reduces the risk of infections and complications (Teo et al. 2016).

Finally, the ability to chemically modify synthetic polymers enhances their functionality for specific medical applications. Polymers can be tailored to incorporate bioactive molecules, improve compatibility with living tissues, or control the release of drugs over time. For example, polyethylene glycol (PEG) is frequently modified to improve the solubility and stability of pharmaceuticals, enabling more effective drug delivery (D'souza and Shegokar 2016). This capacity for customization has driven innovations in fields like regenerative medicine, drug delivery, and minimally invasive procedures, making synthetic polymers indispensable in advancing healthcare technologies.

2.2 Poly(Lactic Acid)

Polylactic acid (PLA) is a biodegradable, thermoplastic polymer derived from renewable resources such as corn starch, sugarcane, and cassava (Fatchurrohman and Muhida 2023). It is one of the most widely used bioplastics globally, valued for its sustainability, versatility, and biodegradability. PLA is composed of lactic acid monomers linked through ester bonds that are polymerized through processes like condensation polymerization or ring-opening polymerization of lactide, a cyclic dimer of lactic acid (Taib et al. 2023) (Fig. 2.3).

2.2.1 Properties of Poly(Lactic Acid)

Poly(lactic acid) exhibits a range of physical properties that make it a versatile and widely used biopolymer. It is transparent or semi-transparent, though its appearance can be modified to suit specific applications. PLA has a low density of approximately 1.25 g/cm^3 (Wang et al. 2007), making it lightweight and suitable for packaging and 3D printing. Thermally, it has a glass transition temperature (T_g) between 55 and 60 °C and a melting point ranging from 150 to 170 °C (Nguyen et al. 2018), indicating limited heat resistance. PLA is mechanically strong, with high tensile strength and stiffness comparable to other polymers, though it is inherently brittle with low impact resistance unless plasticizers or other additives are incorporated (Zhao et al. 2020). These properties, combined with its ease of processing, contribute to its popularity across diverse industries.

PLA has unique chemical properties that contribute to its functionality and environmental appeal. It is a biodegradable polymer that degrades into lactic acid through

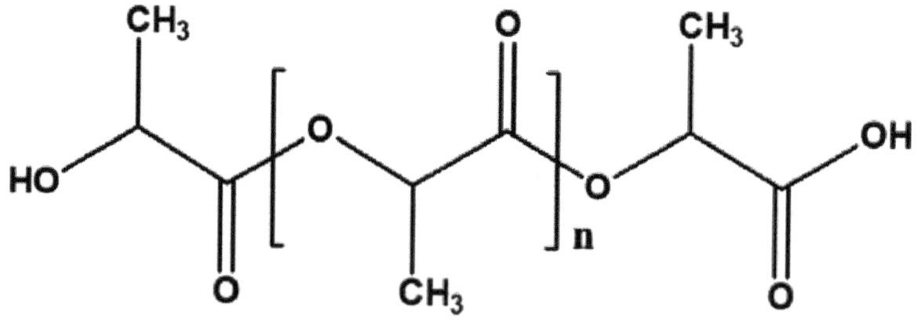

Fig. 2.3 Structure of poly(lactic acid)

hydrolysis or microbial activity, particularly under industrial composting conditions with elevated temperature and humidity (Karamanlioglu et al. 2017; Hughes 2011). PLA is insoluble in water, which makes it suitable for many applications, but it dissolves readily in organic solvents like chloroform, acetone, and dioxane (Pang et al. 2010). The polymer is moderately hydrophobic, with limited water absorption compared to hydrophilic polymers. Chemically, PLA is composed of repeating units of lactic acid, which can exist in different stereoisomeric forms (L-PLA, D-PLA, or racemic mixtures), influencing its crystallinity and overall properties (Palai et al. 2020). These characteristics, combined with its ability to break down into non-toxic by-products, make PLA a prominent choice for sustainable applications.

This polymer exhibits notable optical properties that make it suitable for applications requiring transparency and visual appeal. It is inherently transparent, with a high degree of light transmission. Its refractive index ranges between 1.45 and 1.49, depending on its crystallinity and processing conditions (Bikiaris et al. 2023). PLA's optical properties can be adjusted by controlling its degree of crystallization, where higher crystallinity often reduces transparency and imparts a milky appearance. This tunability allows PLA to be used in both clear and opaque products. Additionally, PLA can exhibit optical activity due to its chiral structure, depending on the proportion of L- and D-isomers in the polymer chain. These optical properties make PLA ideal for packaging, films, and other applications where aesthetics and transparency are critical (Farah et al. 2016).

The degree of crystallinity in PLA plays a significant role in its thermal behaviour; highly crystalline PLA exhibits better thermal resistance and mechanical strength compared to its amorphous form. However, PLA is thermally unstable at elevated temperatures and can degrade, particularly during prolonged exposure to heat (Ivanov et al. 2019). Its thermal conductivity is also low, making it a poor heat conductor (Guo et al. 2019). These properties make PLA suitable for low to moderate temperature applications, but they also necessitate modifications or blending with other materials for use in environments requiring higher thermal resistance.

PLA has limited electrical conductivity, making it primarily an insulating material. Its low dielectric constant and high resistivity are characteristic of thermoplastic polymers, making it suitable for applications where electrical insulation is required. The specific electrical properties of PLA can vary depending on its crystallinity and any additives or fillers incorporated during processing. For instance, blending PLA with conductive materials, such as carbon-based additives or metal powders, can enhance its conductivity for specialized applications like sensors or electromagnetic shielding (Hu et al. 2016). However, in its pure form, PLA's insulating nature, combined with its biodegradability and environmental friendliness, makes it a preferred choice for non-conductive applications in electronics, packaging, and other fields where static dissipation or electrical isolation is needed.

2.2.2 *Synthetic Methods of Poly(Lactic Acid)*

Polylactic acid (PLA) can be synthesized using various methods, each offering distinct advantages and limitations depending on the desired application and production scale. Different synthetic methods are used for PLA synthesis to tailor the polymer's properties, optimize production efficiency, and meet specific application requirements. The choice of method depends on factors such as the desired molecular weight, thermal and mechanical properties, production cost, and environmental considerations. By selecting the appropriate synthesis method, manufacturers can balance technical performance, economic feasibility, and sustainability to meet diverse industrial and environmental needs (Fig. 2.4).

Poly(lactic acid) can be synthesized by direct polycondensation involving the polymerization of lactic acid through a condensation reaction where water is produced as a by-product. The reaction typically occurs at elevated temperatures (150–220 °C) under vacuum or inert gas to facilitate the continuous removal of water and drive the reaction forward. This method is cost-effective and simple, making it attractive for large-scale production. However, it often results in low molecular weight PLA due to the difficulty of completely removing water during the reaction, which limits its mechanical and thermal properties. To address this, chain extenders or coupling agents are sometimes introduced in subsequent steps to increase the molecular weight (Hartmann 1998). Despite its limitations, direct polycondensation remains an important synthetic route for PLA, particularly for applications where high molecular weight is not a critical requirement (Fig. 2.5).

Ring-opening polymerization (ROP) is the most widely used method for synthesizing high molecular weight poly(lactic acid) due to its efficiency and ability to

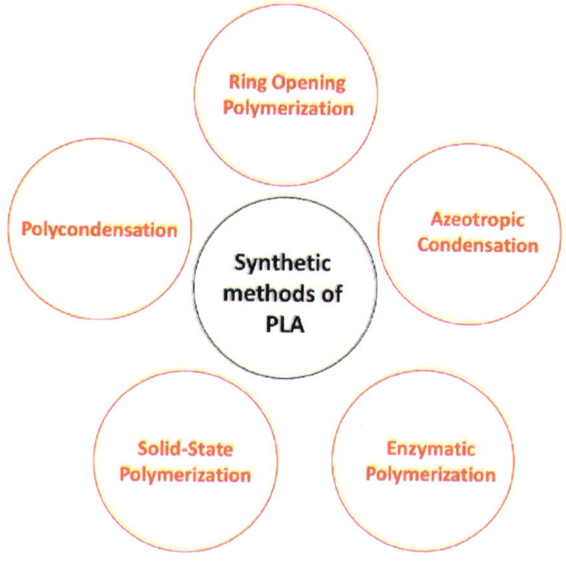

Fig. 2.4 Methods used for the synthesis of polylactic acid

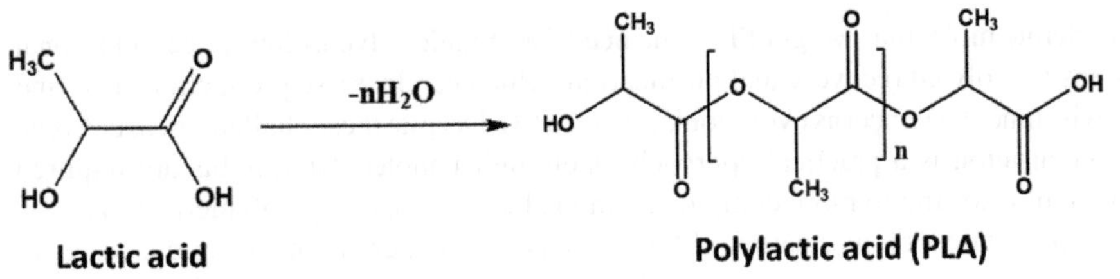

Fig. 2.5 Polycondensation of lactic acid to poly(lactic acid)

produce polymers with controlled properties. This process involves the polymerization of lactide, a cyclic dimer of lactic acid, using catalysts such as tin(II) octoate, zinc lactate, or aluminium-based compounds. The reaction occurs under mild conditions, typically at temperatures between 100 and 180 °C, allowing precise control over the molecular weight and stereochemistry of the PLA (Sarasua et al. 2005). The stereochemistry of the lactide (L-, D-, or racemic) plays a significant role in determining the crystallinity and thermal properties of the resulting polymer (Balla et al. 2021). ROP is preferred for industrial PLA production because it yields high-quality polymers suitable for diverse applications, from biodegradable packaging to medical implants (Fig. 2.6).

Azeotropic condensation is an improved method of synthesizing poly(lactic acid) that enhances the efficiency of direct polycondensation by using an azeotropic solvent to continuously remove water, the by-product of the reaction. In this process, solvents such as toluene or xylene form an azeotrope with water, enabling its removal through distillation as the reaction progresses. This setup helps drive the polymerization equilibrium towards higher molecular weight PLA while reducing the risk of hydrolytic degradation. Typically performed under elevated temperatures and inert conditions, azeotropic condensation allows better control over the polymerization process compared to basic direct condensation. While it is effective for producing

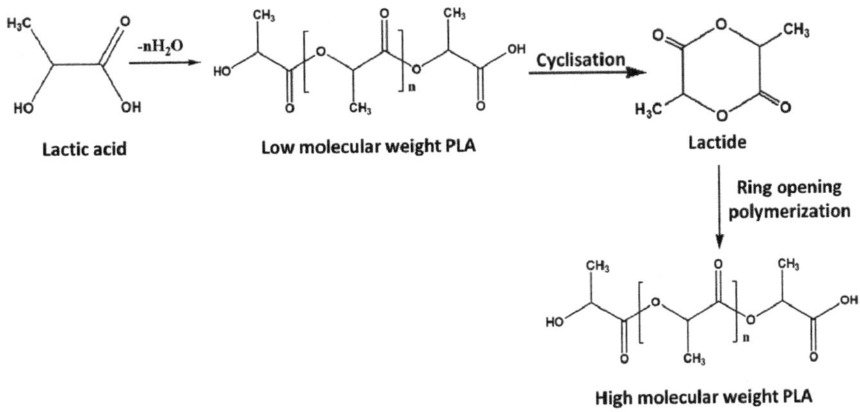

Fig. 2.6 Ring-opening polymerization of lactic acid to poly(lactic acid)

moderate molecular weight PLA, the need for organic solvents introduces additional steps for solvent recovery and purification, which can increase production costs and environmental concerns (Vouyiouka et al. 2013). Despite these challenges, azeotropic condensation is a practical approach when higher molecular weights are required without resorting to more complex methods like ring-opening polymerization.

Solid-state polymerization (SSP) is a post-polymerization technique used to increase the molecular weight of poly(lactic acid) after its initial synthesis via melt or solution polymerization. In this process, low molecular weight PLA is heated below its melting temperature, typically between 90 and 130 °C, under vacuum or an inert gas atmosphere to avoid thermal degradation. The controlled environment allows the polymerization reaction to proceed by removing residual water or other volatile by-products, leading to chain extension and higher molecular weights. SSP is advantageous because it avoids the issues of degradation that can occur at the high temperatures required for melt polymerization, resulting in PLA with improved mechanical and thermal properties. This method is particularly useful for producing PLA suitable for applications requiring high performance, such as durable packaging, fibres, and medical devices (Omay and Guvenilir 2013). However, SSP is more time-intensive compared to other polymerization methods, as the reaction rates are slower in the solid state.

Enzymatic polymerization of lactic acid is an environmentally friendly method for synthesizing polylactic acid (PLA) using enzymes as biocatalysts. This approach offers several advantages over traditional chemical methods, including mild reaction conditions (low temperature and neutral pH), reduced energy consumption, and the absence of toxic metal catalysts, making it ideal for biomedical and eco-friendly applications. Lipases, particularly *Candida antarctica* lipase B (CALB), are commonly used in this process due to their ability to catalyse ester bond formation with high specificity. Enzymatic polymerization typically proceeds via condensation reactions or ring-opening polymerization (ROP) of lactide, enabling control over polymer molecular weight and end-group functionality (Swetha et al. 2023). However, challenges such as lower reaction rates and the higher cost of enzymes compared to chemical catalysts have limited its commercial scalability. Despite these limitations, enzymatic polymerization remains an attractive area of research for sustainable and biocompatible PLA production.

2.2.3 Applications of PLA

Polylactic acid (PLA) is a versatile, biodegradable polymer with a wide range of applications across various industries due to its renewability, biocompatibility, and environmentally friendly nature. The applications of PLA include the following:

- PLA is widely used for making biodegradable containers, cups, bottles, and trays for food and beverages due to its transparency, rigidity, and safety for food contact (Mangaraj et al. 2019).

- Thin PLA films are used for packaging fresh produce, baked goods, and other perishable items, providing an eco-friendly alternative to petroleum-based plastics (Lunt and Shafer 2000).
- It is used in producing biodegradable fibres for clothing, upholstery, and non-woven materials. Its properties, such as low flammability and resistance to UV light, make it suitable for technical textiles (Khouri et al. 2024).
- It is used in sutures, stents, screws, and pins due to its biocompatibility and ability to degrade naturally in the body. It is employed in controlled drug release applications, as it can encapsulate and deliver active compounds over time (Martín-Closas et al. 2017).
- PLA is a preferred material for additive manufacturing (3D printing) due to its ease of processing, low melting point, and ability to produce detailed and precise structures. It is commonly used for prototyping and creating custom objects.
- It is used in casings for gadgets and appliances where biodegradability is a desirable feature.
- PLA films are used to cover soil in farming, helping with moisture retention and weed control while degrading naturally after use (Giammaria et al. 2024).
- PLA is utilized in lightweight, biodegradable interior components such as panels and trims, contributing to the sustainability of automotive production (Ramakrishna et al. 2001).
- PLA's biocompatibility makes it ideal for packaging medications and personal care products, aligning with eco-conscious branding.

By leveraging its environmentally friendly properties and adaptability, PLA continues to find new applications, driving sustainability across multiple sectors.

2.2.4 Limitations of PLA

While polylactic acid (PLA) is a promising biodegradable and renewable polymer, it has several disadvantages that limit its wider adoption in certain industries. One of the primary drawbacks is its brittleness and low impact resistance, which make it unsuitable for applications requiring toughness and flexibility, such as durable goods or high-stress components (Marano et al. 2022). PLA also suffers from a low glass transition temperature (55–65 °C) and poor thermal stability, meaning it cannot withstand high-temperature environments, such as hot food containers or automotive parts, without modifications.

Additionally, PLA has poor barrier properties, limiting its effectiveness in packaging applications where resistance to moisture and gas permeation is crucial. Its slow crystallization rate during processing can lead to dimensional instability, longer manufacturing times, and difficulties in achieving high heat resistance in products (Clarizio and Tatara 2012).

From a mechanical perspective, PLA has low tensile strength and elongation at break, which restrict its use in demanding structural applications without reinforcement or blending (Payne and Jones 2021). Recycling is also a challenge, as PLA requires separate streams from traditional plastics, and contamination with PLA can compromise the quality of recycled materials (Chen et al. 2003).

These limitations can be overcome by blending PLA with other materials which enhances its impact strength and flexibility, making it suitable for more demanding applications. Blending can also reduce costs by combining PLA with inexpensive fillers or recycled materials. This approach not only improves PLA's performance but also broadens its use in industries such as automotive, textiles, and medical devices, while maintaining its sustainability credentials (Nofar et al. 2019).

2.3 Poly(Lactic Acid) Blends

Polylactic acid (PLA) blends are a class of materials created by combining PLA with other polymers or additives to enhance its properties, overcome its limitations, and broaden its application range. While PLA is valued for its biodegradability, renewable nature, and eco-friendliness, it has several drawbacks, such as brittleness, low thermal resistance, slow crystallization rates, and poor barrier properties. By blending PLA with other materials, these shortcomings can be addressed, making PLA more versatile for various industrial applications (Thomas et al. 2014). PLA blends are typically designed to improve mechanical performance, enhance thermal stability, and facilitate processing.

2.3.1 Compatibility of Poly(Lactic Acid) Blends

The compatibility of poly(lactic acid) blends is a critical factor in enhancing its properties for broader applications. Compatibility largely depends on the miscibility of PLA with other polymers, which is influenced by factors such as polarity, molecular weight, and chemical structure (Jiang and Zhang 2012). PLA blends well with certain biodegradable polymers like polyhydroxyalkanoates (PHAs), polycaprolactone (PCL), and polybutylene succinate (PBS), as these materials share similar polar characteristics and biodegradability, resulting in improved toughness, flexibility, and processing behaviour. However, blending PLA with non-polar or hydrophobic polymers like polyethylene (PE), polypropylene (PP), and polystyrene (PS) often leads to immiscibility, characterized by phase separation and poor interfacial adhesion, which can compromise mechanical and thermal properties. To address such challenges, compatibilization strategies are employed, including the use of grafted copolymers (e.g. maleic anhydride or glycidyl methacrylate), plasticizers, or reactive blending to promote interfacial bonding and reduce phase separation. Nanofillers like graphene oxide or silica are also introduced to stabilize the blend and enhance structural integrity. These approaches help tailor PLA blends for specific applications, such as flexible packaging, automotive parts, and biomedical devices, by improving elongation, impact strength, and thermal stability. Importantly, while blending can significantly expand the functionality of PLA, maintaining or enhancing its biodegradability is crucial for ensuring the environmental benefits of the material in sustainable applications.

2.3.2 Types of *Poly(Lactic Acid) Blends*

1. Blending Poly(lactic acid) with Other Biopolymers

Blending poly(lactic acid) with polyhydroxyalkanoates (PHA) creates a versatile biodegradable material that combines the strengths of both polymers. PLA, known for its high rigidity and renewability, often suffers from brittleness and limited thermal stability. PHA, on the other hand, is more flexible, tough, and thermally stable. When blended, PHA enhances the toughness, flexibility, and impact resistance of PLA while maintaining biodegradability and environmental benefits (Taguet et al. 2014).

Blending PLA with starch creates a cost-effective and environmentally friendly material that enhances the biodegradability of PLA while reducing its dependence on petroleum-based components. Starch, being abundant, renewable, and highly biodegradable, serves as an excellent filler or modifier for PLA. However, starch on its own has poor mechanical properties and high moisture sensitivity. When blended, starch can lower the overall material cost and increase the rate of biodegradation, but it may also reduce PLA's mechanical strength and thermal stability. To address these challenges, plasticizers are often added to improve the interfacial adhesion between PLA and starch, resulting in better flexibility, toughness, and processability (Naser et al. 2021).

Blending PLA with cellulose creates a biocomposite that leverages the renewable and biodegradable properties of both materials while enhancing PLA's mechanical performance and thermal stability. Cellulose, derived from plant fibres, serves as a natural reinforcing agent, improving the strength and stiffness of PLA. Additionally, cellulose enhances the material's biodegradability and reduces its environmental impact (Makhijani et al. 2015). However, the hydrophilic nature of cellulose and the hydrophobicity of PLA can lead to poor interfacial adhesion, which may affect the composite's performance which could be overcome by the use of compatibilizers to enhance dispersion and bonding.

2. Blending Poly(lactic acid) with Synthetic Polymers

Blending poly(lactic acid) with polyolefins like polyethylene (PE) and polypropylene (PP) presents an opportunity to combine PLA's biodegradability and renewability with the mechanical toughness and durability of PE and PP. These blends can enhance properties such as impact resistance, flexibility, and thermal stability. However, the significant differences in polarity and chemical structure between PLA and polyolefins often result in poor miscibility and weak interfacial adhesion. To overcome these challenges, grafted copolymers or maleic anhydride are typically added to improve blend homogeneity and interfacial bonding (Gorgun et al. 2024).

Blending polylactic acid with polyurethanes (PU) results in a material that combines the biodegradability and renewability of PLA with the elasticity, toughness, and durability of PU. This blend addresses PLA's inherent brittleness and limited impact resistance, making it suitable for applications requiring flexibility and resilience, such as soft packaging, foams, or biomedical devices. These PLA-PU composites strike a balance between sustainability and enhanced mechanical properties, offering eco-friendly alternatives for various industries while expanding PLA's application potential (Tripathi et al. 2021).

3. Blending Poly(lactic acid) with Elastomers and Plasticizers

Blending polylactic acid with elastomers and plasticizers results in a material with enhanced flexibility, toughness, and impact resistance. Elastomers such as thermoplastic polyurethanes (TPU), polybutadiene, or natural rubber provide elasticity and resilience, making the blend suitable for applications requiring durability and stretchability (Zhang et al. 2020). Plasticizers like polyethylene glycol (PEG), citrate esters, or glycerol reduce the stiffness of PLA by decreasing intermolecular forces, improving ductility and processability (Sun et al. 2021). These blends are ideal for applications like flexible packaging, medical devices, and consumer products, offering an effective balance between mechanical performance and eco-friendliness. Proper plasticizer selection ensures that the material maintains its biodegradability and environmental benefits while expanding its practical applications.

4. Blending Poly(lactic acid) with Nanomaterials

Polylactic acid when blended with nanomaterials creates advanced composites with significantly enhanced properties, expanding PLA's range of applications. Nanomaterials such as nanoclays, carbon nanotubes (CNTs), graphene, or cellulose nanocrystals (CNCs) improve the mechanical strength, thermal stability, and barrier properties of PLA. Additionally, the incorporation of nanomaterials can enhance electrical conductivity, flame retardancy, or biodegradation rates depending on the additive used. Nanoclays can also improve PLA's resistance to UV degradation and enhance its dimensional stability (Farah et al. 2016). These PLA-nanomaterial composites are ideal for high-performance applications, including packaging with improved gas barrier properties, lightweight automotive components, and electronic devices.

5. Blending Poly(lactic acid) with Bio-Based Fillers

Bio-based fillers, such as natural fibres (like flax, hemp, or jute), are renewable and biodegradable, providing an eco-friendly alternative to traditional petroleum-based reinforcements. These fillers improve the tensile strength, stiffness, and impact resistance of PLA, making the material more suitable for a wider range of applications, including packaging, automotive parts, and construction materials (Mastalygina and Aleksanyan 2023). Overall, PLA blends with bio-based fillers not only offer

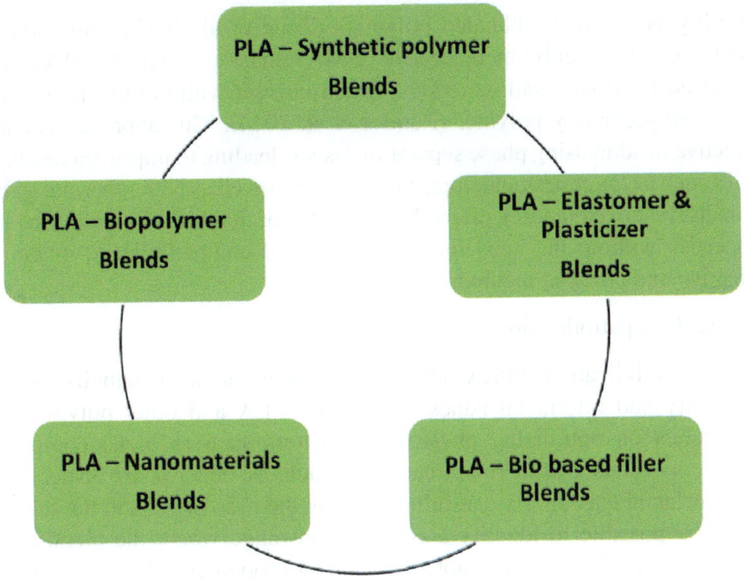

Fig. 2.7 Different types of poly(lactic acid) blends

improved performance but also contribute to reducing plastic waste and promoting sustainability in various industries (Fig. 2.7).

2.3.3 Methods of Compatibilization of Poly(Lactic Acid) Blends

Compatibilization methods for PLA blends are chosen based on the nature of the secondary polymer and the desired application. Reactive compatibilizers, nanofillers, plasticizers, and crosslinking agents are among the most commonly used strategies (Sanivada et al. 2020). The goal is to achieve a homogeneous blend with enhanced mechanical, thermal, and functional properties, while maintaining biodegradability for sustainable applications.

1. Reactive Blending

Reactive blending is a widely employed compatibilization strategy for improving the properties of PLA blends, particularly when combining PLA with immiscible or partially compatible polymers. In reactive blending, specific additives or reactive agents are introduced during the melt processing of PLA and the secondary polymer to facilitate in situ chemical reactions, such as chain extension, branching, or crosslinking. These reactions typically create covalent bonds or strong interactions at the interface, significantly enhancing the interfacial adhesion and phase compatibility between the blended polymers (Zeng et al. 2015). Common reactive agents include glycidyl methacrylate (GMA), maleic anhydride (MA), epoxy-functionalized polymers, and peroxides, which interact with the functional groups in PLA or the secondary polymer (Raquez et al. 2014). This approach is particularly effective in addressing phase separation issues, leading to improved mechanical properties such as toughness and impact resistance, as well as better thermal stability. Additionally, reactive blending offers the advantage of flexibility in modifying material properties without the need for pre-synthesized compatibilizers, making it a cost-effective and efficient method.

2. Physical Compatibilization

Physical compatibilization of PLA blends involves non-chemical methods to improve the miscibility and interfacial adhesion between PLA and other polymers. This approach relies on optimizing physical parameters such as blend composition, processing conditions, and thermal treatment to achieve better phase dispersion and reduced interfacial tension. By carefully selecting the ratio of PLA to the secondary polymer, it is possible to identify a composition range where the blend exhibits improved compatibility and desirable mechanical properties. Thermal processing, including controlled shear forces during melt blending and precise temperature regulation, can help achieve a uniform distribution of polymer phases. Additionally, annealing treatments after blending can promote partial crystallization or improve phase alignment, leading to enhanced interfacial bonding. Though less effective than reactive or chemical methods in creating strong interfacial interactions, physical compatibilization is often combined with other strategies to enhance overall performance. It is a cost-effective and straightforward method, making it suitable for applications where minor enhancements in miscibility are sufficient to meet functional requirements (Mastalygina and Aleksanyan 2023).

3. Crosslinking

Crosslinking is a powerful method for enhancing the properties of PLA by creating a three-dimensional network structure within the material or between PLA and other polymers in blends. This approach involves introducing crosslinking agents, such as peroxides, silanes, or multifunctional monomers, which form covalent bonds between polymer chains during processing. Crosslinking significantly improves the mechanical strength, thermal stability, and chemical resistance of PLA, addressing its inherent brittleness and low heat resistance. For instance, the use of dicumyl peroxide in PLA blends induces chain scission and subsequent crosslinking, leading to improved toughness and durability (Sanivada et al. 2020). In PLA-based blends, crosslinking can also enhance the compatibility between immiscible polymers by reinforcing interfacial adhesion, resulting in better phase dispersion and uniformity. However, careful control of the crosslinking process is essential to avoid excessive gel formation, which can hinder processability. Crosslinked PLA materials are particularly valuable in applications requiring high-performance and durability, such as in automotive parts, packaging, and biomedical implants, where the combination of rigidity and resilience is critical.

4. Hydrolysis and Enzymatic Compatibilization

These are emerging techniques for improving the compatibility and performance of poly(lactic acid) (PLA) in blends with other polymers, especially biodegradable ones. Hydrolysis involves partially breaking down PLA chains into smaller oligomers or introducing hydroxyl and carboxyl groups through controlled chemical or thermal processes. These modifications increase PLA's hydrophilicity and its ability to interact with hydrophilic or polar polymers, enhancing interfacial adhesion and miscibility (Formela et al. 2017). Similarly, enzymatic compatibilization utilizes specific enzymes, such as lipases or cutinases, to catalyse the selective degradation or functionalization of PLA (Olewnik-Kruszkowska et al. 2020). These methods offer an eco-friendly and precise way to modify PLA at the molecular level, paving the way for improved mechanical properties, enhanced processing, and broadened applications in sustainable materials, packaging, and biomedical fields. However, both techniques require careful optimization to balance the extent of modification and maintain the structural integrity of PLA.

5. Annealing

This is a post-processing technique used for compatibilizing poly(lactic acid) blends by improving the physical and thermal interactions between PLA and other polymers. During annealing, the PLA blend is exposed to elevated temperatures, typically just below its melting point, for a controlled duration, allowing molecular chains to reorganize and crystallize. This process enhances the interfacial adhesion between immiscible phases by promoting partial miscibility and reducing interfacial tension. In PLA blends, annealing can lead to improved phase morphology, better dispersion of the secondary polymer, and enhanced mechanical properties such as tensile strength and impact resistance. Additionally, annealing increases the crystallinity of PLA, which boosts its thermal stability and heat resistance, addressing a key limitation of the material. The effectiveness of annealing depends on factors such as temperature, time, and the composition of the blend, which must be carefully optimized to achieve the desired compatibilization without inducing degradation (Tournier et al. 2023).

2.4 Poly(Lactic Acid) Blends in Biomedical Applications

PLA blends allow for the development of materials with improved toughness, elasticity, and controlled degradation rates while maintaining biocompatibility and biodegradability. By combining PLA with certain materials, researchers have enhanced its functionality, enabling the creation of scaffolds, hydrogels, and implantable devices with superior mechanical and biological performance. These tailored blends are paving the way for innovative solutions in regenerative medicine, controlled drug release, and biodegradable medical devices, demonstrating the transformative potential of PLA-based materials in the biomedical field.

2.4.1 Tissue Engineering

PLA blends are at the forefront of advancements in tissue engineering, offering customizable materials that address the specific mechanical, structural, and biological requirements of different tissues (Wach et al. 2018). In scaffold applications, PLA blends allow precise control over porosity, mechanical strength, and degradation rates, enabling scaffolds to closely mimic the structural and functional properties of native tissues (Ogbu and Idumah 2024). For instance, PLA-based blends can be engineered to degrade at a rate synchronized with new tissue formation, ensuring adequate support during the regenerative process without causing harmful by-products (Donate et al. 2020). Additionally, these PLA blends facilitate the incorporation of bioactive agents such as growth factors, peptides, or nanoparticles, which can promote cellular differentiation, proliferation, and angiogenesis (Omigbodun et al. 2024). Overall, PLA blends play a crucial role in tissue engineering by addressing the challenges of regenerating diverse tissues, including bone, cartilage, skin, and cardiovascular structures. Their tunable properties, combined with the ability to integrate bioactive cues, position PLA blends as a cornerstone in the design of next-generation scaffolds for regenerative medicine.

2.4.2 Drug Delivery Systems

Poly(lactic acid) blends are extensively utilized in drug delivery systems due to their ability to control drug release profiles. These blends allow for fine-tuning of the material's degradation rate, hydrophilicity, and mechanical properties, providing tailored drug delivery solutions for a variety of therapeutic needs. For example, poly(lactic acid)/polyethylene glycol blends enhance hydrophilicity and facilitate the encapsulation and sustained release of hydrophilic drugs (Sharma et al. 2021), while poly(lactic acid)/polycaprolactone blends improve flexibility and adjust degradation rates for controlled release over extended periods (Aijaz et al. 2022). In drug delivery, PLA blends are typically used in the form of microspheres, nanoparticles, or implantable devices. These delivery vehicles are designed to protect therapeutic agents from degradation, improve bioavailability, and provide a controlled release over time (Saini et al. 2016). The degradation of PLA-based blends generates lactic acid, a biocompatible by-product that is naturally metabolized in the body, minimizing toxicity risks. By modifying the ratio and properties of the blend components, PLA-based drug delivery systems can be engineered to deliver drugs ranging from small molecules to proteins, nucleic acids, and vaccines (Tyler et al. 2016). These systems are particularly valuable in chronic disease management, cancer therapy, and post-surgical care, where controlled and sustained drug release can significantly improve therapeutic outcomes. The flexibility and tunability of PLA blends ensure their continued relevance in the design of innovative and effective drug delivery platforms.

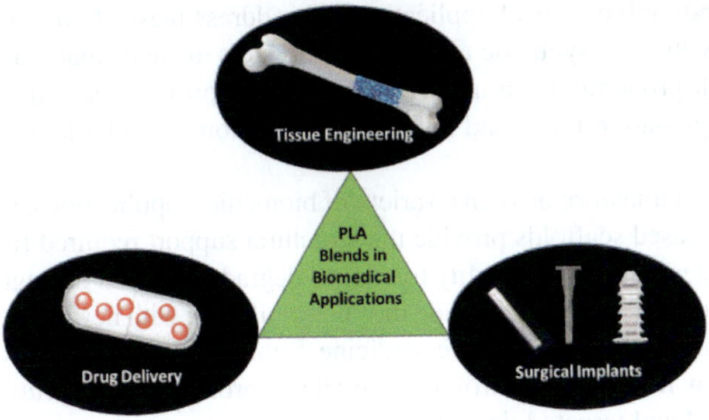

biomedical products.

Fig. 2.8 Biomedical applications of poly(lactic acid) blends

2.4.3 *Surgical Implants*

Poly(lactic acid) blends are increasingly being used in the development of surgical implants due to their enhanced flexibility, impact resistance, and hydrophilicity, while allowing for the fine-tuning of degradation rates and mechanical properties to meet the specific needs of different surgical procedures. In surgical applications, PLA blends are used for screws, pins, plates, and scaffolds that provide temporary mechanical support while the body regenerates natural tissue (Perinelli et al. 2019). The ability to control the degradation rate of PLA blends ensures that the implant maintains structural integrity during the healing process and degrades completely once its function is fulfilled, eliminating the need for secondary surgeries to remove the implant (DeStefano et al. 2020). Additionally, PLA blends can be engineered to incorporate bioactive components, such as growth factors, antibiotics, or anti-inflammatory agents, allowing implants to promote healing, reduce infection risks, and minimize inflammation (Chong et al. 2023). The adaptability of PLA blends ensures their suitability for a wide range of surgical applications, including orthopaedic, dental, and maxillofacial implants (Saini et al. 2016). The biodegradability of PLA blends makes it a highly attractive candidate for creating sustainable and patient-friendly biomedical products (Fig. 2.8).

2.5 Conclusion

Poly(lactic acid) and its blends have emerged as pivotal materials in the development of advanced biomedical applications, driven by their unique combination of biocompatibility, biodegradability, and processability. However, PLA's intrinsic limitations, such as brittleness, low thermal stability, and hydrophobicity, pose challenges for its standalone use in demanding biomedical applications. To address these shortcomings, PLA is combined with other synthetic and natural polymers to create materials with enhanced and tunable properties resulting in products with improved mechanical flexibility, controlled degradation rates, and enhanced interactions with biological systems.

PLA blends have proven transformative in a variety of biomedical applications. In tissue engineering, PLA-based scaffolds provide the structural support required for cell growth and tissue regeneration. The ability to control degradation rates through blend composition ensures that scaffolds degrade in synchronization with new tissue formation, a critical requirement in regenerative medicine. Similarly, in drug delivery systems, PLA blends allow for precise control over the release profiles of therapeutic agents, enabling sustained and targeted delivery of drugs, proteins, or vaccines. In surgical implants, PLA blends offer the dual advantage of mechanical support and resorbability, eliminating the need for secondary surgeries to remove implants after healing.

Looking to the future, PLA and PLA blends are expected to play an even greater role in addressing the challenges of modern medicine. Their ability to incorporate bioactive agents, such as growth factors, antibiotics, or nanoparticles, positions them as key materials for next-generation biomedical products that promote healing, reduce infection, and enhance therapeutic outcomes. Moreover, the environmental benefits of PLA, derived from renewable resources and capable of biodegrading into non-toxic by-products, contribute to their appeal as sustainable materials in a field increasingly focused on eco-friendly solutions.

References

Abasalizadeh F, Moghaddam SV, Alizadeh E, Akbari E, Kashani E, Fazljou SMB, Torbati M, Akbarzadeh A (2020) Alginate-based hydrogels as drug delivery vehicles in cancer treatment and their applications in wound dressing and 3D bioprinting. J Biol Eng 14:1–22

Aijaz MO, Yang SB, Karim MR, Othman MHD, Alnaser IA (2022) Preparation and characterization of poly (lactic acid)/poly (ethylene glycol)-poly (propyl glycol)-poly (ethylene glycol) blended nanofiber membranes for fog collection. Membranes 13(1):32

Akinsemolu AA, Idowu AM, Onyeaka HN (2024) Recycling technologies for biopolymers: current challenges and future directions. Polymers 16(19):2770

Alshangiti DM, El-Damhougy TK, Zaher A, Madani M (2023) Revolutionizing biomedicine: advancements, applications, and prospects of nanocomposite macromolecular carbohydrate-based hydrogel biomaterials: a review. RSC Adv 13(50):35251–35291

Alven S, Aderibigbe BA (2020) Chitosan and cellulose-based hydrogels for wound management. Int J Mol Sci 21(24):9656

Anju S, Prajitha N, Sukanya VS, Mohanan PV (2020) Complicity of degradable polymers in health-care applications. Mater Today Chem 16:100236

Arif U, Haider S, Haider A, Khan N, Alghyamah AA, Jamila N, Khan MI, Almasry WA, Kang IK (2019) Biocompatible polymers and their potential biomedical applications: a review. Curr Pharm Des 25(34):3608–3619

Balla E, Daniilidis V, Karlioti G, Kalamas T, Stefanidou M, Bikiaris ND, Vlachopoulos A, Koumentakou I, Bikiaris DN (2021) Poly (lactic Acid): a versatile biobased polymer for the future with multifunctional properties—from monomer synthesis, polymerization techniques and molecular weight increase to PLA applications. Polymers 13(11):1822

Benatti ACB, Pattaro AF, Rodrigues AA, Xavier MV, Kaasi A, Barbosa MIR, Jardini AL, Maciel Filho R, Kharmandayan P (2019) Bioreabsorbable polymers for tissue engineering: PLA, PGA, and their copolymers. In: Materials for biomedical engineering. Elsevier, pp 83–116

Bhatia S, Bhatia S (2016) Natural polymers vs synthetic polymer. Natural polymer drug delivery systems: nanoparticles, plants, and algae, pp 95–118

Bikiaris ND, Koumentakou I, Samiotaki C, Meimaroglou D, Varytimidou D, Karatza A, Kalantzis Z, Roussou M, Bikiaris RD, Papageorgiou GZ (2023) Recent advances in the investigation of poly (lactic acid)(PLA) nanocomposites: incorporation of various nanofillers and their properties and applications. Polymers 15(5):1196

Bolívar-Monsalve EJ, Alvarez MM, Hosseini S, Espinosa-Hernandez MA, Ceballos-González CF, Sanchez-Dominguez M, Shin SR, Cecen B, Hassan S, Di Maio E, Trujillo-de Santiago G (2021) Engineering bioactive synthetic polymers for biomedical applications: a review with emphasis on tissue engineering and controlled release. Mater Adv 2(14):4447–4478

Causin V, Causin V (2015) Polymers: an overview. Polymers on the crime scene: forensic analysis of polymeric trace evidence, pp 9–52

Chen CC, Chueh JY, Tseng H, Huang HM, Lee SY (2003) Preparation and characterization of biodegradable PLA polymeric blends. Biomaterials 24(7):1167–1173

Chen LH, Xue JF, Zheng ZY, Shuhaidi M, Thu HE, Hussain Z (2018) Hyaluronic acid, an efficient biomacromolecule for treatment of inflammatory skin and joint diseases: a review of recent developments and critical appraisal of preclinical and clinical investigations. Int J Biol Macromol 116:572–584

Chong WJ, Shen S, Li Y, Trinchi A, Simunec DP, Kyratzis IL, Sola A, Wen C (2023) Biodegradable PLA-ZnO nanocomposite biomaterials with antibacterial properties, tissue engineering viability, and enhanced biocompatibility. Smart Mater Manuf 1:100004

Chouhan D, Kaushik S, Arora D (2021) Trends in bio-derived biomaterials in tissue engineering. Biomaterials in tissue engineering and regenerative medicine: from basic concepts to state of the art approaches, pp 163–213

Clarizio SC, Tatara RA (2012) Tensile strength, elongation, hardness, and tensile and flexural moduli of PLA filled with glycerol-plasticized DDGS. J Polym Environ 20:638–646

D'souza AA, Shegokar R (2016) Polyethylene glycol (PEG): a versatile polymer for pharmaceutical applications. Expert Opin Drug Deliv 13(9):1257–1275

DeStefano V, Khan S, Tabada A (2020) Applications of PLA in modern medicine. Eng Regeneration 1:76–87

Donate R, Monzón M, Alemán-Domínguez ME (2020) Additive manufacturing of PLA-based scaffolds intended for bone regeneration and strategies to improve their biological properties. e-Polymers 20(1):571–599

Elkhenany H, Soliman MW, Atta D, El-Badri N (2024) Innovative marine-sourced hydroxyapatite, Chitosan, collagen, and gelatin for eco-friendly bone and cartilage regeneration. J Biomed Mater Res Part A

Farah S, Anderson DG, Langer R (2016) Physical and mechanical properties of PLA, and their functions in widespread applications—a comprehensive review. Adv Drug Deliv Rev 107:367–392

Fatchurrohman N, Muhida R (2023) From corn to Cassava: unveiling PLA origins for sustainable 3D printing. Jurnal Teknologi 13(2):87–93

Formela K, Hejna A, Haponiuk J, Tercjak A (2017) In situ processing of biocomposites via reactive extrusion. In: Biocomposites for high-performance applications. Woodhead Publishing, pp 195–246

Giammaria V, Capretti M, Del Bianco G, Boria S, Santulli C (2024) Application of poly (lactic acid) composites in the automotive sector: a critical review. Polymers 16(21):3059

Gong M, Zhang L, Wan P (2020) Polymer nanocomposite meshes for flexible electronic devices. Prog Polym Sci 107:101279

Gorgun E, Ali A, Islam MS (2024) Biocomposites of poly (lactic acid) and microcrystalline cellulose: influence of the coupling agent on thermomechanical and absorption characteristics. ACS Omega 9(10):11523–11533

Gunatillake P, Mayadunne R, Adhikari R (2006) Recent developments in biodegradable synthetic polymers. Biotechnol Annu Rev 12:301–347

Guo R, Ren Z, Bi H, Xu M, Cai L (2019) Electrical and thermal conductivity of polylactic acid (PLA)-based biocomposites by incorporation of nano-graphite fabricated with fused deposition modeling. Polymers 11(3):549

Hartmann MH (1998) High molecular weight polylactic acid polymers. In: Biopolymers from renewable resources. Springer, Berlin, Heidelberg, pp 367–411

He W, Benson R (2017) Polymeric biomaterials. In: Applied plastics engineering handbook. William Andrew Publishing, pp 145–164

Homaeigohar S, Boccaccini AR (2022) Nature-derived and synthetic additives to poly (ε-caprolactone) nanofibrous systems for biomedicine; an updated overview. Front Chem 9:809676

Hu Y, Daoud WA, Cheuk KKL, Lin CSK (2016) Newly developed techniques on polycondensation, ring-opening polymerization and polymer modification: focus on poly (lactic acid). Materials 9(3):133

Hughes J (2011) The effects of solvent mixture on the thermal and mechanical properties of solvent cast poly-lactic acid (PLA) film

Iñiguez-Moreno M, Ragazzo-Sánchez JA, Calderón-Santoyo M (2021) An extensive review of natural polymers used as coatings for postharvest shelf-life extension: trends and challenges. Polymers 13(19):3271

Ivanov E, Kotsilkova R, Xia H, Chen Y, Donato RK, Donato K, Godoy AP, Di Maio R, Silvestre C, Cimmino S, Angelov V (2019) PLA/Graphene/MWCNT composites with improved electrical and thermal properties suitable for FDM 3D printing applications. Appl Sci 9(6):1209

Jiang L, Zhang J (2012) Biodegradable polymers and polymer blends. In: Handbook of biopolymers and biodegradable plastics: properties, processing and applications, pp 109–128

Karamanlioglu M, Preziosi R, Robson GD (2017) Abiotic and biotic environmental degradation of the bioplastic polymer poly (lactic acid): a review. Polym Degrad Stab 137:122–130

Kenry and Liu B (2018) Recent advances in biodegradable conducting polymers and their biomedical applications. Biomacromolecules 19(6):1783–1803

Khouri NG, Bahú JO, Blanco-Llamero C, Severino P, Concha VO, Souto EB (2024) Polylactic acid (PLA): properties, synthesis, and biomedical applications-a review of the literature. J Mol Struct 138243

Kirillova A, Yeazel TR, Asheghali D, Petersen SR, Dort S, Gall K, Becker ML (2021) Fabrication of biomedical scaffolds using biodegradable polymers. Chem Rev 121(18):11238–11304

Kumar S, Nehra M, Dilbaghi N, Tankeshwar K, Kim KH (2018) Recent advances and remaining challenges for polymeric nanocomposites in healthcare applications. Prog Polym Sci 80:1–38

Laurencin C, Deng M (eds) (2014) Natural and synthetic biomedical polymers. Newnes

Luckachan GE, Pillai CKS (2011) Biodegradable polymers-a review on recent trends and emerging perspectives. J Polym Environ 19:637–676

Lunt J, Shafer AL (2000) Polylactic acid polymers from com. Applications in the textiles industry. J Ind Text 29(3):191–205

Maitz MF (2015) Applications of synthetic polymers in clinical medicine. Biosurface Biotribol 1(3):161–176

Makhijani K, Kumar R, Sharma SK (2015) Biodegradability of blended polymers: a comparison of various properties. Crit Rev Environ Sci Technol 45(16):1801–1825

Malafaya PB, Silva GA, Reis RL (2007) Natural–origin polymers as carriers and scaffolds for biomolecules and cell delivery in tissue engineering applications. Adv Drug Deliv Rev 59(4–5):207–233

Mangaraj S, Mohanty S, Swain S, Yadav A (2019) Development and characterization of commercial biodegradable film from PLA and corn starch for fresh produce packaging. J Packag Technol Res 3:127–140

Marano S, Laudadio E, Minnelli C, Stipa P (2022) Tailoring the barrier properties of PLA: a state-of-the-art review for food packaging applications. Polymers 14(8):1626

Martín-Closas L, Costa J, Pelacho AM (2017) Agronomic effects of biodegradable films on crop and field environment. Soil degradable bioplastics for a sustainable modern agriculture, pp 67–104

Mastalygina EE, Aleksanyan KV (2023) Recent approaches to the plasticization of poly (lactic acid)(PLA)(A review). Polymers 16(1):87

Mohammed ASA, Naveed M, Jost N (2021) Polysaccharides; classification, chemical properties, and future perspective applications in fields of pharmacology and biological medicine (a review of current applications and upcoming potentialities). J Polym Environ 29:2359–2371

Mondal A, Nayak AK, Chakraborty P, Banerjee S, Nandy BC (2023) Natural polymeric nanobiocomposites for anti-cancer drug delivery therapeutics: a recent update. Pharmaceutics 15(8):2064

Naser AZ, Deiab I, Defersha F, Yang S (2021) Expanding poly (lactic acid)(PLA) and polyhydroxyalkanoates (PHAs) applications: a review on modifications and effects. Polymers 13(23):4271

Nayak PL (1999) Biodegradable polymers: opportunities and challenges

Nguyen HTH, Qi P, Rostagno M, Feteha A, Miller SA (2018) The quest for high glass transition temperature bioplastics. J Mater Chem A 6(20):9298–9331

Nofar M, Sacligil D, Carreau PJ, Kamal MR, Heuzey MC (2019) Poly (lactic acid) blends: processing, properties and applications. Int J Biol Macromol 125:307–360

Ogbu JE, Idumah CI (2024) Poly (lactic) acid (PLA) hybrid bionanoarchitectures for tissue engineering. Int J Polym Mater Polym Biomater 1–26

Olewnik-Kruszkowska E, Burkowska-But A, Tarach I, Walczak M, Jakubowska E (2020) Biodegradation of polylactide-based composites with an addition of a compatibilizing agent in different environments. Int Biodeterior Biodegradation 147:104840

Omay D, Guvenilir Y (2013) Synthesis and characterization of poly (d, l-lactic acid) via enzymatic ring opening polymerization by using free and immobilized lipase. Biocatal Biotransform 31(3):132–140

Omigbodun FT, Oladapo BI, Osa-uwagboe N (2024) Exploring the frontier of polylactic acid/hydroxyapatite composites in bone regeneration and their revolutionary biomedical applications—a review. J Reinf Plast Composit 07316844241278045

Palai B, Mohanty S, Nayak SK (2020) Synergistic effect of polylactic acid (PLA) and Poly (butylene succinate-co-adipate)(PBSA) based sustainable, reactive, super toughened eco-composite blown films for flexible packaging applications. Polym Testing 83:106130

Pang X, Zhuang X, Tang Z, Chen X (2010) Polylactic acid (PLA): research, development and industrialization. Biotechnol J 5(11):1125–1136

Payne J, Jones MD (2021) The chemical recycling of polyesters for a circular plastics economy: challenges and emerging opportunities. Chemsuschem 14(19):4041–4070

Perinelli DR, Cespi M, Bonacucina G, Palmieri GF (2019) PEGylated polylactide (PLA) and poly (lactic-co-glycolic acid)(PLGA) copolymers for the design of drug delivery systems. J Pharm Investig 49:443–458

Piggott M (2002) Load bearing fibre composites. Springer Science & Business Media

Prajapati SK, Jain A, Jain A, Jain S (2019) Biodegradable polymers and constructs: a novel approach in drug delivery. Eur Polym J 120:109191

Ramakrishna S, Mayer J, Wintermantel E, Leong KW (2001) Biomedical applications of polymer-composite materials: a review. Compos Sci Technol 61(9):1189–1224

Raquez JM, Ramy-Ratiarison R, Murariu M, Dubois P (2014) Reactive extrusion of PLA-based materials: from synthesis to reactive melt-blending

Saini P, Arora M, Kumar MR (2016) Poly (lactic acid) blends in biomedical applications. Adv Drug Deliv Rev 107:47–59

Saleh Alghamdi S, John S, Roy Choudhury N, Dutta NK (2021) Additive manufacturing of polymer materials: progress, promise and challenges. Polymers 13(5):753

Sanivada UK, Mármol G, Brito FP, Fangueiro R (2020) PLA composites reinforced with flax and jute fibers—a review of recent trends, processing parameters and mechanical properties. Polymers 12(10):2373

Sarasua JR, Rodríguez NL, Arraiza AL, Meaurio E (2005) Stereoselective crystallization and specific interactions in polylactides. Macromolecules 38(20):8362–8371

Satchanska G, Davidova S, Petrov PD (2024) Natural and synthetic polymers for biomedical and environmental applications. Polymers 16(8):1159

Sharma S, Sudhakara P, Singh J, Ilyas RA, Asyraf MRM, Razman MR (2021) Critical review of biodegradable and bioactive polymer composites for bone tissue engineering and drug delivery applications. Polymers 13(16):2623

Song R, Murphy M, Li C, Ting K, Soo C, Zheng Z (2018) Current development of biodegradable polymeric materials for biomedical applications. Drug design, development and therapy, pp 3117–3145

Srivastava A, Yadav T, Sharma S, Nayak A, Kumari AA, Mishra N (2015) Polymers in drug delivery. J Biosci Med 4(1):69–84

Sun M, Huang S, Yu M, Han K (2021) Toughening modification of polylactic acid by thermoplastic silicone polyurethane elastomer. Polymers 13(12):1953

Swetha TA, Bora A, Mohanrasu K, Balaji P, Raja R, Ponnuchamy K, Muthusamy G, Arun A (2023) A comprehensive review on polylactic acid (PLA)–synthesis, processing and application in food packaging. Int J Biol Macromol 234:123715

Taguet A, Cassagnau P, Lopez-Cuesta JM (2014) Structuration, selective dispersion and compatibilizing effect of (nano) fillers in polymer blends. Prog Polym Sci 39(8):1526–1563

Taib NAAB, Rahman MR, Huda D, Kuok KK, Hamdan S, Bakri MKB, Julaihi MRMB, Khan A (2023) A review on poly lactic acid (PLA) as a biodegradable polymer. Polym Bull 80(2):1179–1213

Teo AJ, Mishra A, Park I, Kim YJ, Park WT, Yoon YJ (2016) Polymeric biomaterials for medical implants and devices. ACS Biomater Sci Eng 2(4):454–472

Terzopoulou Z, Zamboulis A, Koumentakou I, Michailidou G, Noordam MJ, Bikiaris DN (2022) Biocompatible synthetic polymers for tissue engineering purposes. Biomacromol 23(5):1841–1863

Thomas S, Grohens Y, Jyotishkumar P (eds) (2014) Characterization of polymer blends: miscibility, morphology and interfaces. Wiley

Tian H, Tang Z, Zhuang X, Chen X, Jing X (2012) Biodegradable synthetic polymers: preparation, functionalization and biomedical application. Prog Polym Sci 37(2):237–280

Tournier V, Duquesne S, Guillamot F, Cramail H, Taton D, Marty A, André I (2023) Enzymes' power for plastics degradation. Chem Rev 123(9):5612–5701

Tripathi N, Misra M, Mohanty AK (2021) Durable polylactic acid (PLA)-based sustainable engineered blends and biocomposites: recent developments, challenges, and opportunities. ACS Eng Au 1(1):7–38

Tyler B, Gullotti D, Mangraviti A, Utsuki T, Brem H (2016) Polylactic acid (PLA) controlled delivery carriers for biomedical applications. Adv Drug Deliv Rev 107:163–175

Vouyiouka S, Theodoulou P, Symeonidou A, Papaspyrides CD, Pfaendner R (2013) Solid state polymerization of poly (lactic acid): some fundamental parameters. Polym Degrad Stab 98(12):2473–2481

Wach RA, Wolszczak P, Adamus-Wlodarczyk A (2018) Enhancement of mechanical properties of FDM-PLA parts via thermal annealing. Macromol Mater Eng 303(9):1800169

Wang X, Kumar V, Li W (2007) Low density sub-critical CO_2-blown solid-state PLA foams. Cell Polym 26(1):11–35

Wang Z, Ye Q, Yu S, Akhavan B (2023) Poly ethylene glycol (PEG)-based hydrogels for drug delivery in cancer therapy: a comprehensive review. Adv Healthcare Mater 12(18):2300105

Yu L, Dean K, Li L (2006) Polymer blends and composites from renewable resources. Prog Polym Sci 31(6):576–602

Zeng JB, Li KA, Du AK (2015) Compatibilization strategies in poly (lactic acid)-based blends. RSC Adv 5(41):32546–32565

Zhang HC, Kang BH, Chen LS, Lu X (2020) Enhancing toughness of poly (lactic acid)/thermoplastic polyurethane blends via increasing interface compatibility by polyurethane elastomer prepolymer and its toughening mechanism. Polym Testing 87:106521

Zhang S, Fang H, Tian H (2024) Recent advances in degradable biomedical polymers for prevention, diagnosis and treatment of diseases. Biomacromolecules

Zhao X, Hu H, Wang X, Yu X, Zhou W, Peng S (2020) Super tough poly (lactic acid) blends: a comprehensive review. RSC Adv 10(22):13316–13368

Chapter 3
Bioactive Glasses/Polylactic Acid for Bone Tissue Engineering

Serra Nur Mutlu and Zehra Durmus

Abstract The fabrication of a suitable scaffold material is one of the major challenges for bone tissue engineering (BTE). Polylactic acid (PLA) is one of the most favorable matrix materials in BTE owing to its biocompatibility and biodegradability. However, PLA suffers from some shortcomings including low degradation rate, low cell adhesion due to its hydrophobic properties, and inflammatory reactions in vivo due to its degradation product, lactic acid. Therefore, incorporating bioactive reinforcements is considered a powerful method to alleviate these issues. This chapter presents a comprehensive study on recent advances in bioactive glasses/polylactic acid (BG/PLA) for BTE, including various methods of production and the evaluation of the scaffolds in terms of porosity, mechanical properties, in vitro and in vivo biocompatibility and bioactivity for BTE applications. The BG/PLA composites combine the advantages of both constituting phases, the mechanical strength of the polymer matrix, ease of processing, and bone growth ability. This chapter aims to empower readers with contemporary knowledge regarding the synthesis techniques of scaffolds, particularly focusing on bioglass and polylactic acid scaffolds. The goal is to establish a foundational understanding of how scaffold design has evolved and been modified, particularly in tissue engineering applications.

Keywords Poly(lactic acid) · Bone tissue engineering · Bioactive glass · Scaffold · Biocompatibility

Abbreviations

ALP assay	Alkaline phosphatase (ALP) activity assay
B_2O_3	Borate

S. N. Mutlu
Faculty of Medicine, Bezmialem Vakif University, Istanbul, Türkiye

Z. Durmus (✉)
Department of Biomedical Engineering, Faculty of Engineering and Natural Sciences, Biruni University, Istanbul, Türkiye
e-mail: zehrad@biruni.edu.tr

P. M. Visakh (ed.), *Engineering Applications of Polymer based Nano Blends*, Advanced Structured Materials 235, https://doi.org/10.1007/978-3-031-90979-5_3

BC	Biocement
BC	Bone cement
BG	Bioglass
BG/PLA	Bioactive glasses/polylactic acid
BG/PLLA	Bioglass/poly(L-lactic acid)
BG-BC	Bioglass-biocement
BGC	Bioactive glass ceramic
BGNPs-PLA	Bioglass nanoparticles-poly(L-lactic acid)
BTE	Bone tissue engineering
CAM assay	Chick embryo chorioallantoic membrane assays
FESEM	Field emission scanning electron microscopy
FTIR	Fourier transform infrared
g-BG/PLLA	Grafted-bioglass/poly(L-lactic acid)
HA	Hydroxyapatite
MBG	Mesoporous bioglass
m-BG	Micron-sized bioglass
MTT	3-(4,5-Dimethylthiazol-2-yl)-2,5-diphenyltetrazolium bromide
n-BG	Nano-sized bioglass
P(L,DL)LA	Poly(L,DL)lactide
P_2O_5	Phosphate
PBG	PLA + BG composite scaffold
PBS	Phosphate buffered saline
PDLLA	Poly-D, L-lactide
PDLLA-BG	Poly-D, L-lactide-bioglass
PLA	Polylactic acid
PLA/MBG	Polylactic acid/mesoporous bioglass
PLA/PMMA	Polylactic acid/polymethylmethacrylate
PLA-BC	Polylactic acid-biocement
PLGA	Poly (D, L-lactide-co-glycolide)
PLGA	Poly(lactide-co-glycolide)
SBF	Simulated body fluid
SEM	Scanning electron microscopy
SiO_2	Silicate
TEM	Transmission electron microscopy
THF	Tetrahydrofuran
TIPS	Thermally induced phase separation
VEGF	Vascular endothelial growth factor
XRD	X-ray diffraction

Objectives

- This chapter summarizes the recent developments in structures, properties, and applications of functional bioactive glasses/polylactic acid for bone tissue engineering.
- Basic definitions, fundamental knowledge, and classifications of bioactive glasses/polylactic acid materials are reported.
- The relevant literature and emerging technologies of functional bioactive glasses/polylactic acid are discussed.
- The overview of scaffold synthesis techniques of bioactive glasses/polylactic acid is exhibited.
- Practical applications such as bone repair and regeneration, scaffold materials, and angiogenesis promotion are exemplified.
- Challenges and current difficulties in manufacturing and applications of functional bioactive glasses/polylactic acid materials are concluded.

3.1 Introduction

Bones and the connective tissue around cartilage are vital components of the human body serving a variety of purposes, such as nourishing and protecting the human body, storing essential nutrition, and producing blood cells (Xue et al. 2022). Millions of individuals worldwide suffer from bone-related conditions, including osteoporosis, bone trauma, orthopedic surgeries, and congenital anomalies. One of the most difficult clinical practices is treating bone defects since they can cause significant harm over long-term therapy periods. Bone tissue engineering seeks novel ways to regenerate disturbed or diseased bone through cells, growth hormones, and biomaterials (Alonzo et al. 2021). The primary objective of BTE, utilizing a mix of cells, scaffolds, and growth factors, is to accelerate the healing process, manage non-union fractures, and reduce donor site morbidity and immune reactions to autografts or allografts. The human body is designed to regenerate various tissues, yet it cannot repair orthopedic tissue, including bone, cartilage, ligament, and tendon. Traditional surgical approaches may decrease distress as well as enhance joint function; however, the new cartilage regenerated by these procedures usually accounts for type I collagen and lacks type II collagen that has been found in hyaline cartilage, resulting in insufficient long-term clinical outcomes (Qiao et al. 2022). However, autograft transplantation is insufficient to treat bone abnormalities since it has several limitations such as donor site morbidity, unexpected immunological responses as well as prolonged hospitalization time (Archunan and Petronis 2021). Thus, researchers have begun looking for artificial alternatives to natural bone-derived implants. They use various techniques to mix cells, biomaterials, and growth factors to produce artificial tissues for bone regeneration (Zhu et al. 2021). BTE is a promising field considering the bone's unique structure, making regeneration difficult. The fundamental scaffold design specifications for BTE have been determined with required different criteria given in Fig. 3.1. These are, in order; *Biocompatibility*: After implanting, the material must allow for cell attachment, proliferation, and differentiation; nevertheless, it must not generate any harmful byproducts or trigger an inflammatory reaction. *Osteoconduction and osteoproduction*: Because they have to do with the material's capacity to adhere to bone and attract cells that surround the implantation site to differentiate into new bone development, osteoconduction and osteoproduction are particularly significant in the context of BTE. *Biodegradability*: If the material is to be employed in a variety of tissue engineering applications, its rate of deterioration must be controlled so that it corresponds to the rate at which new tissue forms at the implant site. *Mechanical competence*: The mechanical characteristics of the scaffold should closely resemble those of the tissue it aims to replace, ensuring adequate support throughout the regeneration process. *Interconnected porous structure*: BTE scaffolds must possess a high porosity of over 80% and a pore size greater than 100 μm. These conditions are optimal for promoting vascularization, allowing cell infiltration, and facilitating tissue growth (Ylänen 2011).

In dentistry, BTE has already succeeded in oral tissue regeneration with regards to dentine, pulp tissue scaffold templates, periodontal membranes, and bone cement

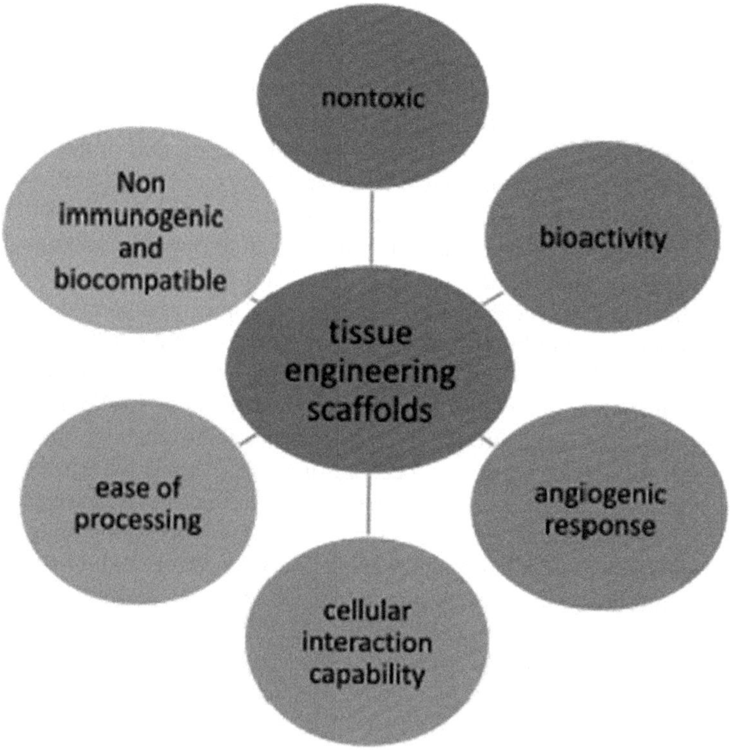

Fig. 3.1 Necessities for bone tissue engineering scaffolds. Adapted from Ylänen (2011)

(Najeeb et al. 2019; Zafar et al. 2015). Dental tissue differs significantly from other tissue ensembles in the body because of structure, organization, and function. Mechanical, thermal, chemical, and microbiological stresses of the human body are all placed on teeth and supporting structures. Dental tissues such as enamel, dentin, pulp, cementum, periodontal ligament, and alveolar bone that support the root are susceptible to injury or gradual degradation due to harmful oral circumstances and microbial infections (Das et al. 2022). As a result, BTE plays an important role in this regard. Dental stem cells, particularly those from the pulp, papilla, follicle, and periodontal ligament, have emerged as promising sources of bone repair (Ercal et al. 2018). Various biomaterials, including bioactive glasses, polylactic acids, and composites, are being studied as scaffolds for BTE (Ercal and Pekozer 2020). Among these types, bioactive glass, one of the accessible materials, is gaining importance as an area with increasing research into it (Skallevold et al. 2019). Bioglasses are classified into three types: silicate (SiO_2), phosphate (P_2O_5), and borate (B_2O_3). These materials are extremely versatile, functioning as porous scaffolds, drug delivery systems, and implant coatings (Kaou et al. 2023). BGs have an amorphous shape that binds to both hard and soft tissues, encouraging new bone growth while dissolving over time (Dash et al. 2023). They have great biological activity, which improves tissue/scaffold/implant interactions and connections. Porous scaffolds promote bone cell adhesion, proliferation, and fast integration with normal bone, with degradation rates corresponding to new bone creation (Dukle et al. 2022). BGs also inhibit bacterial biofilm development and release bioactive ions such as Ag, Ce, Co, Ga, Mg, Se, Sr, and Zn (Schatkoski et al. 2021). During degradation, dissolution and precipitation occur, releasing ions that promote osteogenesis, and angiogenesis, and provide antibacterial and anti-inflammatory properties (Ege et al. 2022; Brauer 2015).

Phosphate-based BGs, which break down quicker than silicate-based ones, are especially useful in physiological conditions since their dissolved particles are already present in the body. Borate-based BGs, known for their fast breakdown and transition into hydroxyapatite (HA), outperform silicate-based glasses regarding solubility and biological applications (Abodunrin et al. 2023). To improve bioactivity, BGs have been doped with bioactive ions. Further research is required to enhance understanding the clinical applications of these elements and drugs. Polylactic acid is a potential biopolymer for biomedical applications, especially BTE, considering its biocompatibility, biodegradability, and mechanical strength (Singhvi et al. 2019). The scientist Scheele from Switzerland originally discovered lactic acid in the 1700s, and it was initially used in medicine to treat canine mandibular fractures (Tyler et al. 2016). Lactic acid is the building block of the synthetic PLA polymer. PLA may be produced via direct polycondensation or ring-opening polymerization of lactic acid, derived from renewable resources (Lasprilla et al. 2012). Despite its benefits, PLA has drawbacks such as a slow decomposition rate, hydrophobicity, and potential inflammatory responses (Tajbakhsh and Hajiali 2017). To overcome these obstacles, researchers are looking at PLA mixes and composites with other materials such as BGs to improve their specific applications (Khouri et al. 2024).

BG and PLA composites have noteworthy synergistic qualities in biomedical applications. BGs, known for their osteoconductive properties, and PLA, known for their biodegradability, have drawn a considerable amount of attention in the field of biomaterials because they have the potential to improve both mechanical and biological efficiency when combined (Tinco et al. 2020), BGs may be included in PLA using a variety of processes, including solvent casting, melt mixing, and 3D printing. Solvent casting disintegrates PLA in an organic solvent, mixing it with bioactive glass particles and casting the solution into a mold (Shearer et al. 2023). Melt blending, on the other hand, is the process of melting PLA and blending it with bioactive glass particles before molding it (Bento et al. 2021). 3D printing provides a more complicated method, permitting the production of complex, porous elements with customized features (Gentile et al. 2014). These composites combine BG's osteoconductive characteristics with PLA's biodegradability, potentially improving mechanical and biological performance (Pawlik et al. 2020). BGs stimulate bone cell adhesion and development, whereas biodegradable PLA provides a suitable carrier and structure for tissue regeneration. Recent studies examined how BG composition and polymer mix ratios affect surface qualities, mechanical performance, and biological response. These composites are promising for many orthopedic and dental applications, including tissue engineering scaffolds and regenerative medicine.

3.2 Bioactive Glasses

BGs are highly specialized materials in tissue engineering and medical applications, including bone and soft tissue treatment (Negut and Ristoscu 2023). They have a special capacity to interact with bone and soft tissues, stimulating new bone development while progressively disintegrating (Cole et al. 2020). Bioactive glasses are highly appealing materials for creating scaffolds aimed at bone regeneration because of their adaptable properties and flexible composition. A key characteristic that makes BGs ideal for BTE is their capacity to promote revascularization, facilitate osteoblast adhesion, increase enzyme activity, and support the differentiation of mesenchymal stem cells and osteoprogenitor cells (Kaur 2017). Larry Hench and his colleagues invented BGs in the late 1960s based on the SiO_2–Na_2O–CaO–P_2O_5 system (Hench 2006). Their findings emerged from research aimed at developing materials that could bind with living bone tissue, which resulted in the invention of 45S5 Bioglass®, a silicate-containing glass (Huang et al. 2006). BGs are important in BTE because they promote bone regeneration and repair. When a bioglass material is implanted in the body, it undergoes a two-step process. The initial step involves particular surface reactions, particularly with simulated body fluid (SBF). In the second step, HA like layer forms, which facilitates interactions between hard and soft tissues. Their bioactivity allows for the creation of that HA layer that closely matches the mineral phase of bone, resulting in strong connections between the glass and bone tissue (Baino et al. 2018). The term bioactive glass refers to glass that must be compatible with surrounding living tissues and should be non-mutagenic, non-carcinogenic, and non-antigenic to prevent any negative effects on cells. A crucial requirement for any biomaterial, including bioactive glasses, is their resorbability; this means that

Table 3.1 Composition of 45S5 Bioglass® (Hench 2006)

Compound	Percentage (wt%)
SiO_2	45.0
CaO	24.5
Na_2O	24.5
P_2O_5	6.0

after implantation, the biomaterials should gradually degrade over time, allowing for their replacement by the body's natural tissue. Resorbable biomaterials need to maintain a stable interface throughout the degradation process, and their resorption rates should align to facilitate effective tissue repair. BGs garner significant interest primarily because they can exhibit a wide range of property changes with only minor adjustments to their chemical composition (Table 3.1).

BGs consist primarily of Na_2O, SiO_2, CaO, and P_2O_5 components in varying ratios, produced from silicate, phosphate, and borate origins. The ratios of these components can be modified to customize the glass for specific purposes. Thus, BG compositions vary, resulting in different characteristics. For example, silicate-based BGs are recognized for their strong bonding with bone (Baino and Fiume 2019). Still, borate and phosphate-based BGs have higher degradation rates, making them suited for applications requiring rapid tissue regeneration (Fu et al. 2010; Brito et al. 2017). BGs' bioactivity is determined by their contact with physiological fluids, which produces a HA layer on their surface (Menci et al. 2019). This process involves dissolving the glass, releasing calcium and phosphate ions, and precipitation of HA, which attaches to bone tissue (Samudrala et al. 2021). In biological circumstances, BGs go through various degradation and precipitation cycles. The release of ions from the glass, such as Si, Ca, and P, triggers biological reactions that promote bone repair and regeneration. The development of a HA layer is a vital feature of BG bioactivity. This layer occurs due to ion exchange between the BG and physiological fluids, resulting in a connection that connects the glass to bone tissue. In vitro research reveals that BG surfaces stimulate cell proliferation and differentiation, while in vivo studies confirm their capacity to promote bone repair and integration with the surrounding tissue (Fiume et al. 2021).

BGs act as a framework for new bone development, which fosters *osteoconduction.* Certain BG compositions containing Mg, Sr, F, Mn, Fe, Zn, and Co can activate progenitor cells and growth factors, and stimulate *osteoinduction.* BGs have strength and toughness that resemble natural bone, and their mechanical qualities are impacted by their composition. BGs are seen as biocompatible, however, conducting any clinical applications is complicated since the risk of an immunological reaction is always possible (Kargozar et al. 2018). Their degradation products are considered non-toxic and may even promote cellular function. The decomposition of BGs causes the release of bioactive ions, thereby helping in tissue repair. These ions are essential for the proper functioning of the human body. The most common elements are Mg, Zn, Sr, Cu, and Mn. Several studies have been performed to produce bioactive artificial replacements for bones that may be positioned directly to the wounded area and generate osteogenesis on-site. Because of their remarkable, varied, and flexible qualities (Aalto-Setälä et al. 2021), BGs are promising candidates for the treatment of hard tissue/bone injuries. It is also critical in tissue engineering that the BGs adhere properly to hard and soft living tissues (Hench and Jones 2015). Silicate-based BGs are frequently used in bone regeneration in the form of particles. This scaffold carries compressive strength comparable to cancellous bone and may be used for surgical applications. Eriksson et al. further confirmed this kind of scaffold was capable of supporting 3D bone ingrowth in rabbit femurs, meaning load-bearing bones (Eriksson et al. 2021). In addition, detailed studies on borate-based BGs show that these materials can be utilized for in vivo bone repair, without any side effects (Yao et al. 2007; Liu et al. 2015).

Since bioactive borate glasses have higher degradation rates, they are more eligible for both hard and soft tissue regeneration than silicate-based ones (Sengupta et al. 2021; Thyparambil et al. 2020). The main purpose of BGs in dentistry is to reload gaps in injured bones and teeth (bone fillers or bone grafts) or in osteoplasty. The classic, originally developed bioglass (45S5 Bioglass®) has been widely utilized in orthopedics and dentistry since 1985, though its capacity to produce apatite is relatively slow. However, increasing the phosphate concentration in BGs led to quicker apatite formation and better integration (Li et al. 2021; O'Donnell et al. 2009) than the original one. Some dental conditions (e.g., periodontitis, peri-implantitis, maxillofacial abnormalities) often result in bone loss or defects. These disorders significantly reduce the patient's quality of life. Bone grafts, both autologous and allogeneic, are widely utilized therapy options.

3.3 Preparation of Bioactive Glasses/Polylactic Blends

Understanding the properties and characteristics of materials is essential for their use. All glasses including BGs, whether made through melt quenching, sol–gel processing, or other methods, share two key traits: a glass structure and a gradual change in properties when transitioning between solid (glass) and liquid (melt) states. Glass is defined as "an amorphous solid lacking long-range atomic order," exhibiting a transformation where liquid properties gradually shift to those of the solid state. Key rules of glass formation and the impact of glass structure on its properties are crucial in the glass-forming process. Over the years, interest in BG compositions and morphologies has increased along with the development of new synthesis and production methods. The primary drawback of BGs is their low fracture toughness and brittleness. In BTE applications, BGs are frequently combined with biodegradable polymers to optimize mechanical and biological performance. Despite the inherent disadvantages of poor cellular affinity and low bone bioactivity for implantable biomedical applications, synthetic polymers form suitable composite structures for BTE with BGs because they possess many beneficial properties such as low stiffness, moldability, degradability, and drug encapsulation capacity. Different processes were developed to fabricate polymer-bioactive glass scaffolds/composites with interpenetrating network microstructure in a synergistic gain (Boccaccini et al. 2016). The clustering of nanoscale reinforcements allows for a more uniform distribution of the nanofillers within the polymer matrix, leading to improved mechanical properties. The contribution of nano BGs into bioresorbable polymers can change the degradation behavior of the polymer. Among polymer-bioactive glass scaffolds/composites, PLA composites are known for their high osteoconductivity, osteoinductivity, and biodegradability properties. Moreover, these composites have attracted interest in load-bearing applications because they combine bioactive behavior with mechanical strength and processability. In support of this, Liu et al. prepared bioglass-poly(L-lactic acid) (PLLA) composites with two different techniques and compared their mechanical strengths. First, they grafted low molecular weight PLLA onto the surface of BG nanoparticles via the coupling of hexamethylene diisocyanate, used toluene and dibutyltin dilaurate as the solvent and catalyst in this reaction respectively, and called this composite as g-BG/PLLA. Secondly, they suspended BG and PLLA homogeneously in chloroform with an ultrasonicator and precipitated in ethanol called BG/PLLA and dried under a vacuum. They reported that when the filler content was low, g-BG/PLLA composites showed higher tensile strength than BG/PLLA composites or pure PLLA. Similarly, porous nanocomposite scaffolds of PLLA containing different quantities of bioactive glass ceramic (BGC) nanoparticles (SiO_2:CaO:P_2O_5 $\cong$ 55:40:5 (mol)) were prepared by a different method, thermally induced phase separation (TIPS). Dioxane was used as the solvent for PLLA (Hong et al. 2008). In a similar approach, Maquet et al. prepared highly porous composites scaffolds of poly-D,L-lactide (PDLLA), and poly(lactide-co-glycolide) (PLGA) containing different amounts (10, 25, and 50 wt%) of bioactive glass (45S5 Bioglass) by TIPS and subsequent solvent sublimation (Maquet et al. 2004).

Blaker and Maquet et al. also synthesized in another study the foam formations of PDLLA and PDLLA/Bioglass → filled composite prepared by TIPS and subsequent solvent sublimation (Blaker et al. 2005). Another TIPS synthesis was performed by Sultan et al. with a composite filament of PLA and 45S5 BG to produce 3D print scaffolds for bone tissue regeneration given in Fig. 3.2 (Sultan et al. 2022). They printed various 3D scaffolds with PLA as well as with PLA-BG composite filament in the form of cylinders, discs, and cubes.

Gerhardt et al. manufactured BGNPs-PLA membranes and scaffolds, fabricating both composite membranes by solvent casting and scaffolds by particulate leaching methods. They examined the angiogenic properties of micron-sized (m-BG) and nano-sized (n-BG) filled PDLLA composites. Based on cell culture studies that assessed the secretion of vascular endothelial growth factor (VEGF) by human fibroblasts interacting with composite films of various wt%, they developed porous 3D composite scaffolds optimized for BG filler content to promote an angiogenic response. In vivo vascularization of these scaffolds was evaluated and high porosity (81–93%), permeability ($k = 5.4–8.6 \times 10^{-9}$ m^2), and compressive strength values (0.4–1.6 MPa) were observed, all comparable to trabecular bone. In composite films containing 20 wt% m-BG or n-BG, human fibroblasts produced five times more VEGF than those on pure PDLLA films. After eight weeks of implantation, scaffolds containing m-BG and n-BG were well infiltrated with newly formed tissue and

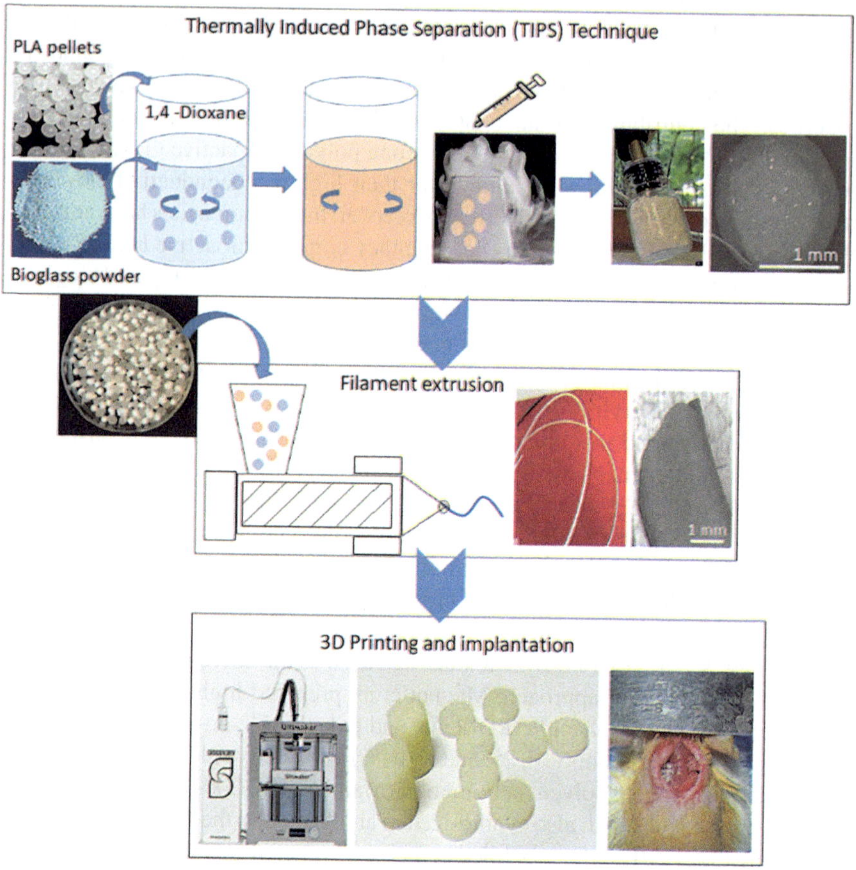

Fig. 3.2 PLA–bioglass scaffold processing and performance evaluation. Reproduced with permission from Sultan et al. (2022) Copyright 2022, MDPI

showed greater vascularization and a higher percentage of blood vessels to tissue (11.6–15.1%) compared to PDLLA scaffolds (8.5%). This study indicates the potential for regenerating hard-soft tissue defects and enhancing bone formation due to increased vascularization in the constructs (Gerhardt et al. 2011). As another strategy, Pant et al. have used pneumatic extrusion-based advanced 3D bioprinting to fabricate PLA and PLA/BG composite scaffolds given in Fig. 3.3. The preparation of PLA bio-ink involves the dissolution of PLA pellets in an appropriate amount of solvent chloroform and evaporating slowly to obtain 3D bio-ink based on neat PLA. On the other hand, the PLA/BG composite scaffold was fabricated by mixing an appropriate amount of BG into PLA bio-ink prepared using the same procedure with the ratio of PLA: BG maintained to be 30:70 to mimic the composition of natural. The orthogonal layer configuration of 0/90° was used for fabrication of 3D-printed PLA and PLA/BG scaffolds, as it was reported to offer better mechanical properties than displaced layer mode (Pant et al. 2022).

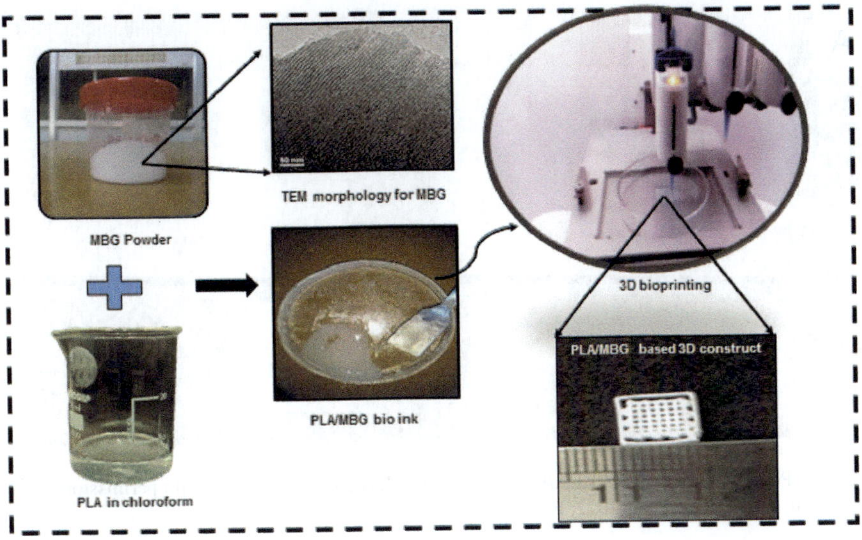

Fig. 3.3 Schematic illustration for the procedure involved in fabricating 3D-printed PLA/MBG composite scaffolds, reproduced with permission from Pant et al. (2022) Copyright 2022, Elsevier

Composite scaffolds were also prepared using 3D printing by Ding et al. in a study. The silica gel mold was prepared through 3D printing mold pouring. The mold had a hole of 1 cm wide, 1 cm long, and 2 mm high, which was fit for the printing stent in Fig. 3.4. The bone cement (BC) was prepared after the printing stent was placed, added to each stent, and extruded into the gap structure of the stent. The samples were obtained after drying. The samples were divided into 4 groups to compare material characterizations (Ding et al. 2022).

Moreover, two-dimensional round scaffolds of each PLA–BG composite were printed by Chaigneau et al. using fused filament fabrication on a 3D printer in another study (Cichos et al. 2023). Sample geometries of a flat disk consisting of two layers were designed and preprocessed with computer-aided design software. The synthesis of composite with a multistep process was achieved (Chaigneau et al. 2023). First, the starting materials of BG are weighed, mixed, and crushed in an agate bowl. The mixture is then heated up to 850 °C to allow de-carbonization. Finally, the mixture is heated at 1350 °C for 45 min to reach the melting and fining step. The resulting liquid is poured onto a stainless-steel support and then cooled in an ambient laboratory atmosphere. The BG is then ground using a mechanical grinder with an agate bowl. The resulting powder is finally sieved to a particle size smaller than 100 μm and checked by laser granulometry measurements. The polymer/glass composite precursor is then obtained by mixing the as-prepared glass powder with the PLA Resinex powder. In a different paper, La Torre et al. performed a method to synthesize BG in two steps. In the first step, tetraethyl silicate and nitric acid were mixed at room temperature. Then, triethyl phosphate and calcium nitrate tetrahydrate were added at intervals. The BG was obtained in this step. In the second step, the material

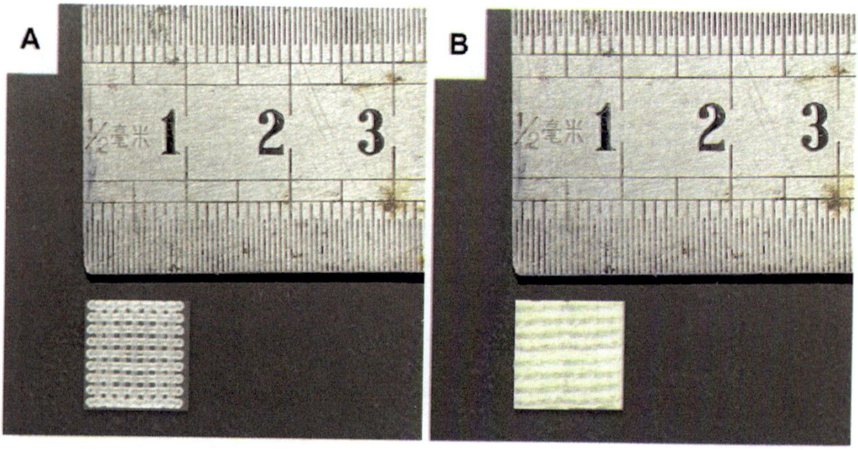

Fig. 3.4 Photos of samples **a** PLLA/BG and **b** PLLA/BG-BC. Reproduced with permission from Ding et al. (2022) Copyright 2022, Frontiers

obtained as a gel form was mixed with the PLA obtained in two weight percentages, namely BG30-PLA70 and BG70-PLA30, employing dissolvent propanone. The composite biomaterial was first kept in a sealed glass jar at room temperature for ten days and later at 70 °C for three days. Thermal treatment was performed at 120 °C for two days (Carbajal-De la Torre et al. 2022). A melt-blending method was used by Lee et al. using a kneader at 190C for 10 min, resulting in BG/PLLA composite pellets. The volume ratios of BG powders in the composites were set to 10 and 30 vol%. The volumes of PLLA and BG powders were calculated from their density. The densities were measured by Archimedes' method using acetone and water as immersion fluid for BG and PLLA, respectively, at 25C ($n = 3$) (Lee et al. 2019). Vergnol et al. also used a poly(L,DL)lactide (P(L,DL)LA) with a 70 L/30DL ratio and 45S5 BioglassVR particles produced from scratch with high purity SiO_2, Na_2CO_3, $CaCO_3$, and P_2O_5 powders. The powders were weighed, mixed, and melted in a platinum crucible at 1400 °C with a de-carbonization step at 950 °C. The melt was then quenched in water, ground by planetary milling in ethanol to a fine powder and sieved to obtain particles with a mean size of 3.5 mm (particle size distribution of 1–15 mm). Composite granules with 20, 30, and 50 wt% (i.e., 11, 17, and 30 vol%, respectively) of BioglassVR particles. Briefly, the BioglassVR particles were first put in suspension in acetone (100 g/500 mL), then the polymer was added progressively and mixed to reach the desired bioactive glass/polymer ratio. The suspension was then precipitated in water. The precipitate was milled and sieved to obtain granules of a mean diameter of 1 mm. Polymer and composite granules were molded by injection (145C, 150 bars) into plates or cylinders with 20, 30, and 50 wt% of BioglassVR (Vergnol et al. 2016).

In another example, by Han et al., a homogeneous solution was first prepared by dissolving a known amount of BG precursor (0.00/0.15/0.30 g) in THF under ultrasound treatment. After complete dissolving, 0.45 g of PLA was added and then, the solution was loaded into a plastic syringe equipped with a blunt-ended stainless-steel needle (12-gauge), which was connected to a high-voltage (20 kV) supply. A piece of grounded flat aluminum foil was placed 20 cm far from the needle tip to collect the fibers. These hybrid fibers were then dried at 25 °C under vacuum (Han et al. 2014). Kim et al. developed herein a novel nanocomposite made up of bioactive glass in a nanofibrous form and a degradable synthetic polymer, PLA. The glass nanofiber with a bioactive composition was generated via an electrospinning process with an average diameter of *320 nm. The nanofiber was homogenized with PLA solution at various concentrations (up to 35% nanofiber), followed by drying and thermal pressing to produce dense nanocomposites seen in Fig. 3.5 (Kim et al. 2008). Alternatively, by Han et al., the macro/mesoporous bioactive glasses/polylactic acid nanofibers have been successfully synthesized via electrospun method followed by acid treatment processing (Han et al. 2014).

In a present study, poly (D, L-lactide-co-glycolide) (PLGA) and BG45S5/PLGA fibers were investigated using human oral bacteria specified with mass spectrometry by Hild et al. BG45S5 nanoparticles were prepared by flame spray synthesis. For nanocomposite scaffolds, an electrospinning solution contained 6.5 wt% PLGA (referred to the total solution weight) and 5.0 wt% Tween 20 (referred to as the PLGA weight) was prepared. Electrospun scaffolds containing 30 wt% BG45S5 nanoparticles were produced by first dispersing the corresponding amount of BG45S5 nanoparticles in a solution of THF and dimethylformamide by ultrasonication (Hild et al. 2013). Liu et al. fabricated a nano-bioceramic incorporated lysozyme PLGA microsphere for sustained protein release formulation. First, the nano-bioceramics were prepared by sol–gel method, second, the lysozyme PLGA microsphere incorporated with nano-bioceramics was fabricated by the S/W/O/W double emulsion

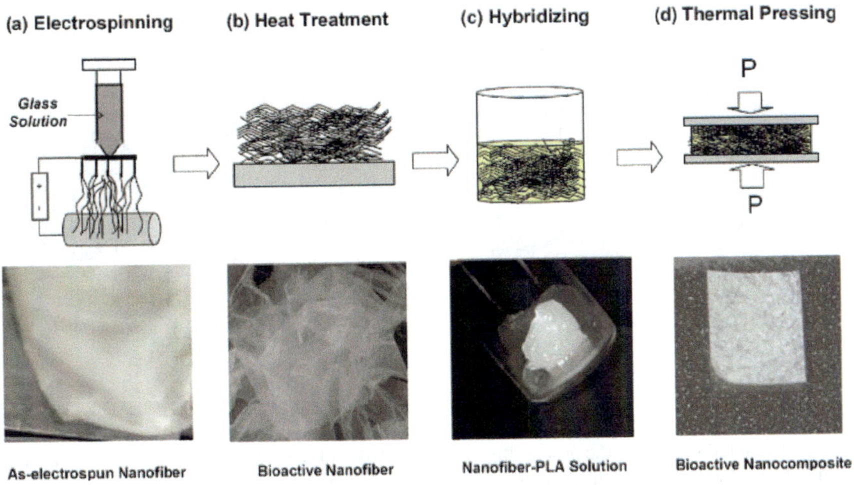

Fig. 3.5 Schematic illustration of the processings introduced to fabricate bioactive glass nanofiber and its nanocomposite with PLA. Below shown are optical images of the materials produced at each processing stage. Reproduced with permission from (Kim et al. 2008) Copyright 2022, Wiley

solvent evaporation method (Liu et al. 2013). Differing from the mentioned studies, Xiao et al. fabricated BG scaffolds with the geometry of a beam and microstructure by finite element modeling simulations using a robotic deposition technique. PLA sheets were prepared by dissolving as-received pellets in $CHCl_3$, by casting method. Then to bond a PLA layer to the top and bottom surface of the bioactive glass beams, one surface of a PLA sheet was wetted with $CHCl_3$ and applied to one surface of the scaffold by applied pressure, and the system was allowed to dry. The process was repeated to bond a PLA layer to the opposite surface of the BG scaffold (Xiao et al. 2017). Carrasco et al. designed hybrid scaffolds given in Fig. 3.6. that are based on PLA and PLA/PMMA with 75/25, 50/50, and 25/75 weight ratios and functionalized with 10 wt.% of bioglass nanoparticles were developed using an electrospinning technique with a chloroform/dimethylformamide mixture in a 9:1 ratio for bone tissue engineering applications (Álvarez-Carrasco et al. 2024).

PLA/BG structures produced with all the above-mentioned production methods show some improvements in mechanical and barrier properties due to the synergistic effect of the combination of these two structures, suggesting that it can be a good replacement for mineral reinforcements such as silica. This enhancement is attributed to the unique interactions between the bioglass and polylactic acid, which not only bolster the overall strength and durability of the scaffolds but also improve their resistance to environmental factors. Furthermore, the incorporation of BG into PLA matrices can enhance bioactivity, promoting better cellular responses and integration in tissue engineering applications. As a result, these composite scaffolds represent a promising alternative to traditional mineral reinforcements, offering both mechanical advantages and improved biological compatibility.

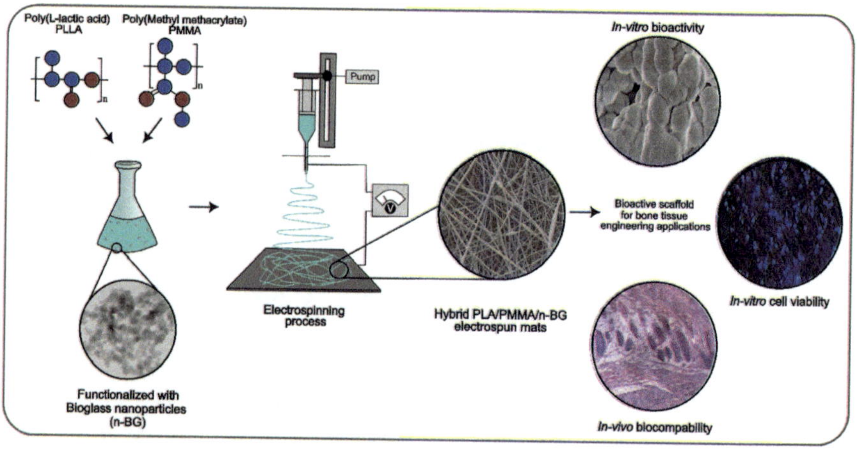

Fig. 3.6 Depicted scheme of the processing of designed hybrid scaffolds that are based on PLA and PLA/PMMA. Reproduced with permission from Álvarez-Carrasco et al. (2024) Copyright 2024, MDPI

3.4 Characterization of Bioactive Glasses/Polylactic Blends

3.4.1 Structural Characterization

The technique of X-ray diffraction (XRD) is widely used to investigate atomic spacing and crystal structure. In the context of PLA/BG composites, XRD is employed to identify the crystalline phases present within the material. The random orientation of powdered components allows for scanning across a range of 2θ angles to capture all possible lattice diffraction patterns. This analysis can reveal significant information about the crystallinity of the BG and its interaction with PLA, including the formation of HA layers when immersed in SBF. The resulting diffraction patterns provide insights into the structural integrity and bioactivity of the composites, confirming their potential for applications in BTE. Fourier Transform Infrared (FTIR) Spectroscopy is a crucial technique for analyzing the chemical changes in BG and PLA composites. It provides both qualitative and quantitative characterization of chemical modifications and degradation products. The formation of various functional groups is significantly influenced by the chemical structure of the polymer. In the context of PLA/BG composites, FTIR analysis can reveal important insights into the interaction between PLA and BG, including the presence of characteristic absorption peaks associated with C–C, C–O, and C=O bonds. These peaks indicate the structural integrity of the polymer matrix and its compatibility with BG. Additionally, FTIR can monitor changes over time, such as the formation of HA during immersion in SBF, which is critical for assessing bioactivity and potential applications in tissue engineering. Overall, FTIR serves as an essential tool for understanding the chemical interactions and stability of PLA/BG composites in biomedical applications.

3.4.2 Morphological Characterization

Morphological analysis is essential for understanding the structure and surface characteristics of BG and PLA composites. Techniques such as scanning electron microscopy (SEM), transmission electron microscopy (TEM), and field emission scanning electron microscopy (FESEM) are commonly employed to characterize the morphologies of these materials. Using SEM, researchers can observe the surface topography and texture of the BG/PLA composites, revealing insights into the distribution of bioglass particles within the PLA matrix. TEM allows for a closer examination of the internal structure at the nanoscale, providing information about the crystalline phases and any interfaces formed between the BG and PLA. FESEM, with its enhanced resolution, offers detailed images that help identify fine structural features and surface defects. Through these morphological studies, we gain a comprehensive understanding of the physical changes that occur during the fabrication and modification processes of BG/PLA composites. This knowledge is crucial for optimizing material properties for applications in tissue engineering and regenerative medicine, ensuring that these composites meet the necessary criteria for effective biological performance.

3.4.3 *In Vitro* and *In Vivo* Analysis

In vitro and in vivo analyses are crucial for evaluating the performance of BG and PLA composites in biomedical applications. For in vitro analysis, various studies have demonstrated the biocompatibility of PLA/BG composites, highlighting their potential for BTE further supporting the composite's potential for promoting vascularization in tissue engineering applications. Also the in vivo performance of PLA/BG composites is assessed through animal models to evaluate their integration and response within biological systems. Studies have shown that these composites can form a HA layer when implanted, which is indicative of bioactivity and compatibility with bone tissue. Furthermore, investigations into the tissue response to these materials have demonstrated minimal inflammatory reactions and effective bone regeneration capabilities. The combination of favorable in vitro results with positive in vivo responses underscores the potential of PLA/BG composites as effective scaffolds for bone repair and regeneration. Overall, both in vitro and in vivo analyses are essential for validating the efficacy of BG/PLA composites in clinical applications, ensuring they meet the necessary biological and mechanical requirements for successful tissue engineering outcomes.

3.4.4 *Mechanical Analysis*

Tensile strength is defined as the highest stress that the composites can withstand when being tested. To determine whether a material has enough mechanical strength for its intended uses, its tensile strength—a mechanical property—is crucial in several applications such as BTE, the strength of the material is very important to understand whether the mechanical properties including tensile strength should have sufficient mechanical strength. Successful incorporation of polymeric matrices has contributed to the mechanical properties of BG. With compression test in deformation modes, which is one of the most important mechanical tests, the elastoplastic nature of the scaffolds is confirmed by Pant et al. for PLA and polylactic acid/mesoporous bioglass (PLA/MBG) composite scaffolds and showed that both the PLA and PLA/MBG scaffolds display an elastic and plastic region. It is determined that PLA's ultimate stress value is 77 MPa. The ultimate stress/compressive strength of PLA/MBG composite scaffolds rises remarkably to 91.5 MPa following the addition of MBG (mesoporous bioglass) to the PLA matrix. PLA has an elastic modulus of 3.71 GPa and a yield strength of 64.4 MPa, respectively. Ding et al. reviewed 3D printing polylactic acid polymer-bioactive glass loaded with bone cement for bone defects in weight-bearing areas and indicated that for a bone defect in a weight-bearing area, long-term stability could be obtained by bone integration through this frame-filling structure They reported bone cement contribution on the compressive strength of PLA and BG was about 20 MPa. The values of PLA-BC and BG-BC reached about 80 MPa, which was similar to that of PBG, indicating that the addition of BC significantly improved the compressive strength of the scaffolds. Cao et al. claim that the PLA/MBG composite made by solution casting had a considerable improvement in compressive strength and water contact angle, going from 26 to 61 MPa and 84° to 34°, respectively (Cao et al. 2015). Chaigneau et al. prepared mechanical specimens of PLA/glass or PLA/BG composites by Fused Filament Fabrication (FFF) 3D printing (Chaigneau et al. 2023). But they observed that this composite presents some problems including a lower mechanical strength than the two compounds alone, probably due to the ionic salting-out induced by the BG. Then, they aimed to counter this phenomenon by coating the BG particles with a thin layer of gold. Mechanical tests revealed that the properties of the composites are influenced by several factors, including the amount and shape of the mineral filler. Notably, substituting glass beads with BG led to a significant reduction in tensile strength due to PLA chain degradation from ionic release. Conversely, incorporating gold particles at the PLA-BG interface limited ionic release, thereby enhancing mechanical tensile properties. While the addition of gold slightly slowed HA formation in SBF, it did not inhibit it. This study suggests that adding gold particles to the BG surface in PLA-BG composites can improve mechanical properties without adversely affecting chemical reactivity. In the mechanical measurements of 3D-printed scaffolds produced from PLA and 45S5BG components by TIPS synthesis mentioned above by Sultan et al., they found that an 80% increase in the compressive strength with the addition of BG particles. They reported that the compressive strengths of the PLA and PLA-BG scaffolds were 8 ± 2 MPa and 14 ± 2 MPa, respectively. The modulus was not affected significantly by the addition of BG (PLA 2.8 ± 0.2 and PLA-BG 2.3 ± 0.2 GPa).

3.5 Bioactive Glasses/Polylactic Blends for Bone Tissue Engineering

Blends of BG and PLA are emerging as key materials in the field of BTE due to their unique properties that facilitate bone regeneration. Bioactive glasses, particularly Bioglass® (45S5), release beneficial ions such as calcium, magnesium, and strontium, which enhance osteoblast proliferation and support new bone formation. Recent developments have focused on optimizing fabrication techniques for these materials, including electrospinning and 3D printing, to create scaffolds with controlled porosity and anisotropic structures that mimic natural bone architecture. For example, oriented fibermats made of BG/PLA composites have been developed to support cell alignment and proliferation, which are essential for restoring both the quality and quantity of bone tissue. These scaffolds provide both a supportive framework for cell attachment and facilitate the gradual release of ions that stimulate cellular activities critical for bone healing. In addition, the combination of these materials addresses the significant challenges in treating critical-sized bone defects without the need for autografts or allografts, thereby reducing donor site morbidity and improving patient outcomes.

For example, Gerhardt et al. evaluated in vivo vascularization of PDLLA-BG (poly-D, L-lactide-bioglass) scaffolds and high porosity (81–93%), permeability ($k = 5.4–8.6 \times 10^{-9} m^2$), and compressive strength values (0.4–1.6 MPa) observed, all comparable to trabecular bone. In composite films containing 20 wt% m-BG or n-BG, human fibroblasts produced five times more VEGF than those on pure PDLLA films. After eight weeks of implantation, scaffolds containing m-BG and n-BG were well infiltrated with newly formed tissue and showed greater vascularization and a higher percentage of blood vessels to tissue (11.6–15.1%) compared to PDLLA scaffolds (8.5%). This study indicates the potential for regenerating hard-soft tissue defects and enhancing bone formation due to increased vascularization in the constructs. Pant et al. assessed in vitro bioactivity on 3D-printed PLA/MBG composite scaffold even during the initial stages of SBF treatment, thus indicating the faster bioactivity. With further progression in the SBF treatment till 14 days, apatite coating was observed to cover the entire surface of PLA/MBG composite scaffold. The formation of hydroxyapatite on the 3D-printed PLA/MBG composite scaffold is confirmed by the Ca/P ratio of 1.66, which follows the natural bone mineral. On the other hand, formation of apatite crystals is rarely observed on the 3D-printed PLA scaffolds indicating the slower bioactivity characteristics. The wettability is found to be improved for 3D-printed PLA/MBG composite scaffold as observed by water contact angle analysis, thus indicating the impartment of hydrophilic nature for the scaffold. In the case of in vitro biological assessment, PLA/MBG composite formulation (30:70) proposed in the current work showed positive response in terms of MTT and ALP assay. Remarkable improvement in the expression level for six important osteoblast markers is observed in the case of 3D-printed PLA/MBG composite scaffolds, which indicates the prospective feature exhibited by the formulation in terms of BTE. The mixed matrix composite scaffold based on PLA/MBG composite (30:70) formulation mimicking the natural bone composition could be seen as a potential scaffold for bone tissue engineering application, which is evidenced by the rapid HA formation ability complemented with osteogenic characteristics. In a present study, Cichos et al. showed that the 3D-printed PLA–BG, especially at a BG concentration of 20%, induces angiogenesis in vitro with endothelial cells as well as in ovo, as demonstrated in the CAM assay. This work demonstrated the angiogenic efficiency of a 3D-printed PLA-BG composite material for bone tissue engineering. De la Torre et al. examined their PLA/BG materials' potential as BTE scaffolds by in vitro testing using Hank's and SBF solutions, confirming the bioactivity of the composites by their ability to form HA on the surfaces and their adequate biodegradation when immersed in PBS after 21 days of immersion. Both properties were confirmed by SEM and FTIR characterization. The electrochemical evaluation of the scaffolds in Hank's saline solution and SBF, both samples showed activation mechanisms at the early stages, followed by pseudo-passivation or current limited behavior due to the physical characteristics of the bioactive glass, which suggests that improvements in the formation of HA nucleation consequently allow redox reactions at the surface of the coating. At Lee et al. study, the oriented BG/PLLA composite fibermats enabled cell alignment along the fibers and promoted cell proliferation due to the released ions from the BGs. The degree of fiber orientation was successfully controlled by modulating the content of the BGs and fiber diameters. The cell proliferation was significantly upregulated by the release of Mg^{2+} and Sr^{2+} ions from the BGs. The cell orientation was determined by the cell recognition of the fiber orientation and adherence along single fibers or the formation of cell branches protruding across multiple fibers. The obtained results not only indicated that the fabricated composites could control the cytoskeletal arrangement of osteoblasts along the fiber direction but also, that the controlled release of ions successfully improved osteoblast proliferation. These findings can lead to the development of innovative multifunctional biomaterials suitable for tissue regeneration treatments by the optimization of the structural properties of the scaffolds and inorganic ion element release.

P(L,DL)LA-BioglassVR 45S5 composite associates advantages of both constituting phases, the ease of processing and the mechanical strength of the polymer matrix, and the bone growth ability promoted by the BioglassVR. The addition of BioglassVR accelerates the degradation of the composite in vitro medium. Biocompatibility of the P(L,DL)LA-BioglassVR 45S5 (30 wt%) composite was confirmed both in vitro and in vivo. Implantation studies have revealed that the presence of BioglassVR accelerates bone integration. The kinetics of in vitro and in vivo degradation assays were not similar, highlighting the limits of the correlation between these two types of tests due to the different environments and the limited relevance of in vitro tests to simulate the actual in vivo degradation rate. The in vitro tests can be considered as a preliminary study before the in vivo study to determine the degradation of material or not. A complementary in vivo study at long time of implantation and with a quantitative histomorphometric analysis would be interesting to define the degradation kinetics of the composite and to precise the new bone formation. All these results suggest that the P(L,DL)LA-BioglassVR composite is a promising material for bone fixation devices applications. The synthesized materials display the macroporous (10 μm) and mesoporous (5 nm) structure, which benefits the drug/nutriment transport and tissue growth. Rapid HA-inducing mineralization after immersing in SBF solution indicates high bioactivity of the samples. Moreover, the component affects the hydrophilic–hydrophobic mechanics as well as the cell attachment property of nanofibers that are very important for practical application. With the novel hierarchical porous structure, and suitable wettability and bioactivity, these materials exhibit potential applications in bone substitution or regeneration. Sultan et al. processed and demonstrated the performance of 3D printing of a porous and biocompatible material, specifically PLA-BG composites, for bone tissue engineering. The strategy was to take advantage of the (1) biocompatibility of the PLA and BG in tissue engineering, (2) potential of the TIPS method to develop PLA-BG composites with homogeneous dispersion and distribution of BG in a PLA matrix, and (3) 3D printability of PLA to develop a porous scaffold with precise control of the pore size, microstructure, scaffold geometry, and mechanical properties. PLA and PLA-BG cylindrical and disc samples with pore size distributions of 150–500 μm were 3D printed. The microscopy and tomography studies revealed well-dispersed bioglass that favorably affected the compressive strength and promoted interaction with the biological environment during the in vitro and in vivo studies. The in vitro cytotoxicity analysis of PLA-BG scaffolds using L929 mouse fibroblast cells confirmed the non-lethal interaction of the scaffold with the cells. Furthermore, morphological analysis (H&E staining) post implantation during the in vivo studies showed an elevated level of macrophages and active fibroblasts populated around the PLA-BG scaffolds that were actively involved in collagen extracellular matrix synthesis, indicating the excellent potential of these scaffolds in bone tissue engineering. Further pre-clinical studies using this scaffold to evaluate efficient bone regeneration and the biochemical- and molecular-level interactions will be performed, which will have therapeutic significance in orthopedics and dentistry.

3.6 Conclusion

Bioactive glass/polylactic acid scaffolds are at the forefront of innovations in tissue engineering due to their biocompatibility and ability to promote cell growth. This chapter emphasizes several key synthesis techniques that have gained prominence such as Sol–Gel derived scaffolds which exhibit a hierarchical pore architecture, essential for mimicking natural tissue and facilitating cell ingrowth, and the customization of scaffold architecture with the integration of 3D printing technology of PLA-BG composites showcasing their potential as implant materials with tunable pore structures and mechanical properties suitable for bone tissue regeneration or mesoporous scaffold design technique focuses on creating scaffolds with specific pore sizes that enhance cellular interactions and nutrient flow. The mesoporous structure is particularly beneficial for drug delivery systems within tissue engineering frameworks.

In conclusion, this chapter underscores the importance of understanding various scaffold synthesis techniques and their implications in developing effective tissue engineering solutions. By focusing on BG/PLA composites, readers will gain insights into cutting-edge methodologies that are paving the way for advanced clinical applications in regenerative medicine.

References

Aalto-Setälä L, Uppstu P, Sinitsyna P, Lindfors NC, Hupa L (2021) Dissolution of amorphous S53P4 glass scaffolds in dynamic in vitro conditions. Materials 14(17)

Abodunrin OD, El Mabrouk K, Bricha M (2023) A review on borate bioactive glasses (BBG): effect of doping elements, degradation, and applications. J Mater Chem B 11(5):955–973. https://doi.org/10.1039/D2TB02505A

Alonzo M, Primo FA, Kumar SA, Mudloff JA, Dominguez E, Fregoso G, Ortiz N, Weiss WM, Joddar B (2021) Bone tissue engineering techniques, advances and scaffolds for treatment of bone defects. Curr Opin Biomed Eng. https://doi.org/10.1016/j.cobme.2020.100248

Álvarez-Carrasco F, Varela P, Sarabia-Vallejos MA, García-Herrera C, Saavedra M, Zapata PA, Zárate-Triviño D, Martínez JJ, Canales DA (2024) Development of bioactive hybrid poly(lactic acid)/poly(methyl methacrylate) (PLA/PMMA) electrospun fibers functionalized with bioglass nanoparticles for bone tissue engineering applications. Int J Mol Sci 25(13)

Archunan MW, Petronis S (2021) Bone grafts in trauma and orthopaedics. Cureus 13(9):e17705. https://doi.org/10.7759/cureus.17705

Baino F, Fiume E (2019) Elastic mechanical properties of 45S5-based bioactive glass–ceramic scaffolds. Materials 12(19)

Baino F, Fiume E, Miola M, Verné E (2018) Bioactive sol-gel glasses: processing, properties, and applications. Int J Appl Ceram Technol 15(4):841–860. https://doi.org/10.1111/ijac.12873

Bento R, Gaddam A, Ferreira JMF (2021) Sol–gel synthesis and characterization of a quaternary bioglass for bone regeneration and tissue engineering. Materials 14(16)

Blaker JJ, Maquet V, Jérôme R, Boccaccini AR, Nazhat SN (2005) Mechanical properties of highly porous PDLLA/Bioglass® composite foams as scaffolds for bone tissue engineering. Acta Biomaterialia 1(6):643–652. https://doi.org/10.1016/j.actbio.2005.07.003

Boccaccini AR, Brauer DS, Hupa L (2016) Bioactive glasses: fundamentals, technology and applications. In: The Royal Society of Chemistry, pp 0. https://doi.org/10.1039/9781782622017-FP001

Brauer DS (2015) Bioactive glasses—structure and properties. Angewandte Chemie Int Edition 54(14):4160–4181. https://doi.org/10.1002/anie.201405310

Brito AF, Antunes B, dos Santos F, Fernandes HR, Ferreira JMF (2017) Osteogenic capacity of alkali-free bioactive glasses. In vitro studies. J Biomed Mater Res Part B Appl Biomater 105(8):2360–2365. https://doi.org/10.1002/jbm.b.33771

Cao L, Weng W, Chen X, Ding Y, Yan Y, Li H, Zhao H, Shin J-W, Wei J, Ji F, Su J (2015) Development of degradable and bioactive composite as bone implants by incorporation of mesoporous bioglass into poly(l-lactide). Compos Part B Eng 77:454–461. https://doi.org/10.1016/j.compositesb.2015.03.034

Carbajal-De la Torre G, Zurita-Méndez NN, Ballesteros-Almanza MD, Ortiz-Ortiz J, Estévez M, Espinosa-Medina MA (2022) Characterization and evaluation of composite biomaterial bioactive glass–polylactic acid for bone tissue engineering applications. Polymers 14(15)

Chaigneau L, Perrot A, Brezulier D, Coulon J-F, Chevire F, Lebullenger R (2023) Bioresorbable polylactic acid (PLA) and bioactive glasses (BG) composite: influence of gold coated of BG powder on mechanical properties and chemical reactivity. J Mech Behav Biomed Mater 138:105571. https://doi.org/10.1016/j.jmbbm.2022.105571

Cichos S, Schätzlein E, Wiesmann-Imilowski N, Blaeser A, Henrich D, Frank J, Drees P, Gercek E, Ritz U (2023) A new 3D-printed polylactic acid-bioglass composite for bone tissue engineering induces angiogenesis in vitro and in ovo. IJB 9(5). https://doi.org/10.18063/ijb.751

Cole KA, Funk GA, Rahaman MN, McIff TE (2020) Mechanical and degradation properties of poly(methyl methacrylate) cement/borate bioactive glass composites. J Biomed Mater Res Part B Appl Biomater 108(7):2765–2775. https://doi.org/10.1002/jbm.b.34606

Das EC, Komath M, Anil Kumar PR (2022) Chapter 21—dental tissue engineering. In: Sharma CP, Chandy T, Thomas V, Thankam FG (eds) Tissue engineering. Academic Press, pp 493–529. https://doi.org/10.1016/B978-0-12-824064-9.00021-6

Dash PA, Mohanty S, Nayak SK (2023) A review on bioactive glass, its modifications and applications in healthcare sectors. J Non-Cryst Solids 614:122404. https://doi.org/10.1016/j.jnoncrysol.2023.122404

Ding Y, Liu X, Zhang J, Lv Z, Meng X, Yuan Z, Long T, Wang Y (2022) 3D printing polylactic acid polymer-bioactive glass loaded with bone cement for bone defect in weight-bearing area [Original Research]. Front Bioeng Biotechnol 10. https://www.frontiersin.org/journals/bioengineering-and-biotechnology/articles/10.3389/fbioe.2022.947521

Dukle A, Murugan D, Nathanael AJ, Rangasamy L, Oh T-H (2022) Can 3D-printed bioactive glasses be the future of bone tissue engineering? Polymers 14(8)

Ege D, Zheng K, Boccaccini AR (2022) Borate bioactive glasses (BBG): bone regeneration, wound healing applications, and future directions. ACS Appl Bio Mater 5(8):3608–3622. https://doi.org/10.1021/acsabm.2c00384

Ercal P, Pekozer GG (2020) A current overview of scaffold-based bone regeneration strategies with dental stem cells. In: Turksen K (ed) Cell biology and translational medicine, volume 9: stem cell-based therapeutic approaches in disease. Springer International Publishing, pp 61–85. https://doi.org/10.1007/5584_2020_505

Ercal P, Pekozer GG, Kose GT (2018) Dental stem cells in bone tissue engineering: current overview and challenges. Cell biology and translational medicine, vol 3: stem cells. Bio-materials and tissue engineering 1107:113–127. https://doi.org/10.1007/5584_2018_171

Eriksson E, Björkenheim R, Strömberg G, Ainola M, Uppstu P, Aalto-Setälä L, Leino VM, Hupa L, Pajarinen J, Lindfors NC (2021) S53P4 bioactive glass scaffolds induce BMP expression and integrative bone formation in a critical-sized diaphysis defect treated with a single-staged induced membrane technique. Acta Biomaterialia 126:463–476. https://doi.org/10.1016/j.actbio.2021.03.035

Fiume E, Tulyaganov DU, Akbarov A, Ziyadullaeva N, Cochis A, Scalia AC, Rimondini L, Verné E, Baino F (2021) Biological evaluation of a new sodium-potassium silico-phosphate glass for bone regeneration: in vitro and in vivo studies. Materials 14(16)

Fu Q, Rahaman MN, Fu H, Liu X (2010) Silicate, borosilicate, and borate bioactive glass scaffolds with controllable degradation rate for bone tissue engineering applications. I. Preparation and in vitro degradation. J Biomed Mater Res Part A 95A(1):164–171. https://doi.org/10.1002/jbm.a.32824

Gentile P, Chiono V, Carmagnola I, Hatton PV (2014) An overview of poly(lactic-co-glycolic) acid (PLGA)-based biomaterials for bone tissue engineering. Int J Mol Sci 15(3):3640–3659

Gerhardt L-C, Widdows KL, Erol MM, Burch CW, Sanz-Herrera JA, Ochoa I, Stämpfli R, Roqan IS, Gabe S, Ansari T, Boccaccini AR (2011) The pro-angiogenic properties of multi-functional bioactive glass composite scaffolds. Biomaterials 32(17):4096–4108. https://doi.org/10.1016/j.biomaterials.2011.02.032

Han X, Wang D, Chen X, Lin H, Qu F (2014) One-pot synthesis of macro-mesoporous bioactive glasses/polylactic acid for bone tissue engineering. Mater Sci Eng C 43:367–374. https://doi.org/10.1016/j.msec.2014.07.042

Hench LL (2006) The story of Bioglass®. J Mater Sci Mater Med 17(11):967–978. https://doi.org/10.1007/s10856-006-0432-z

Hench LL, Jones JR (2015) Bioactive glasses: frontiers and challenges [review]. Front Bioeng Biotechnol 3. https://www.frontiersin.org/journals/bioengineering-and-biotechnology/articles/10.3389/fbioe.2015.00194

Hild N, Tawakoli PN, Halter JG, Sauer B, Buchalla W, Stark WJ, Mohn D (2013) pH-dependent antibacterial effects on oral microorganisms through pure PLGA implants and composites with nanosized bioactive glass. Acta Biomaterialia 9(11):9118–9125. https://doi.org/10.1016/j.actbio.2013.06.030

Hong Z, Reis RL, Mano JF (2008) Preparation and in vitro characterization of scaffolds of poly(l-lactic acid) containing bioactive glass ceramic nanoparticles. Acta Biomaterialia 4(5):1297–1306. https://doi.org/10.1016/j.actbio.2008.03.007

Huang W, Day DE, Kittiratanapiboon K, Rahaman MN (2006) Kinetics and mechanisms of the conversion of silicate (45S5), borate, and borosilicate glasses to hydroxyapatite in dilute phosphate solutions. J Mater Sci Mater Med 17(7):583–596. https://doi.org/10.1007/s10856-006-9220-z

Kaou MH, Furkó M, Balázsi K, Balázsi C (2023) Advanced bioactive glasses: the newest achievements and breakthroughs in the area. Nanomaterials 13(16)

Kargozar S, Baino F, Hamzehlou S, Hill RG, Mozafari M (2018) Bioactive glasses entering the mainstream. Drug Discov Today 23(10):1700–1704. https://doi.org/10.1016/j.drudis.2018.05.027

Kaur G (2017) Bioactive glasses: potential biomaterials for future therapy. Bioact Glasses Potential Biomater Future Ther. https://doi.org/10.1007/978-3-319-45716-1

Khouri NG, Bahú JO, Blanco-Llamero C, Severino P, Concha VOC, Souto EB (2024) Polylactic acid (PLA): properties, synthesis, and biomedical applications—a review of the literature. J Mol Struct 1309:138243. https://doi.org/10.1016/j.molstruc.2024.138243

Kim H-W, Lee H-H, Chun G-S (2008) Bioactivity and osteoblast responses of novel biomedical nanocomposites of bioactive glass nanofiber filled poly(lactic acid). J Biomed Mater Res Part A 85A(3):651–663. https://doi.org/10.1002/jbm.a.31339

Lasprilla AJR, Martinez GAR, Lunelli BH, Jardini AL, Filho RM (2012) Poly-lactic acid synthesis for application in biomedical devices—a review. Biotechnol Adv 30(1):321–328. https://doi.org/10.1016/j.biotechadv.2011.06.019

Lee S, Matsugaki A, Kasuga T, Nakano T (2019) Development of bifunctional oriented bioactive glass/poly(lactic acid) composite scaffolds to control osteoblast alignment and proliferation. J Biomed Mater Res Part A 107(5):1031–1041. https://doi.org/10.1002/jbm.a.36619

Li Y, Chen L, Chen X, Hill R, Zou S, Wang M, Liu Y, Wang J, Chen X (2021) High phosphate content in bioactive glasses promotes osteogenesis in vitro and in vivo. Dent Mater 37(2):272–283. https://doi.org/10.1016/j.dental.2020.11.017

Liu X, Rahaman MN, Day DE (2013) Conversion of melt-derived microfibrous borate (13–93B3) and silicate (45S5) bioactive glass in a simulated body fluid. J Mater Sci Mater Med 24(3):583–595. https://doi.org/10.1007/s10856-012-4831-z

Liu H, Shi S, Cao J, Ji L, He Y, Xi J (2015) Preparation and evaluation of a novel bioactive glass/lysozyme/PLGA composite microsphere (1520–5762 (Electronic))

Maquet V, Boccaccini AR, Pravata L, Notingher I, Jérôme R (2004) Porous poly(α-hydroxyacid)/Bioglass® composite scaffolds for bone tissue engineering. I: preparation and in vitro characterisation. Biomaterials 25(18):4185–4194. https://doi.org/10.1016/j.biomaterials.2003.10.082

Menci PF, Mari A, Charbonneau C, Lefebvre L-P, De Nardo L (2019) Aging of bioactive glass-based foams: effects on structure, properties, and bioactivity. Materials 12(9)

Najeeb S, Khurshid Z, Ghabbani H, Zafar MS, Sefat F (2019) Nano glass ionomer cement: modification for biodental applications. Adv Dent Biomater. https://doi.org/10.1016/B978-0-08-102476-8.00010-4

Negut I, Ristoscu C (2023) Bioactive glasses for soft and hard tissue healing applications—a short review. Appl Sci 13(10)

O'Donnell MD, Watts SJ, Hill RG, Law RV (2009) The effect of phosphate content on the bioactivity of soda-lime-phosphosilicate glasses. J Mater Sci Mater Med 20(8):1611–1618. https://doi.org/10.1007/s10856-009-3732-2

Pant S, Thomas S, Loganathan S, Valapa RB (2022) 3D bioprinted poly(lactic acid)/mesoporous bioactive glass based biomimetic scaffold with rapid apatite crystallization and in-vitro Cytocompatability for bone tissue engineering. Int J Biol Macromol 217:979–997. https://doi.org/10.1016/j.ijbiomac.2022.07.202

Pawlik J, Dziadek M, Cholewa-Kowalska K, Osyczka AM (2020) The synergistic effect of combining the bioactive glasses with polymer blends on biological and material properties. J Am Ceram Soc 103(8):4558–4572. https://doi.org/10.1111/jace.17131

Qiao K, Xu L, Tang J, Wang Q, Lim KS, Hooper G, Woodfield TBF, Liu G, Tian K, Zhang W, Cui X (2022) The advances in nanomedicine for bone and cartilage repair. J Nanobiotechnology 20(1):141. https://doi.org/10.1186/s12951-022-01342-8
Samudrala RK, Patel S, Penugurthi V, Manavathi B, Azeem A (2021) In vitro studies of B_2O_3–SiO_2–Na_2O–CaO–ZnO bioactive glass system. J Non Cryst Solids 574(121164):121164. https://doi.org/10.1016/j.jnoncrysol.2021.121164
Schatkoski VM, Larissa do Amaral Montanheiro T, Canuto de Menezes BR, Pereira RM, Rodrigues KF, Ribas RG, Morais da Silva D, Thim GP (2021) Current advances concerning the most cited metal ions doped bioceramics and silicate-based bioactive glasses for bone tissue engineering. Ceram Int 47(3):2999–3012. https://doi.org/10.1016/j.ceramint.2020.09.213
Sengupta S, Michalek M, Liverani L, Švančárek P, Boccaccini AR, Galusek D (2021) Preparation and characterization of sintered bioactive borate glass tape. Mater Lett 282:128843. https://doi.org/10.1016/j.matlet.2020.128843
Shearer A, Montazerian M, Mauro JC (2023) Modern definition of bioactive glasses and glass-ceramics. J Non-Cryst Solids 608:122228. https://doi.org/10.1016/j.jnoncrysol.2023.122228
Singhvi MS, Zinjarde SS, Gokhale DV (2019) Polylactic acid: synthesis and biomedical applications. J Appl Microbiol 127(6):1612–1626. https://doi.org/10.1111/jam.14290
Skallevold HE, Rokaya D, Khurshid Z, Zafar MS (2019) Bioactive glass applications in dentistry. Int J Mol Sci 20(23). https://doi.org/10.3390/ijms20235960
Sultan S, Thomas N, Varghese M, Dalvi Y, Joy S, Hall S, Mathew AP (2022) The design of 3D-printed polylactic acid–bioglass composite scaffold: a potential implant material for bone tissue engineering. Molecules 27(21)
Tajbakhsh S, Hajiali F (2017) A comprehensive study on the fabrication and properties of biocomposites of poly(lactic acid)/ceramics for bone tissue engineering. Mater Sci Eng C 70:897–912. https://doi.org/10.1016/j.msec.2016.09.008
Thyparambil NJ, Gutgesell LC, Hurley CC, Flowers LE, Day DE, Semon JA (2020) Adult stem cell response to doped bioactive borate glass. J Mater Sci Mater Med 31(2):13. https://doi.org/10.1007/s10856-019-6353-4
Di Tinco R, Sergi R, Bertani G, Pisciotta A, Bellucci D, Carnevale G, Cannillo V, Bertoni L (2020) Effects of a novel bioactive glass composition on biological properties of human dental pulp stem cells. Materials 13(18)
Tyler B, Gullotti D, Mangraviti A, Utsuki T, Brem H (2016) Polylactic acid (PLA) controlled delivery carriers for biomedical applications. Adv Drug Delivery Rev 107:163–175. https://doi.org/10.1016/j.addr.2016.06.018
Vergnol G, Ginsac N, Rivory P, Meille S, Chenal J-M, Balvay S, Chevalier J, Hartmann DJ (2016) In vitro and in vivo evaluation of a polylactic acid-bioactive glass composite for bone fixation devices. J Biomed Mater Res Part B Appl Biomater 104(1):180–191. https://doi.org/10.1002/jbm.b.33364
Xiao W, Zaeem MA, Li G, Bal BS, Rahaman MN (2017) Tough and strong porous bioactive glass-PLA composites for structural bone repair (0022–2461 (Print))
Xue N, Ding X, Huang R, Jiang R, Huang H, Pan X, Min W, Chen J, Duan JA, Liu P, Wang Y (2022) Bone tissue engineering in the treatment of bone defects. Pharmaceuticals (Basel) 15(7). https://doi.org/10.3390/ph15070879
Yao A, Wang D, Huang W, Fu Q, Rahaman MN, Day DE (2007) In vitro bioactive characteristics of borate-based glasses with controllable degradation behavior. J Am Ceram Soc 90(1):303–306. https://doi.org/10.1111/j.1551-2916.2006.01358.x
Ylänen HO (2011) Bioactive glasses materials, properties and applications introduction. Bioactive glasses: materials, properties and applications, Xiii–Xiv. https://doi.org/10.1533/9780857093318
Zafar MS, Khurshid Z, Almas K (2015) Oral tissue engineering progress and challenges. Tissue Eng Regenerative Med 12(6):387–397. https://doi.org/10.1007/s13770-015-0030-6
Zhu G, Zhang T, Chen M, Yao K, Huang X, Zhang B, Li Y, Liu J, Wang Y, Zhao Z (2021) Bone physiological microenvironment and healing mechanism: basis for future bone-tissue engineering scaffolds. Bioact Mater 6(11):4110–4140. https://doi.org/10.1016/j.bioactmat.2021.03.043

Chapter 4
Rubber-Based Nanoblends: Biomedical Applications

Patrick Ehi Imoisili

Abstract Rubber-based nanoblends are a novel family of materials that combine the improved mechanical, thermal, and barrier qualities of nanoscale fillers with the natural elasticity, flexibility, and resilience of rubber. The manufacture, characteristics, and uses of rubber nanoblends are examined in this chapter, focusing on how they can get beyond the conventional drawbacks of rubber materials. To achieve better performance, it explores the use of several nanofillers, including carbon nanotubes, graphene, silica, and clay, which are distributed across natural and synthetic rubber matrices. This chapter emphasizes the importance of advanced processes such as melt blending, solution casting, in situ polymerization, blend morphology, interfacial adhesion, and filler dispersion in influencing the material's overall qualities. The discussion of rubber nanoblends' sustainability issues and environmental effects also includes a look at biodegradable nanofillers, recycling methods, and renewable rubber sources. These materials' revolutionary potential is demonstrated by their applications in the automotive, aerospace, medicinal, and energy sectors. Furthermore, the chapter outlines potential avenues for future research that will maximize multifunctional qualities, lessen the environmental effects, and open up new applications for rubber-based nanoblends in high-performance settings.

Keywords Rubber-based nanoblends · Rubber elasticity · Nanoblends · Nanofillers · Nanocomposites · Mechanical properties

4.1 Introduction to Rubber

Rubber is a robust and flexible substance necessary for everyday living and modern industries. Rubber, well-known for its excellent elasticity, resilience, and water resistance, has a long history and a substantial economic impact (Sisanth et al. 2017; Rodgers and Waddell 2005; Song 2017). Its discovery and advancement changed

P. E. Imoisili (✉)
Mechanical Engineering Science Deoartment, University of Johannesburg, Johannesburg, South Africa
e-mail: pimoisili@uj.ac.za

P. M. Visakh (ed.), *Engineering Applications of Polymer based Nano Blends*, Advanced Structured Materials 235, https://doi.org/10.1007/978-3-031-90979-5_4

the automotive and medical sectors, and it is still essential in various uses today, from tires to medical equipment. Rubber comes in two main varieties: natural and synthetic. Each has unique qualities and uses (Rodgers and Waddell 2005; Song 2017).

4.1.1 Historical Background of Rubber

The history of rubber begins centuries ago with South American indigenous groups who made impermeable shoes and garments out of the milky latex of Hevea brasiliensis plants in the Amazon Basin. When British chemist Joseph Priestley found that rubber could "rub out" pencil lines in 1770, the term "rubber" was born. The need for rubber increased in the nineteenth century as industrialization spread, particularly after Charles Goodyear discovered vulcanization in 1839. Vulcanization transformed rubber, a chemical process that improves flexibility, heat resistance, and durability (Ikeda et al. 2018; Kohjiya and Ikeda 2021; Kauffman 2001). This made rubber appropriate for more robust uses, such as tires and machinery parts. The creation of massive rubber plantations in Southeast Asia was sparked by the exponential surge in demand for rubber caused by the early 20th-century car boom. This demand brought about significant economic opportunities, but it also resulted in problems with labor exploitation and deforestation, which still affect the rubber sector today (Kohjiya and Ikeda 2021; Kauffman 2001; Guerra et al. 2021).

4.1.2 Types of Rubber

Rubber is vital to many industries because of its flexibility, durability, and resilience. It is available in synthetic and natural forms, each with special qualities that make it appropriate for a given use. There are two primary types of rubber: natural and synthetic. Based on their distinct qualities, each type is suitable for a specific use (Rodgers and Waddell 2005; Song 2017; Ikeda et al. 2018; Kohjiya and Ikeda 2021; Kauffman 2001).

4.1.3 Natural Rubber

Natural rubber (NR) is a multipurpose, organic polymer that is mostly made from the latex of the tree Hevea brasiliensis, also known as the rubber tree. Its remarkable qualities—such as strong elasticity, tensile strength, and resilience—have led to its use for ages. Consumer goods, manufacturing, automotive, and medical are just a few of the industries that depend on this substance. Let us take a closer look at NR's properties, manufacturing methods, benefits, drawbacks, and uses (Guerra et al. 2021;

Mooibroek and Cornish 2000; Venkatachalam et al. 2013; Bokobza 2018). NR is primarily made of cis-1,4-polyisoprene molecules, which give it inherent flexibility and elasticity. Its molecular structure allows the material to stretch significantly under force and then return to its original shape, a property known as elasticity. NR also has excellent tensile strength, resilience, and flexibility, which make it suitable for applications where durability and flexibility are paramount. NR's exceptional elasticity, which enables it to expand considerably without breaking and then regain its original shape, and its resistance to abrasion, which makes it perfect for high-wear items like tires, are two of its key characteristics. A further factor in its endurance is its high tensile strength, which allows it to tolerate significant strain. NR's comparatively low heat conductivity also makes it an excellent thermal insulator (Mooibroek and Cornish 2000; Venkatachalam et al. 2013; Bokobza 2018; Cruz-Morales et al. 2023).

Harvesting latex from rubber trees is the first step in the manufacturing of NR. Tappers cut into the tree bark, releasing a thick, milky substance called latex (Fig. 4.1). After collecting, the latex is subjected to coagulation, which thickens it into a solid state by treating it with acids. After that comes the processing step, in which the coagulated latex is dried and rolled into sheets. The rubber is then frequently heated with sulfur to vulcanize it, a technique that Charles Goodyear discovered. By creating cross-links between polymer chains, vulcanization improves the rubber's flexibility, resilience to temperature fluctuations, and durability (Kohjiya and Ikeda 2021; Kauffman 2001; Guerra et al. 2021; Mooibroek and Cornish 2000; Venkatachalam et al. 2013; Bokobza 2018; Cruz-Morales et al. 2023).

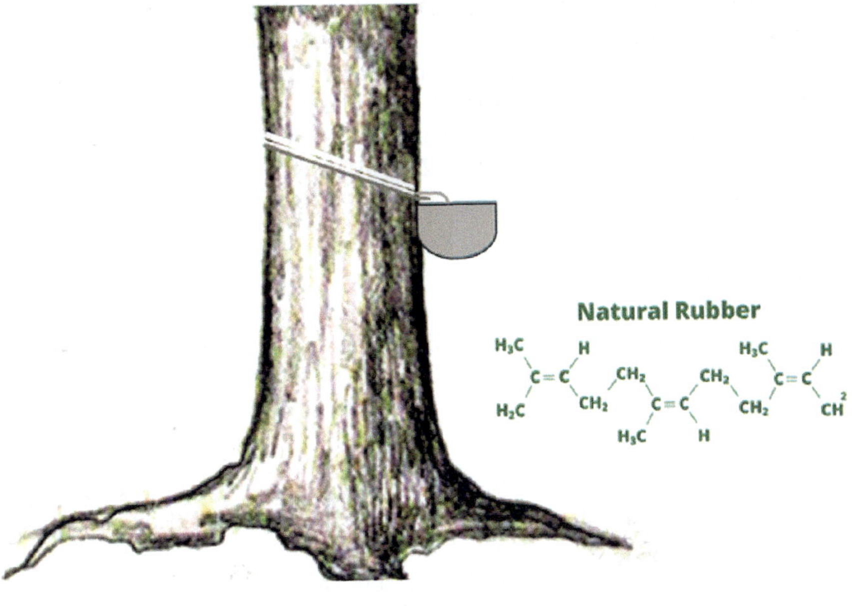

Fig. 4.1 Schematic of the natural rubber tree

Advantages of NR

- **High Elasticity and Flexibility**: NR is perfect for products that move a lot since it can be stretched and bent without breaking or cracking (Venkatachalam et al. 2013; Bokobza 2018).
- **Good Resilience and Shock Absorption**: It is appropriate for shock-absorbing applications since it rapidly regains its shape following compression (Mooibroek and Cornish 2000; Venkatachalam et al. 2013; Bokobza 2018; Cruz-Morales et al. 2023).
- **Tensile Strength and Durability**: NR can tolerate a lot of stress without tearing because of its exceptional tensile strength (Cruz-Morales et al. 2023; Roy et al. 2021).
- **Abrasion Resistance**: Because of its strong abrasion resistance, it is long-lasting for uses like conveyor belts and tires (Bokobza 2018; Cruz-Morales et al. 2023; Ansari et al. 2021; Roy et al. 2021).

Disadvantages of NR

Although NR has numerous benefits, it also has certain drawbacks.

- **Poor Resistance to Heat, Ozone, and UV Exposure**: NR's outdoor applications may be limited due to its degradation in hot, sunny, or ozone-rich environments (Cruz-Morales et al. 2023; Ansari et al. 2021; Roy et al. 2021).
- **Susceptibility to Oil and Chemical Degradation**: NR is susceptible to chemical and oil degradation when it comes into touch with oils, greases, and certain chemicals (Roy et al. 2021; Sethulekshmi et al. 2022).
- **Limited Temperature Resistance**: NR functions best in a moderate temperature range and may become less elastic in extremely cold or deteriorate in really hot temperatures (Cruz-Morales et al. 2023; Ansari et al. 2021; Roy et al. 2021; Sethulekshmi et al. 2022).

Applications of NR

NR is widely employed in many industries because of its qualities, especially in applications with crucial durability, flexibility, and shock absorption (Roy et al. 2021; Sethulekshmi et al. 2022; Kurian and Mathew 2011). Among the principal applications are:

- **Tires for cars**: One of the most significant applications for NR is in tire manufacturing, where its resistance and longevity are crucial (Roy et al. 2021; Sethulekshmi et al. 2022; Kurian and Mathew 2011).
- **Footwear**: Because NR is comfortable, flexible, and non-slip, it is frequently used in shoe soles (Roy et al. 2021; Kurian and Mathew 2011).

- **Industrial Belts and Hoses**: NR is perfect for conveyor belts, hoses, and other equipment parts because of its strength and flexibility (Ansari et al. 2021; Roy et al. 2021; Sethulekshmi et al. 2022; Kurian and Mathew 2011).
- **Medical Equipment**: NR is strong and flexible, and it is frequently used to make latex gloves, catheters, and other medical equipment (Sethulekshmi et al. 2022; Kurian and Mathew 2011; Singha et al. 2019).
- **Shock Absorbers and Seals**: It is appropriate for gaskets, seals, and other shock-absorbing items due to their capacity to absorb shock and vibration (Sethulekshmi et al. 2022; Kurian and Mathew 2011; Singha et al. 2019).

4.1.4 Synthetic Rubbers

A class of elastomers known as synthetic rubbers is created artificially to mimic or enhance the qualities of NR. Synthetic rubbers were first created to address NR shortages, particularly during times of war, and were first introduced in the early twentieth century. Due to their adaptability and variety of qualities that may be customized for specific uses, they are now essential to many different sectors (Shah et al. 2013; Boon et al. 2022). Every variety of synthetic rubber has unique characteristics that make it appropriate for particular commercial, consumer, or industrial applications. The primary synthetic rubber kinds' history, varieties, traits, and uses are examined below, offering insight into their crucial function in contemporary technology and production (Kurian and Mathew 2011; Singha et al. 2019; Shah et al. 2013; Boon et al. 2022).

Historical Development of Synthetic Rubbers

The concept of synthetic rubber was first proposed in the 1900s, and it saw significant development during World War I when NR supplies were restricted or stopped. To replace NR, particularly for military uses, early synthetic rubber varieties, including styrene-butadiene Rubber, were created in Germany and USA. By improving elasticity and durability through cross-linking with sulfur, Charles Goodyear's revolutionary 19th-century discovery of vulcanization also paved the way for advancements in natural and synthetic rubbers. The development of polymer science throughout time has produced a vast variety of synthetic rubbers, each with unique qualities designed for uses (Kauffman 2001; Shah et al. 2013; Boon et al. 2022; Norton 2021).

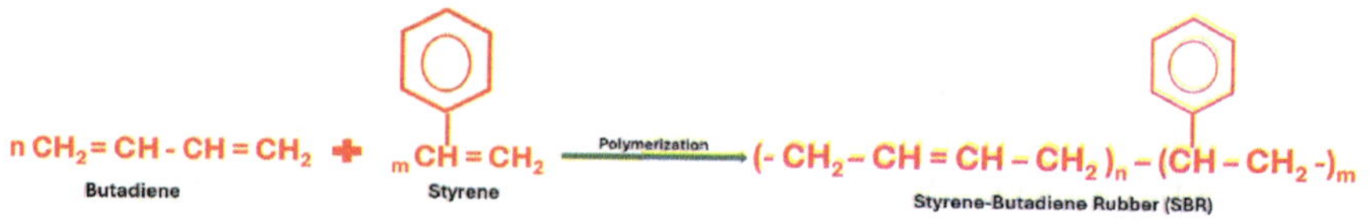

Fig. 4.2 Schematic of SBR production

4.1.5 Key Types of Synthetic Rubber

Styrene-Butadiene Rubber

One of the most popular synthetic rubbers is styrene-butadiene rubber (SBR), which is made by copolymerizing butadiene and styrene (Fig. 4.2). It is similar to NR in many ways, including flexibility and durability, but it is better at resisting abrasion, which makes it appropriate for high-wear applications (Norton 2021; Dhanorkar et al. 2021). Since SBR offers superior road traction and lifespan, it is frequently used in tire manufacture. It is frequently used with NR to balance price and performance in tire formulations. Conveyor belts, shoe soles, and other molded rubber products frequently use SBR due to its strength and flexibility (Boon et al. 2022; Norton 2021; Dhanorkar et al. 2021; Yang et al. 2021). Applications of SBR include gaskets, hoses, conveyor belts, shoe bottoms, and car tires.

Nitrile Rubber (NBR)

Synthetic rubber made by polymerizing butadiene and acrylonitrile is called nitrile rubber or Buna-N (Fig. 4.3). Due to its remarkable resistance to chemicals, oils, and fuels, it is highly prized and perfect for usage in the automotive and aerospace sectors. NBR is a popular option for protective gloves and seals in chemical and oil handling applications due to its resistance to severe environments. Manufacturers can tailor NBR goods to meet needs by varying the acrylonitrile in the material to improve its flexibility or oil resistance (Abdullah et al. 2022; Zou et al. 2021). They are used in the production of gloves for handling chemicals and fuel, gaskets, oil seals, and fuel hoses (Zou et al. 2021; Bouaziz et al. 2020).

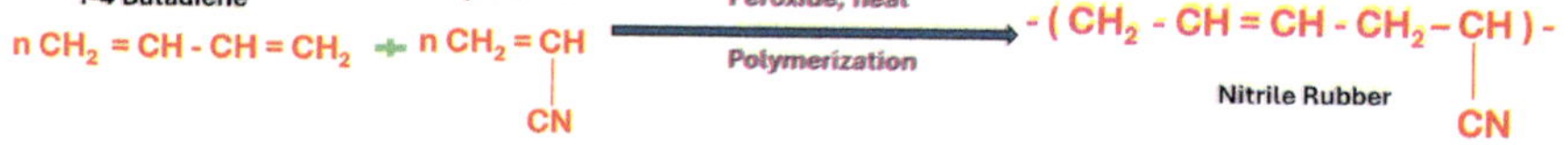

Fig. 4.3 Schematic of NBR production

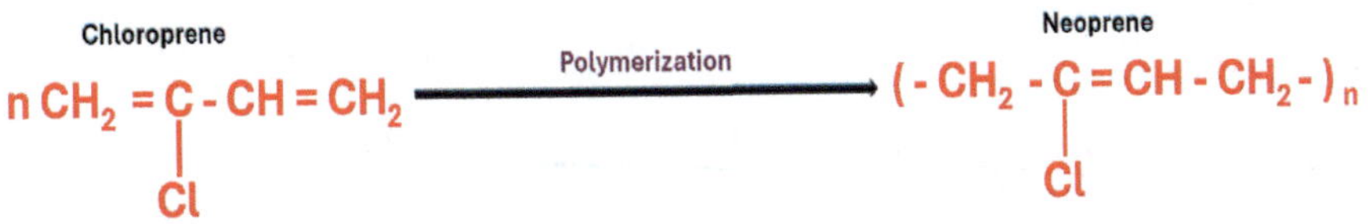

Fig. 4.4 Schematic of neoprene production

Neoprene (Polychloroprene)

Renowned for its remarkable resistance to heat, chemicals, oils, and weather, it is a multipurpose synthetic rubber called neoprene. It is produced by the polymerizing the chloroprene monomers called neoprene (Fig. 4.4). Neoprene is a common material in sectors that need long-lasting rubber that can survive harsh conditions because of its strength and stability (Bouaziz et al. 2020). It is utilized in various products, such as insulation, industrial hoses, and wetsuits. In addition to its use in the electronics, automotive, and construction sectors, neoprene is especially well-suited for maritime applications due to its insulating qualities, flexibility, and resilience to water (Bouaziz et al. 2020; Alazhary et al. 2024). Hoses, insulation materials, automotive components, wetsuits, and weather-resistant seals are some of the applications of neoprene rubber (Bouaziz et al. 2020; Alazhary et al. 2024).

Butyl (Isobutylene) Rubber

Isobutylene and a trace amount of isoprene copolymerize to form butyl rubber (IIR) (Fig. 4.5), which has unique qualities like superior gas impermeability. Because of this feature, it is especially useful in applications that need to retain air, such as tire liners and inner tubes (Sharma et al. 2021; Behera et al. 2022). Low moisture absorption, elasticity at low temperatures, and strong chemical resistance are further qualities of butyl rubber. Because it is airtight, it is commonly utilized in protective gloves, sealing applications, and medicinal bottle stoppers. For more rigorous uses, halogenated butyl rubber—a modified form with improved aging and chemical resistance is also offered (Sharma et al. 2021; Behera et al. 2022). Applications of IIR include, protective gloves, inner tubes, tire liners, pharmaceutical closures, and sealing applications (Shah et al. 2013; Sharma et al. 2021; Behera et al. 2022).

Ethylene Propylene Diene Monomer

- Synthetic rubber made of ethylene propylene diene monomer (EPDM) (Fig. 4.6) is renowned for its exceptional resistance to weathering, heat, and UV radiation. It is frequently used in outdoor applications like weather seals, roofing membranes, and auto parts (Hu et al. 2023). EPDM is an excellent material for outdoor construction because of its high resistance to sunshine and ozone, and it

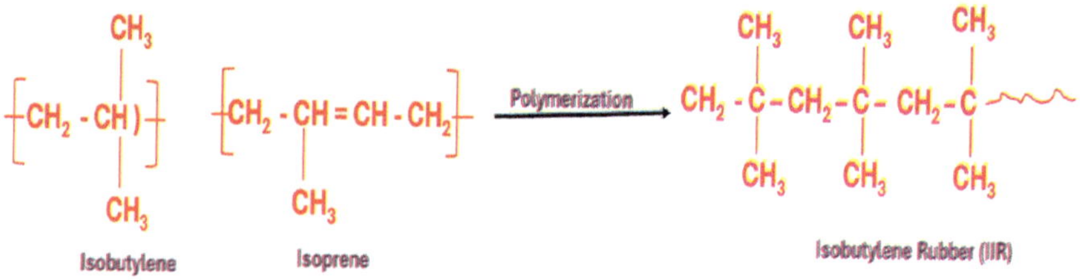

Fig. 4.5 Schematic of Isobutylene Rubber Production

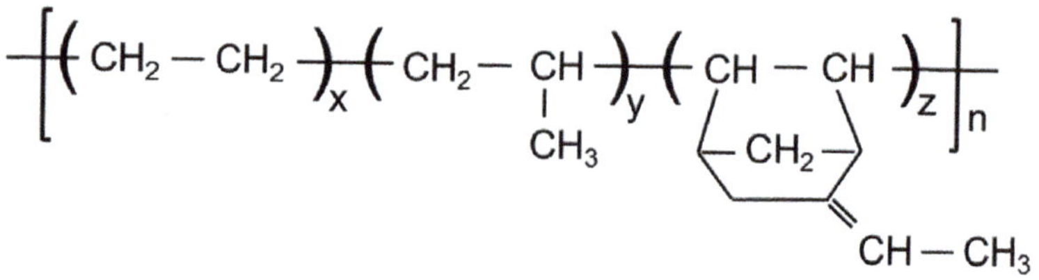

Fig. 4.6 Structure of ethylene propylene diene monomer

may be used in cooling system hoses and vehicle weather stripping because of its flexibility and durability across a wide temperature range (Costa et al. 2024). EPDM is also water and chemical resistant, which makes it appropriate for industrial settings where mild chemicals and moisture are present (Hu et al. 2023; Costa et al. 2024). EPDM finds applications in hoses, weatherproofing materials, vehicle weather seals, and roofing membranes (Boon et al. 2022; Hu et al. 2023; Costa et al. 2024).

Silicone Rubber

Among synthetic rubbers, silicone rubber is unique because it is highly flexible and resistant to temperature, staying stable in temperatures ranging from −55 °C to over 200 °C. Because of its exceptional resistance to UV light, chemicals, and water, it can be used in high-precision and high-temperature applications (Han et al. 2022). The automobile business uses silicone rubber for gaskets and seals that must endure high temperatures, the medical industry uses it extensively for products like implants and tubing, and the kitchen uses it in baking molds. Electrical insulation, cookware, medical equipment, vehicle seals, and aerospace parts are some of the applications of silicon rubber (Han et al. 2022; Kumar et al. 2021) (Fig. 4.7).

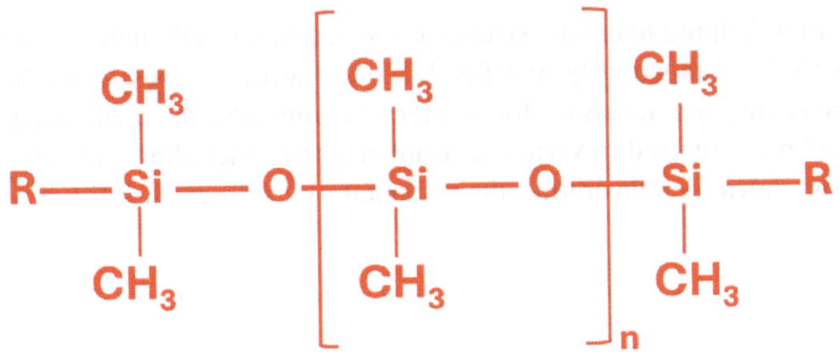

Fig. 4.7 Structure of silicone rubber

Advantages of Synthetic Rubber

Compared to NR, synthetic rubber has several benefits that make it better suited for particular uses (Shah et al. 2013; Boon et al. 2022; Norton 2021; Dhanorkar et al. 2021; Yang et al. 2021).

- **Increased Chemical Resistance**: Many synthetic rubbers, including neoprene and NBR, have a high level of resistance to fuels, lubricants, and chemicals, which makes them perfect for use in industrial and automotive settings (Boon et al. 2022; Norton 2021; Dhanorkar et al. 2021).
- **Temperature Stability**: Rubbers that can withstand a wide range of temperatures, such as silicone and EPDM, can be used in extremely hot and cold environments (Shah et al. 2013; Boon et al. 2022; Norton 2021).
- **Increased Durability and Longevity**: Synthetic rubbers often have a longer lifespan in demanding conditions because they are more resilient to weathering, UV radiation, and ozone deterioration (Boon et al. 2022; Norton 2021; Dhanorkar et al. 2021; Yang et al. 2021; Abdullah et al. 2022).
- **Customization**: By modifying their chemical makeup, synthetic rubbers can be designed to satisfy particular property needs, providing more design flexibility (Norton 2021; Dhanorkar et al. 2021; Yang et al. 2021; Abdullah et al. 2022; Zou et al. 2021; Bouaziz et al. 2020; Alazhary et al. 2024; Sharma et al. 2021; Behera et al. 2022).

Applications and Impact of Synthetic Rubbers on Industry

Numerous industries have transformed thanks to synthetic rubbers, which offer materials that satisfy specific needs for flexibility, durability, and resilience to harsh environments. The automotive industry depends on synthetic rubbers for tires, hoses, and seals that must withstand high temperatures and oil exposure (Shah et al. 2013; Boon et al. 2022; Norton 2021; Dhanorkar et al. 2021; Yang et al. 2021; Abdullah et al. 2022). EPDM is frequently used in roofing and waterproofing applications in construction, helping to create weather-resistant, energy-efficient structures. Silicone rubber's biocompatibility and flexibility are advantageous to the healthcare sector because they are necessary for prosthetics, implants, and medical equipment. Synthetic rubbers are used in various consumer goods, including wetsuits, culinary utensils, and shoes, demonstrating their adaptability (Hu et al. 2023; Costa et al. 2024).

4.2 Introduction to Rubber Blends

Rubber's elasticity, resistance, and durability make it one of the most adaptable materials in many industries. However, certain synthetic or NR varieties could not have the particular qualities needed for a given use. Rubber blending has become a crucial material engineering technique to overcome these constraints (Vayyaprontavida Kaliyathan et al. 2020; Sidkey et al. 2002). Composite materials known as rubber blends are made by mixing two or more varieties of rubber to produce a unique combination of qualities. Manufacturers can customize materials to maximize performance, satisfy budgetary constraints, and even enhance sustainability by combining rubbers (Kaliyathan et al. 2020). This session thoroughly explains how rubber blending improves rubber's usefulness and adaptability in various sectors by examining the types, applications, advantages, processes, and purposes of rubber mixes (Vayyaprontavida Kaliyathan et al. 2020; Sidkey et al. 2002; Kaliyathan et al. 2020).

4.2.1 Purpose of Rubber Blending

Utilizing rubber blending, engineers may use the most outstanding qualities of each variety of rubber to create materials with precisely balanced attributes. Enhancing material performance is rubber mixes' primary goal, enabling products to satisfy the unique requirements of many applications. For instance, NR has superior tensile strength and suppleness, but it is not resistant to chemicals, oils, or extremely high or low temperatures. By combining NR with synthetic rubber, such as SBR or nitrile rubber (NBR), these shortcomings are made up for, resulting in a composite that is more elastic and has better chemical resistance (Vayyaprontavida Kaliyathan et al. 2020; Sidkey et al. 2002; Kaliyathan et al. 2020). Blending is another useful tactic for controlling production costs without compromising output. Although they cost a lot, some synthetic rubbers have advantageous properties like resistance to heat or chemicals. These can be blended with less expensive varieties, such as NR, to create a balanced material at a reduced price. Additionally, as environmental responsibility and sustainability get more attention, rubber blends that improve durability, use fewer resources, and produce less waste have been developed (Sidkey et al. 2002; Kaliyathan et al. 2020).

The possibilities of rubber materials have been transformed by rubber blending, which allows producers to produce compounds that are precisely tailored to the needs of contemporary applications. Mixing natural and synthetic rubbers can improve strength, flexibility, durability, and resistance to chemicals and environmental factors (Venkatachalam et al. 2013; Bokobza 2018; Cruz-Morales et al. 2023; Ansari et al. 2021; Roy et al. 2021; Sethulekshmi et al. 2022; Kurian and Mathew 2011; Singha et al. 2019; Shah et al. 2013; Boon et al. 2022; Norton 2021; Dhanorkar et al. 2021; Yang et al. 2021). Because of their adaptability, rubber mixes are essential to various sectors, including consumer goods, healthcare, and the automotive and aerospace industries. Rubber blending will become increasingly crucial as material science develops to create high-performance, sustainable materials that satisfy changing societal and technological demands (Vayyaprontavida Kaliyathan et al. 2020; Sidkey et al. 2002; Kaliyathan et al. 2020).

4.2.2 Types of Rubber Blends

The choice of blend depends on the desired properties, such as flexibility, durability, thermal resistance, or chemical resilience. Rubber blends are widely used in applications ranging from automotive and aerospace components to medical devices and consumer goods, demonstrating their versatility and importance in modern material science. Numerous rubber blends with distinct qualities suited for uses have been made possible by the wide range of natural and synthetic rubbers that are currently accessible (Vayyaprontavida Kaliyathan et al. 2020; Sidkey et al. 2002; Kaliyathan et al. 2020). Table 4.1 outlines the commercially available rubber blends and the type of polymer involved.

4.2.3 Benefits of Rubber Blend

Rubber blending improves rubber's applicability for a variety of applications by providing several advantages that address both technical and financial factors:

- **Improved Mechanical Properties**: Better mechanical properties: Rubber blends give producers the perfect mix of durability, strength, flexibility, and resistance to wear. SBR can improve NR's abrasion resistance, for instance, making it a more durable material for high-wear applications (Sidkey et al. 2002; Kaliyathan et al. 2020).
- **Enhanced Resistance to Chemicals and Environment**: Synthetic rubbers, such as NBR and EPDM, provide durability against weathering, chemicals, oils, and ultraviolet radiation. Combining them with other synthetic or NR makes the material more adaptable and resilient to adverse conditions (Dhanorkar et al. 2021; Zou et al. 2021; Kaliyathan et al. 2020).

Table 4.1 Rubber blends, polymers, and applications (Tang et al. 2024; Barghamadi et al. 2020; Lee et al. 2021; Khanra et al. 2020; Waddell and Tsou 2004)

Rubber blends	Type of polymers used	
Natural rubber (NR)/Styrene-butadiene rubber (SBR)	Natural rubber	SBR
Nitrile rubber (NBR)/Polyvinyl chloride (PVC)	Nitrile rubber	PVC
Ethylene propylene diene monomer (EPDM)/ Natural rubber (NR)	EPDM	Natural rubber
Chloroprene rubber (CR)/Natural rubber (NR	Chloroprene rubber	Natural rubber
Silicone rubber/Fluoro-elastomers	Silicone rubber	Fluoro-elastomers
Thermoplastic elastomer (TPE)/ Polypropylene (PP)	Thermoplastic elastomer	Polypropylene
Butyl rubber/Polyisobutylene	Butyl rubber	Polyisobutylene
Natural rubber (NR)/Polybutadiene rubber (BR)	Natural rubber	Polybutadiene rubber
Styrene-butadiene rubber (SBR)/Polyvinyl acetate (PVA)	SBR	Polyvinyl acetate
Silicone rubber/Natural rubber	Silicone rubber	Natural rubber

- **Cost optimization**: Specialty rubbers are typically more expensive than NR and some synthetic rubbers, such as SBR. Manufacturers can lower material prices while maintaining performance criteria by blending them with more costly rubbers, reducing the cost of products for producers and customers (Dhanorkar et al. 2021; Yang et al. 2021; Abdullah et al. 2022; Zou et al. 2021; Sidkey et al. 2002).
- **Customization and Flexibility**: Rubber blending enables producers to modify materials to meet the requirements of a given application, including elasticity, tensile strength, and temperature resistance. Blends can be customized to create products with specific properties appropriate for consumer goods, automotive, aerospace, and healthcare industries (Vayyaprontavida Kaliyathan et al. 2020; Sidkey et al. 2002; Kaliyathan et al. 2020).
- **Sustainability**: By extending product life, rubber blends that improve durability and resistance to deterioration help reduce waste and the need for frequent replacements. This has positive environmental effects because it saves raw materials and reduces the carbon footprint linked to the manufacture and disposal of rubber (Dhanorkar et al. 2021; Sidkey et al. 2002; Kaliyathan et al. 2020).

4.2.4 Rubber Blending Techniques

In creating innovative materials, rubber blending is an essential step that allows several kinds of rubber to be combined or fillers, nanoparticles, or additives to be added for specific qualities. The blending procedure greatly influences the material's ultimate performance properties, such as its mechanical strength, elasticity, thermal stability, and chemical resistance. This section examines the most widely used methods for blending rubber, emphasizing their benefits, uses, and guiding principles (Vayyaprontavida Kaliyathan et al. 2020). The development of goods that satisfy specific performance standards across industries is made possible by rubber blending procedures, which are at the core of material innovation. These procedures guarantee exact control over the material's properties, ranging from mechanical mixing for conventional rubber blends to more sophisticated approaches like dynamic vulcanization and nanoparticle dispersion (Sidkey et al. 2002; Whba et al. 2024; Mohite et al. 2022). The compatibility of rubber kinds, the intended use, and the production scale all influence the selection of the blending procedure. Rubber blending is changing due to ongoing technological breakthroughs to produce materials that are more effective and long-lasting and environmentally friendly (Vayyaprontavida Kaliyathan et al. 2020; Whba et al. 2024).

Mechanical Blending

Mixing different types of rubber or adding fillers to the rubber matrix by using mechanical forces is known as mechanical blending. This method is popular for creating rubber mixtures since it is easy to use and efficient at creating uniformity (Whba et al. 2024; Mohite et al. 2022; Saad and El-Sabbagh 2001).

- **Internal Mixing**: Done with a Banbury mixer or another type of internal mixer. Significant shear forces are applied during the operation to evenly distribute rubber polymers, fillers, and additives. Ideally suited for manufacturing large-scale blends with reliable qualities, particularly for industrial uses like tires for cars (Yue et al. 2023).
- **Two-Roll Milling**: Rubber compounds are mixed using two counter-rotating wheels in a process known as two-roll milling. Ideal for blending or fine-tuning formulas on a laboratory scale improves control over the blending process, although it is less effective when produced on a large scale (Yu et al. 2020).

Solutions Blending

Solution blending creates a homogenous solution by dissolving the rubbers and additives in a common solvent. After that, the solvent is removed, leaving behind a consistent mixture which provides exact control over rubber interactions and nanoparticle dispersion. It is especially good at mixing rubbers with drastically varying polarity or viscosity, and prevents mechanical shear, which can cause delicate rubber materials to deteriorate. Applications include sophisticated coatings, medical devices, and high-performance composites (Dorigato 2021).

Blending Latex

Mixing two or more rubber varieties in their latex state (aqueous dispersions) is known as latex blending. When it comes to attaining molecular consistency, this approach is perfect. Applications of this method includes coatings, gloves, and balloons (Boratto et al. 2020).

- **Coagulation Blending**: Rubber lattices are combined and then coagulated using salts or acids to create a solid blend. This method guarantees superior homogeneity and compatibility of the combined materials (Mekonnen et al. 2019).
- **Direct Mixing**: This method involves mixing two or more latex solutions directly, followed by casting or coagulation. This makes it possible to create blends with microstructures (Phumnok et al. 2022).

Reactive Blending

Reactive blending creates graft copolymers or cross-linked networks by causing chemical reactions throughout the mixing process (Sahakaro et al. 2007). This enhances the compatibility of different kinds of rubber, including polar and non-polar ones. They require adding compatibilizers, like maleic anhydride, to encourage chemical connection between rubber phases. Improves the blend's thermal and mechanical stability. Applications of this technique include high-stress settings such as aircraft parts, gaskets, and seals (Mekonnen et al. 2019; Phumnok et al. 2022; Sahakaro et al. 2007).

Dynamic Vulcanization

Blending and vulcanization (cross-linking) are combined in a single phase via dynamic vulcanization. When one rubber phase is combined with another rubber or plastic phase, it vulcanizes dynamically (Karger-Kocsis 2019; Babu and Naskar 2011). This process produces thermoplastic elastomers (TPEs), which combine the processability of plastics with the elasticity of rubber. It enhances phase dispersion and compatibility, resulting in better mechanical qualities. Applications of this blending techniques include vibration-dampening elements, flexible seals, and automotive parts (Karger-Kocsis 2019; Babu and Naskar 2011).

Melt Blending

Rubbers are heated to respective softening points and blended under carefully regulated shear and temperature conditions in a process known as melt blending. They are used frequently for elastomers made of thermoplastic and ensures that fillers and rubber phases are distributed uniformly without the use of solvents. They are economical for large-scale manufacturing while having little effect on the environment. Flexible pipelines, shoe soles, and industrial belts are a few examples of its applications (Cantamessa et al. 2022).

Nanoparticle Dispersion Techniques

Achieving consistent dispersion of nanoparticles is essential for optimizing the performance of polymer-based nanoblends (Albdiry 2024; Imoisili et al. 2020a). Specialized blending processes guarantee uniform dispersion of nanoparticles such as silica, graphene, or carbon nanotubes. Applications include thermal insulation, sophisticated coatings, and high-performance tires (Imoisili et al. 2020a; Bheemappa and Gurumurthy 2022).

- **High-Shear Mixing**: Provides a homogeneous blend by breaking up agglomerates of nanoparticles (Imoisili et al. 2020b).
- **Ultrasound-Assisted Blending**: This method equally distributes nanoparticles throughout the rubber matrix by using ultrasonic vibrations (Buruga and Kalathi 2018).
- **Sol–gel Processing**: Sol–gel processing ensures molecular-level dispersion by synthesizing nanoparticles in situ within the rubber matrix (Miloskovska et al. 2015).

4.3 Rubber Blends Applications

Rubber blends are extensively utilized in many industries because of their unique qualities, which combine the benefits of several rubbers to satisfy specific application needs. By combining natural and synthetic rubbers, producers can produce materials with the ideal ratio of flexibility, toughness, and resistance to weathering, chemicals, and temperature fluctuations (Vayyaprontavida Kaliyathan et al. 2020; Rodgers and Halasa 2011; Popescu et al. 2017). A previous study claims that producers can enhance material qualities for specific industrial applications by combining different types of rubber (Song 2017; Ikeda et al. 2018; Kohjiya and Ikeda 2021; Kauffman 2001; Guerra et al. 2021; Mooibroek and Cornish 2000; Venkatachalam et al. 2013; Bokobza 2018; Cruz-Morales et al. 2023; Ansari et al. 2021; Roy et al. 2021; Sethulekshmi et al. 2022; Kurian and Mathew 2011; Singha et al. 2019; Shah et al. 2013; Boon et al. 2022). For example, adding NR to SBR increases durability and abrasion resistance, making this blend perfect for tire manufacturing while keeping costs down. Conveyor belts, shoes, and seals where solid and long-lasting materials are needed also employ these blends. Here are some critical applications of rubber blends across various sectors (Popescu et al. 2017; Nirmal Ghosh et al. 2017; Xiao et al. 2021; Ramarad et al. 2015).

4.3.1 *Automotive Industry Applications of Rubber Blends*

One of the biggest markets for rubber blends is the automotive sector, as rubber is essential to the manufacturing of many vehicle parts that need to have particular qualities like flexibility, durability, resistance to wear and tear, and the capacity to endure harsh environments. In the automotive sector, rubber blends combine the advantages of synthetic and NR to satisfy the high-performance requirements of contemporary automobiles. These blends contribute to the overall efficiency and durability of the vehicle while providing optimal performance in various applications, guaranteeing the safety and comfort of the occupants (Ramarad et al. 2015; Zanchet et al. 2012).

Rubber blends are essential to the automobile industry because they give components in different vehicle systems vital qualities including flexibility, durability, heat resistance, and wear resistance. Rubber mixtures improve vehicle performance, comfort, and safety in everything from tires and seals to hoses and suspension components. In addition to improving longevity and lowering environmental impact, the ability to blend rubber compositions allows automakers to satisfy the ever-tougher specifications of contemporary automobiles. Rubber mixes will continue to play a crucial part in creating high-performance, reasonably priced, and environmentally friendly automobiles as automotive technology develops (Ramarad et al. 2015; Zanchet et al. 2012; Begum et al. 2020).

Applications of Rubber Blends in Automotive Tires

One of the most important uses of rubber blends in the automotive sector is tires. In addition to being reasonably priced, the rubber used to make tires needs to have exceptional elasticity, traction, durability, and resistance to abrasion. Tire manufacture often uses a combination of NR and synthetic rubbers such as butadiene rubber (BR) and SBR. A tire's resistance to stretching, compression, and repeated road impacts is improved by NR exceptional tensile strength, elasticity, and resilience (Venkatachalam et al. 2013; Ansari et al. 2021; Kurian and Mathew 2011). All-weather tires can benefit from SBR, which improves the tire's traction on damp conditions and provides exceptional wear resistance. High-performance tires employ BR because it increases wear resistance and offers a better trade-off between longevity and performance. This blend of rubbers reduces rolling resistance and increases fuel efficiency while allowing tires to withstand high stress, offer the best grip, and function effectively in a variety of road conditions (Yang et al. 2021; Abdullah et al. 2022; Zou et al. 2021; Bouaziz et al. 2020; Alazhary et al. 2024).

Applications of Rubber Blends in Seals and Gaskets

Rubber mixes are frequently utilized in the production of gaskets and seals for automotive applications. These parts are essential for keeping fluids and gasses from leaking and for preserving vehicle performance, particularly in engines, transmissions, and other crucial systems. Although NBR has exceptional resistance to lubricants, fuels, and other automotive fluids, it is frequently used for gaskets and seals. For engine seals, fuel hoses, and transmission gaskets, NBR is perfect since it can be combined with other rubbers to increase their flexibility, heat resistance, and compression set qualities (Dahi Taleghani and Santos 2023; Periyasamy et al. 2024). To create seals that can resist water and anti-freeze fluids, as well as harsh weather, UV rays, and ozone depletion, EPDM is frequently combined with NBR. Vehicles are guaranteed to retain their integrity throughout a variety of operating situations, including high temperatures, the presence of chemicals, and severe environmental stressors, thanks to blended rubbers in gaskets and seals (Hu et al. 2023; Costa et al. 2024; Barghamadi et al. 2020; Lee et al. 2021; Periyasamy et al. 2024).

Applications of Rubber Blends in Hoses and Belts

Rubber hoses and belts are necessary for power transmission, fluid flow regulation, and system integrity maintenance across a variety of engine components in automobiles. For hoses and belts to be strong, flexible, and impervious to heat and chemicals, rubber mixtures are essential. Radiator hoses, coolant hoses, and air intake hoses frequently use EPDM due to its exceptional resistance to steam, high temperatures, and automotive fluids (Hu et al. 2023; Costa et al. 2024; Prakash et al. 2024). Blending EPDM with other rubbers, such as SBR, increases its durability and flexibility. Belts frequently employ neoprene because of its strength and resistance to abrasion, weather, and oil. Belts in the engine compartment, often exposed to heat and hard circumstances, are a good fit for it because of its resilience to ozone and aging. To minimize the need for frequent repairs and guarantee that vital systems like cooling, fuel, and air intake operate as intended, rubber blends for hoses and belts help ensure constant performance in vehicle systems (Periyasamy et al. 2024; Prakash et al. 2024; Dhanasekar et al. 2023).

Applications of Rubber Blends in Bushings and Mounts

Vehicle suspension systems use rubber bushings and mounts to absorb shock, lessen vibration and noise, and give a smooth ride. Rubber blends with superior flexibility, durability, and vibration-damping qualities are needed for these parts. Automotive bushings frequently use mixtures of NR and PU blends due to its strength, high load-bearing capacity, and wear resistance (Singha et al. 2019; Madhavan et al. 2024). The combination improves ride comfort and vehicle handling by offering the best possible impact resistance and vibration damping. Given their exceptional thermal stability and resistance to deterioration at high temperatures, silicone rubber blends are also utilized in some high-performance applications. Bushings and mounts made of rubber blends that effectively absorb shocks and lessen the effects of vibrations increase handling, ride quality, and suspension system longevity (Han et al. 2022; Kumar et al. 2021; Khanra et al. 2020; Madhavan et al. 2024).

Applications of Rubber Blends in Window Seals and Weather Stripping

Weather stripping and window seals are crucial for keeping cars waterproof and airtight, reducing wind noise, and shielding the interior from the elements. The qualities that these parts require to stay pliable, strong, and weatherproof over time are supplied by rubber mixtures (Li et al. 2024). The most popular material for vehicle weather stripping is EPDM because of its superior resistance to ozone, UV radiation, and weather conditions like rain, snow, and extremely high or low temperatures. EPDM can perform better in automotive applications and have better processing properties when blended with other elastomers. It can be mixed with EPDM to provide better resilience to higher temperatures, which makes it perfect for seals used in hot or abrasive environments like those around engines or exhaust systems (Hu et al. 2023; Costa et al. 2024; Barghamadi et al. 2020; Sahakaro et al. 2007). Rubber blends used in weather stripping, and window seals keep out dust, wetness, and noise while acting as an efficient barrier against inclement weather, ensuring that cars maintain their energy efficiency, comfort, and safety (Zou et al. 2021; Vayyaprontavida Kaliyathan et al. 2020; Rodgers and Halasa 2011; Wang et al. 2011).

Applications of Rubber Blends in Suspension and Anti-Vibration Products

Suspension parts including strut mounts, shock absorber bushings, and anti-vibration pads are made from rubber blends to lessen the effect of vibrations and absorb shock. These elements are essential for enhancing the vehicle's overall performance, stability, and ride quality. Suspension bushings use both natural and synthetic rubbers (such SBR or nitrile rubber) to give them the flexibility and shock absorption they need (Dhanorkar et al. 2021; Prakash et al. 2024). The mixture lessens noise and guarantees a better ride by absorbing road shocks and vibrations. Considering its superior damping qualities, fatigue resilience, and adaptability to a range of temperatures, EPDM is frequently combined with other rubbers to create anti-vibration pads and suspension components. These rubber mixtures improve vehicle stability, minimize vibration, and reduce noise, all of which contribute to a comfortable driving experience (Hu et al. 2023; Costa et al. 2024).

Applications of Rubber Blends in Interior Components

Rubber blends are frequently utilized in dashboard components, trim, floor mats, and other interior components. Rubber materials that are both very comfortable and esthetically pleasing must be able to withstand chemicals, UV light, and wear and tear. Floor mats and trim pieces are made from thermoplastic rubber (TPR) mixes, which combine rubber's elasticity with plastics' processability (Sethulekshmi et al. 2022; Joseph Jebaraj and Rajendran 2024). These materials can be easily molded into various shapes and are incredibly resilient to chemicals. Soft-touch interior components like knobs, buttons, and handles are frequently made of silicone and PVC blends due to their exceptional resistance to aging and environmental deterioration. Rubber mixtures in car interiors guarantee enduring function and enhance the cabin's general comfort and visual appeal (Kurian and Mathew 2011; Singha et al. 2019; Kumar et al. 2021).

4.3.2 Applications of Rubber Blends in Aerospace and Aviation

Rubber is essential to the aerospace and aviation sectors because of its unique qualities, which include flexibility, durability, resistance to high temperatures, and the capacity to withstand challenging climatic conditions. Rubber blends are designed to satisfy the demanding requirements of the aircraft industry by blending several kinds of natural and synthetic rubbers. Applications of rubber mixes in aerospace and aviation include vital parts that enhance the longevity, performance, and safety of aircraft, such as vibration dampening and sealing systems (Kumar et al. 2021; Popescu et al. 2017). Rubber blends are crucial to the aerospace and aviation sectors because they provide materials that satisfy the strict specifications for efficiency, performance, and safety during flight. Rubber mixes are essential for ensuring that airplanes operate dependably and securely in harsh environments, whether in fuel lines, vibration-damping parts, sealing systems, or fire protection. Manufacturers can customize materials to fulfill strict industrial standards for performance and safety while also offering the required resistance to chemicals, wear, and temperature extremes by fusing the special qualities of natural and synthetic rubbers. Rubber mixes will continue to be essential to creating safer, more effective, and longer-lasting aircraft as the aerospace industry develops (Popescu et al. 2017; Sahu et al. 2024).

Seals and O-Rings

O-rings and seals are essential for aircraft systems because they keep fluids and gasses from leaking and guarantee airtight, waterproof conditions in very pressured settings. These components must be manufactured using rubber blends to satisfy the aerospace industry's strict criteria for temperature, pressure, and chemical resistance. O-rings and seals in aviation frequently use fluoro-elastomers (like Viton) and silicone-based mixes because of their exceptional resilience to high temperatures, chemicals, and fuels. These materials are necessary to maintain the integrity of the aircraft's engines, hydraulic systems, and fuel systems, as contamination or fuel leaks could result in catastrophic failure (Sahu et al. 2024; Białecki et al. 2021). Blends of nitrile rubber (NBR) and ethylene propylene diene monomer (EPDM) are used for applications that demand superior elasticity and weather resistance. They provide excellent performance in sealing applications for airframe components exposed to the outside environment. These rubber blends guarantee the seamless functioning of the aircraft's internal systems, preserve pressure integrity, and lower the possibility of fluid leaks (Sahu et al. 2024; Białecki et al. 2021; Liu et al. 2021; Valentini et al. 2022).

Applications of Rubber Blends in Fuel and Hydraulic Hoses

Flexible hoses are frequently needed in aerospace and aviation systems to move hydraulic fluids, fuel, and oil. Rubber mixes offer the flexibility and durability required for these hoses, which must endure high pressures, exposure to fuels and oils, and extremely high or low temperatures. To make hoses resistant to oils, fuels, and hydraulic fluids, nitrile rubber (NBR) is commonly mixed with other materials like PVC or SBR. These hoses must withstand high temperatures, pressure changes, and chemical exposure because they are utilized in fuel systems, engines, and hydraulic circuits (Dhanorkar et al. 2021; Veloso et al. 2024). Fluoro-elastomers are occasionally utilized in applications that call for strong thermal stability and chemical resistance, guaranteeing that hoses in vital systems—like the engine or auxiliary power units—continue functioning even in harsh operating circumstances. To ensure the safety and lifespan of aviation systems, these hoses are made of rubber blends that provide flexibility while simultaneously preserving their resistance to cuts, abrasions, and degradation from harsh chemicals or high-pressure fluids (Sahu et al. 2024; Białecki et al. 2021; Liu et al. 2021).

Applications of Rubber Blends in Vibration and Noise Control

Reducing vibration and managing noise is essential for passenger comfort and aircraft equipment longevity in aerospace applications. Vibration-damping parts like mounts, bushings, and isolators frequently use rubber blends to absorb and lessen undesired vibrations from airframes, engines, and other mechanical systems (Veloso et al. 2024; Crocker 2018). Neoprene and NR are frequently utilized as vibration dampers because of their exceptional flexibility, durability, and capacity to tolerate the enormous forces produced by aircraft engines and structures. Manufacturers can enhance the rubber's vibration-damping properties and its performance across a wide temperature range by combining NR with synthetic elastomers. Blends of silicone rubber are frequently utilized to isolate vibrations, especially in aircraft applications where low compression set and superior temperature stability are necessary. Silicone is perfect for use in components subjected to significant temperature fluctuations during flight because of its capacity to retain flexibility in extremely high or low temperatures. By lowering noise levels, rubber blends used in vibration dampening increase passenger and crew comfort and shield delicate mechanical and avionic systems from the detrimental effects of vibration (Crocker 2018; Geethamma et al. 2014; Sołoducho et al. 2021).

Applications of Rubber Blends in Thermal Insulation and Fire Protection

During flight and ground operations, aircraft are subjected to drastic temperature changes in the atmosphere. Rubber mixes are utilized in fire safety and thermal insulation applications to safeguard delicate parts and guarantee the security of both passengers and workers. The perfect material for thermal insulation is silicone rubber because of its exceptional resistance to hot and low temperatures. When combined with other materials like fiberglass or ceramic, silicone-based mixes can effectively insulate hot and cold areas, including electrical harnesses, exhaust systems, and engine components (Ilina et al. 2022). Gaskets and seals exposed to high temperatures, as those in turbine engines or exhaust systems, use blends based on fluoro-elastomers because of their exceptional tolerance to high temperatures. Under extreme heat stress, these materials continue to function as sealants and retain their integrity. Aircraft utilize fire-resistant rubber mixtures to ensure safety in a fire, frequently containing intumescent additives that expand when heated. These materials contain fires and prevent them from spreading throughout the airplane, protecting passengers and vital components (Ilina et al. 2022; Basuli et al. 2012; Rahman et al. 2023).

Applications of Rubber Blends in Aircraft Interiors

Aircraft interiors are also built and designed using rubber mixtures, which enhance comfort, toughness, and safety. These blends must adhere to stringent fire, smoke, and toxicity (FST) regulations to protect passengers and offer long-term resilience under the harsh circumstances of an airplane. Thermoplastic elastomer (TPE) blends are utilized in the interior for floors, sealants, and ornamental components (Parveez et al. 2022). Regarding achieving strict safety and regulatory standards for minimal smoke and toxicity, TPE offers flexibility, durability, and ease of processing. These blends provide superior protection against UV radiation, ozone, and severe weather, guaranteeing that seals will stay intact and work appropriately for the duration of the aircraft's life. Manufacturers can lower maintenance costs, increase passenger comfort, and adhere to stringent safety and environmental rules by utilizing rubber blends in airplane interiors (Sahu et al. 2024; Parveez et al. 2022; Srivastava and Sharma 2023).

Applications of Rubber Blends in Aircraft Window and Windshield Seals

Seals on windows and windshields are essential parts of airplanes because they keep the interior pressurized and weatherproof. Rubber mixtures are utilized to make resilient, flexible seals that can tolerate changes in pressure, temperature, and UV rays while in flight. EPDM rubber's exceptional weathering durability, low-temperature flexibility, and capacity to tolerate UV and ozone exposure make it a popular choice for window and windshield seals (Begum et al. 2020). Blending EPDM with other elastomers like silicone improves its performance in harsh environments. Applications needing a low compression set and exceptional high-temperature resistance also use silicone mixes, which guarantee that seals in the structural elements of the aircraft continue to function as intended over time. These rubber mixtures guarantee that windows and windshields stay firmly in place during flight, help maintain cabin pressure, and stop water intrusion (Srivastava and Sharma 2023; Pandey et al. 2024).

Applications of Rubber Blends in Aircraft Landing Gear and Tires

High speeds during takeoff and landing and significant weight during ground operations subject an aircraft's landing gear and tires to tremendous forces. These parts use rubber mixtures to give them the strength, flexibility, and resilience to the required temperature changes and abrasion. Aircraft tires frequently employ butadiene rubber (BR) because of its exceptional resilience and wear resistance. It gives better traction, stability, and longevity when combined with other elastomers, like SBR and NR. High-performance airplane tires occasionally use silicone-based blends due to their exceptional suppleness and tolerance to high temperatures, which provide safe landings even in the most adverse circumstances. The airplane can fly, land, and take off securely thanks to these rubber blends, increasing the landing gear and tires' dependability and safety (Pandey et al. 2024; Indriasari et al. 2021; Valentini and Lopez-Manchado 2020).

4.4 Conclusion

Rubber-based nanoblends, which use nanotechnology to improve the qualities of conventional rubber blends, have become a ground-breaking field of material science in recent years. Nanoblends, which are created by integrating nanoparticles into rubber matrices, have exceptional properties in terms of strength, flexibility, thermal stability, and chemical resistance. This makes them perfect for high-performance applications in a variety of industries, such as electronics, medical devices, automotive, and aerospace. The chapter examined the synthesis, processing techniques, and many uses of these cutting-edge materials, highlighting the substantial influence that nanoparticle compatibility, dispersion, and selection have on the blend's overall performance.

The constraints of traditional rubber materials may be overcome by nanoblends, which offer specialized qualities for certain uses. They are particularly useful in settings that require resilience to harsh conditions, like high temperatures, chemical exposure, and high mechanical loads, due to their improved mechanical qualities. Nevertheless, there are still issues that can prevent large-scale usage, such as controlling production costs and attaining the best possible nanoparticle dispersion.

Rubber-based nanoblends have a bright future thanks to continuous research into sustainable alternatives, novel nanoparticle types, and processing technique refinement, which will guarantee production that is ecologically benign. A major breakthrough in material innovation, rubber-based nanoblends are predicted to transform a number of industries with their unparalleled performance and adaptability as nanotechnology continues to advance.

References

Abdullah NAS, Abdullah FF, Sufian AH, Abidin ANSZ, Jamaludin AS, Razali MNM (2022) Effect of degradation by temperature onto nitrile rubber elastomer mechanical properties. Mater Today: Proc 48:1941–1946

Alazhary S, Shaafaey M, Bahrololoumi A, Dargazany R (2024) Investigating the effects of sequential aging temperature profiles on the response of neoprene rubber. J Polym Res 31(4):102

Albdiry M (2024) Effect of melt blending processing on mechanical properties of polymer nanocomposites: a review. Polym Bull 81(7):5793–5821

Ansari AH, Jakarni FM, Muniandy R, Hassim S, Elahi Z (2021) Natural rubber as a renewable and sustainable bio-modifier for pavement applications: a review. J Clean Prod 289:125727

Babu RR, Naskar K (2011) Recent developments on thermoplastic elastomers by dynamic vulcanization. Adv Rubber Composites 219–247

Barghamadi M, Ghoreishy MHR, Karrabi M, Mohammadian-Gezaz S (2020) Investigation on the kinetics of cure reaction of acrylonitrile–butadiene rubber (NBR)/polyvinyl chloride (PVC)/graphene nanocomposite using various models. J Appl Polym Sci 137(18):48632

Basuli U, Jose J, Lee RH, Yoo YH, Jeong KU, Ahn JH, Nah C (2012) Properties and degradation of the gasket component of a proton exchange membrane fuel cell—a review. J Nanosci Nanotechnol 12(10):7641–7657

Begum SA, Rane AV, Kanny K (2020) Applications of compatibilized polymer blends in automobile industry. In: Compatibilization of polymer blends. Elsevier, pp 563–593

Behera PK, Kumar A, Mohanty S, Gupta VK (2022) Overview on post-polymerization functionalization of butyl rubber and properties. Ind Eng Chem Res 61(46):16910–16923

Bheemappa S, Gurumurthy H (2022) Recent advances in fabrication and characterization of nanofiller filled epoxy nanocomposites. Trends Fabr Polym Polym Compos 2022:1–40

Białecki T, Sitkiewicz A, Giemza B, Sarnecki J, Skolniak M, Gawron B (2021) Compatibility of different automotive elastomers in paraffinic diesel fuel. Appl Sci 11(23):11312

Bokobza L (2018) Natural rubber nanocomposites: a review. Nanomaterials 9(1):12

Boon ZH, Teo YY, Ang DTC (2022) Recent development of biodegradable synthetic rubbers and bio-based rubbers using sustainable materials from biological sources. RSC Adv 12(52):34028–34052

Boratto MH, Nozella NL, Ramos RA, Silva RA, Graeff CF (2020) Flexible conductive blend of natural rubber latex with PEDOT: PSS. APL Mater 8(12)

Bouaziz R, Truffault L, Borisov R, Ovalle C, Laiarinandrasana L, Miquelard-Garnier G, Fayolle B (2020) Elastic properties of polychloroprene rubbers in tension and compression during ageing. Polymers 12(10):2354
Buruga K, Kalathi JT (2018) A facile synthesis of halloysite nanotubes based polymer nanocomposites for glass coating application. J Alloy Compd 735:1807–1817
Cantamessa F, Damonte G, Monticelli O, Arrigo R, Fina A (2022) Thermoreversible cross-linked rubber prepared via melt blending and its nanocomposites. ACS Appl Polym Mater 4(7):4796–4807
Costa NL, Hiranobe CT, Cardim HP, Dognani G, Sanchez JC, Carvalho JAJ, Torres GB, Paim LL, Pinto LF, Cardim GP et al (2024) A review of EPDM (ethylene propylene diene monomer) rubber-based nanocomposites: properties and progress. Polymers 16:1720
Crocker MJ (2018) Principles and methods of noise control. In: Noise and noise control. CRC Press, pp 1–49
Cruz-Morales JA, Gutiérrez-Flores C, Zárate-Saldaña D, Burelo M, García-Ortega H, Gutiérrez S (2023) Synthetic polyisoprene rubber as a mimic of natural rubber: recent advances on synthesis, nanocomposites, and applications. Polymers 15(20):4074
Dahi Taleghani A, Santos L (2023) Elastomeric sealing assemblies. Wellbore Integrity: From Theory to Practice, 165–193
Dhanasekar S, Baskar S, Vishvanathperumal S (2023) Halloysite nanotubes effect on cure and mechanical properties of EPDM/NBR nanocomposites. J Inorg Organomet Polym Mater 33(10):3208–3220
Dhanorkar RJ, Mohanty S, Gupta VK (2021) Synthesis of functionalized styrene butadiene rubber and its applications in SBR–silica composites for high performance tire applications. Ind Eng Chem Res 60(12):4517–4535
Dorigato A (2021) Recycling of polymer blends. Adv Indus Eng Polym Res 4(2):53–69
Geethamma VG, Asaletha R, Kalarikkal N, Thomas S (2014) Vibration and sound damping in polymers. Resonance 19:821–833
Guerra NB, Pegorin GSA, Boratto MH, de Barros NR, de Oliveira Graeff CF, Herculano RD (2021) Biomedical applications of natural rubber latex from the rubber tree Hevea brasiliensis. Mater Sci Eng, C 126:112126
Han R, Li Y, Zhu Q, Niu K (2022) Research on the preparation and thermal stability of silicone rubber composites: a review. Composites Part c: Open Access 8:100249
Hu Q, Chen Q, Song P, Gong X, Chen J, Zhao Y (2023) Performance of thermal-oxidative aging on the structure and properties of ethylene propylene diene monomer (EPDM) vulcanizates. Polymers 15(10):2329
Ikeda Y, Kato A, Kohjiya S, Nakajima Y (2018) Rubber science, vol 10. Springer, Singapore, pp 978–981
Ilina M, Selivanov O, Kurochkin I (2022) The development of polymer fire-resistant thermal insulation coating filled with anthropogenic waste. In: Technological advancements in construction: selected papers. Springer International Publishing, pp 117–125
Imoisili PE, Ukoba KO, Jen T-C (2020a) Green technology extraction and characterisation of silica nanoparticles from palm kernel shell ash via sol–gel. J Mater Res Technol 9(1):307–313
Imoisili PE, Ukoba KO, Jen T-C (2020b) Synthesis and characterization of amorphous mesoporous silica from palm kernel shell ash. Boletín De La Sociedad Española De Cerámica y Vidrio. 59:159–164
Indriasari, Noordermeer J, Dierkes W (2021) Incorporation of oligomeric hydrocarbon resins for improving the properties of aircraft tire retreads. Appl Sci 11(21):9834
Joseph Jebaraj JJS, Rajendran R (2024) An experimental study of the properties of carbon fiber/epoxy composites mixed with rubber granules. Int J Polym Sci 2024(1):5555592
Kaliyathan AV, Rane AV, Huskic M, Kunaver M, Kalarikkal N, Rouxel D, Thomas S (2020) Carbon black distribution in natural rubber/butadiene rubber blend composites: distribution driven by morphology. Compos Sci Technol 200:108484

Karger-Kocsis J (2019) Thermoplastic rubbers via dynamic vulcanization. In: Polymer blends and alloys. Routledge, pp 125–154
Kauffman GB (2001) Charles Goodyear (1800–1860), American Inventor, on the Bicentennial of His Birth. Chem Educ 6(1):50–54
Khanra S, Ganguly D, Ghorai SK, Goswami D, Chattopadhyay S (2020) The synergistic effect of fluorosilicone and silica towards the compatibilization of silicone rubber and fluoroelastomer based high performance blend. J Polym Res 27:1–17
Kohjiya S, Ikeda Y (2021) A short history of natural rubber research. In: Chemistry, manufacture, and applications of natural rubber. Woodhead Publishing, pp 407–427
Kumar V, Alam MN, Manikkavel A, Song M, Lee DJ, Park SS (2021) Silicone rubber composites reinforced by carbon nanofillers and their hybrids for various applications: a review. Polymers 13(14):2322
Kurian T, Mathew NM (2011) Natural rubber: production, properties and applications. In: Biopolymer and biomedical environment applications, pp 403–436
Lee SH, Park SY, Chung KH, Jang KS (2021) Phlogopite-reinforced natural rubber (NR)/ethylene-propylene-diene monomer rubber (EPDM) composites with aminosilane compatibilizer. Polymers 13(14):2318
Li Y, Yang J, Heris SZ, Zhang Y, Zhang X, Chen J (2024) Pyrolysis and combustion characteristics of typical sealing materials for exterior building windows. Fuel 360:130546
Liu M, Cataldi P, Young RJ, Papageorgiou DG, Kinloch IA (2021) High-performance fluoroelastomer-graphene nanocomposites for advanced sealing applications. Compos Sci Technol 202:108592
Madhavan MK, Nambaithodi V, Krishnamoorthy A, Palanisamy S (2024) Hemicellulose rubber composites and rubber bionanocomposites. In: Rubber based bionanocomposites: characterisation. Springer Nature Singapore, Singapore, pp 105–127
Mekonnen TH, Ah-Leung T, Hojabr S, Berry R (2019) Investigation of the co-coagulation of natural rubber latex and cellulose nanocrystals aqueous dispersion. Colloids Surf, A 583:123949
Miloskovska E, Friedrichs C, Hristova-Bogaerds D, Persenair O, van Duin M, Hansen MR, de With G (2015) Chemical mapping of silica prepared via sol–gel reaction in rubber nanocomposites. Macromolecules 48(4):1093–1103
Mohite AS, Rajpurkar YD, More AP (2022) Bridging the gap between rubbers and plastics: a review on thermoplastic polyolefin elastomers. Polym Bull 79(2):1309–1343
Mooibroek H, Cornish K (2000) Alternative sources of natural rubber. Appl Microbiol Biotechnol 53:355–365
Nirmal Ghosh OS, Gayathri S, Sudhakara P, Misra SK, Jayaramudu J (2017) Natural rubber nanoblends: preparation, characterization and applications. In: Rubber nano blends: preparation, characterization and applications, pp 15–65
Norton MG (2021) Rubber—the material of possibilities. Ten materials that shaped our world. Springer International Publishing, Cham, pp 125–144
Pandey SK, Mishra S, Ghosh S, Rohan R, Maji PK (2024) Self-healing polymers for aviation applications and their impact on circular economy. Polym Eng Sci 64(3):951–987
Parveez B, Kittur MI, Badruddin IA, Kamangar S, Hussien M, Umarfarooq MA (2022) Scientific advancements in composite materials for aircraft applications: a review. Polymers 14(22):5007
Periyasamy M, Mubasshir AA, Kodra S, Chandramouli S, Campbell R, Kazmer DO, Mead JL (2024) From formulation to application: effects of plasticizer on the printability of fluoro elastomer compounds and additive manufacturing of specialized seals. Micromachines 15(5):622
Phumnok E, Khongprom P, Ratanawilai S (2022) Preparation of natural rubber composites with high silica contents using a wet mixing process. ACS Omega 7(10):8364–8376
Popescu RC, Popescu D, Grumezescu AM (2017) Applications of rubber-based blends. In: Recent developments in polymer macro, micro and nano blends. Woodhead Publishing, pp 75–109
Prakash PC, Srinivasan D, Navaneethakrishnan V, Vishvanathperumal S (2024) Effect of modified nanographene oxide loading on the swelling and compression set behavior of EPDM/SBR nano-composites. J Inorg Organomet Polym Mater 34(2):593–610

Rahman MZ, Rahman M, Mahbub T, Ashiquzzaman M, Sagadevan S, Hoque ME (2023) Advanced biopolymers for automobile and aviation engineering applications. J Polym Res 30(3):106
Ramarad S, Khalid M, Ratnam CT, Chuah AL, Rashmi W (2015) Waste tire rubber in polymer blends: a review on the evolution, properties and future. Prog Mater Sci 72:100–140
Rodgers B, Halasa A (2011) Compounding and processing of rubber/rubber blends. Encyclopedia of Polymer Blends, 163–206
Rodgers B, Waddell W (2005) The science of rubber compounding. In: Science and technology of rubber. Academic Press, pp 401–454
Roy K, Debnath SC, Pongwisuthiruchte A, Potiyaraj P (2021) Recent advances of natural fibers based green rubber composites: properties, current status, and future perspectives. J Appl Polym Sci 138(35):50866
Saad AL, El-Sabbagh S (2001) Compatibility studies on some polymer blend systems by electrical and mechanical techniques. J Appl Polym Sci 79(1):60–71
Sahakaro K, Naskar N, Datta RN, Noordermeer JW (2007) Reactive blending, reinforcement and curing of NR/BR/EPDM compounds for tire sidewall applications. Rubber Chem Technol 80(1):115–138
Sahu BB, Moharana S, Behera PK (2024) Elastomeric-based composite materials for engineering applications. In: Polymer composites: fundamentals and applications. Springer Nature Singapore, Singapore, pp 329–355
Sethulekshmi AS, Saritha A, Joseph K (2022) A comprehensive review on the recent advancements in natural rubber nanocomposites. Int J Biol Macromol 194:819–842
Shah AA, Hasan F, Shah Z, Kanwal N, Zeb S (2013) Biodegradation of natural and synthetic rubbers: a review. Int Biodeterior Biodegradation 83:145–157
Sharma RK, Mohanty S, Gupta V (2021) Advances in butyl rubber synthesis via cationic polymerization: an overview. Polym Int 70(9):1165–1175
Sidkey MA, Yehia AA, Abd El Malak NA, Gaafar MS (2002) Compatibility studies on some rubber blend systems by ultrasonic techniques. Mater Chem Phys 74(1):23–32
Singha NR, Mahapatra M, Karmakar M, Chattopadhyay PK (2019) Processing, characterization and application of natural rubber based environmentally friendly polymer composites. In: Sustainable polymer composites and nanocomposites, pp 855–897
Sisanth KS, Thomas MG, Abraham J, Thomas S (2017) General introduction to rubber compounding. In: Progress in rubber nanocomposites. Woodhead Publishing, pp 1–39
Sołoducho J, Zając D, Spychalska K, Baluta S, Cabaj J (2021) Conducting silicone-based polymers and their application. Molecules 26(7):2012
Song K (2017) Micro-and nano-fillers used in the rubber industry. In: Progress in rubber nanocomposites. Woodhead Publishing, pp 41–80
Srivastava A, Sharma S (2023) Recent advances in experimental and molecular dynamics study of graphene-oxide/natural rubber composites: a review. J Reinf Plast Compos 42(3–4):110–130
Tang S, Li Z, Sun W, Liu Y, Wang J, Wang X, Lin J (2024) Natural rubber/styrene–butadiene rubber blend composites potentially applied in damping bearings. Polymers 16(13):1945
Valentini L, Lopez-Manchado MA (2020) Classification of rubbers and components for harsh environmental systems. In: High-performance elastomeric materials reinforced by nano-carbons. Elsevier, pp 1–14
Valentini F, Dorigato A, Fambri L, Bersani M, Grigiante M, Pegoretti A (2022) Production and characterization of novel EPDM/NBR panels with paraffin for potential thermal energy storage applications. Therm Sci Eng Progress 32:101309
Vayyaprontavida Kaliyathan A, Varghese KM, Nair AS, Thomas S (2020) Rubber–rubber blends: a critical review. Progress Rubber, Plast Recycl Technol 36(3):196–242
Veloso V, Pinto E, Santiago M, Bortoli B, Ribeiro W, Polkowski R (2024) Potential application of rubber-graphene compounds in the automotive parts (No. 2023-36-0028). SAE Technical Paper
Venkatachalam P, Geetha N, Sangeetha P, Thulaseedharan A (2013) Natural rubber producing plants: an overview. African J Biotechnol 12(12)

Waddell WH, Tsou AH (2004) Butyl rubbers. In: Rubber compounding. CRC Press, pp 142–171

Wang JJ, Lee J, Woo CS, Kim BK, Lee SB (2011) An experimental study and finite element analysis of weatherstrip. Int J Precis Eng Manuf 12:97–104

Whba R, Su'ait MS, Whba F, Sahinbay S, Altin S, Ahmad A (2024) Intrinsic challenges and strategic approaches for enhancing the potential of natural rubber and its derivatives: a review. Int J Biol Macromol 133796

Xiao X, Wang J, Cai D, Lou L, Xiao F (2021) A novel application of thermoplastic polyurethane/waste rubber powder blend for waterproof seal layer in high-speed railway. Transport Geotech 27:100503

Yang Q, Yu S, Zhong H, Liu T, Yao E, Zhang Y, Zou H, Du W (2021) Gas products generation mechanism during co-pyrolysis of styrene-butadiene rubber and natural rubber. J Hazard Mater 401:123302

Yu P, Jin G, Wang W, Wang R, Guo B, Zhang L (2020) Dispersing carbon dots in non-polar rubber by slurry compounding and in situ compatibilizing. Composites Communications 22:100429

Yue S, Liu Y, Geng J, Hua J (2023) High vinyl polybutadiene rubber/polypropylene thermoplastic elastomer blends: optimization of internal mixing process parameters and screening of processing methods. J Appl Polym Sci 140(20):e53845

Zanchet A, Carli LN, Giovanela M, Brandalise RN, Crespo JS (2012) Use of styrene butadiene rubber industrial waste devulcanized by microwave in rubber composites for automotive application. Mater des 39:437–443

Zou H, Jing Y, Tu J, Shi X, Kadlcak J, Yong Z, Liu G (2021) Investigation on the mechanical properties and oil resistance of sulfur cured nitrile rubber/hydrogenated nitrile butadiene rubber blends. Polym Eng Sci 61(12):3050–3059

Chapter 5
Polymer/Rubber-Based Nanoblends

Patrick Ehi Imoisili

Abstract This chapter explores the development, properties, and applications of polymer/rubber-based nanoblends, a class of materials that combines the elasticity and flexibility of rubber with the strength and durability of polymers, enhanced further by nanoscale fillers. The integration of nanofillers, such as carbon nanotubes, silica, and graphene, offers significant improvements in mechanical strength, thermal stability, abrasion resistance, and chemical resilience, making these materials highly adaptable for demanding applications in industries including automotive, aerospace, healthcare, and electronics. Key challenges in compatibility between polymers and rubber phases are addressed through the use of compatibilizers and advanced blending techniques to achieve optimal interfacial bonding and stability. Furthermore, the chapter discusses recent advancements toward sustainability, highlighting the potential of bio-based and recyclable polymers in nanoblends as industries increasingly prioritize eco-friendly solutions. By analyzing the synthesis, customization, and application potential of polymer/rubber-based nanoblends, this chapter provides a comprehensive overview of their current and future role as versatile, high-performance materials capable of meeting both technical and environmental demands.

Keywords Polymer/rubber nanoblends · Nanocomposites · Nanofillers · Mechanical properties · Bio-based polymers

5.1 Introduction to Polymers and Rubber

Rubber and polymers have developed into essential materials in our world, serving as the building blocks for many goods that influence daily living and propel industrial advancement. Due to their adaptability, durability, and versatility, polymers and rubber are essential components in everything from the tires on our cars to the

P. E. Imoisili (✉)
Mechanical Engineering Science Department, University of Johannesburg, Johannesburg, South Africa
e-mail: pimoisili@uj.ac.za

P. M. Visakh (ed.), *Engineering Applications of Polymer based Nano Blends*, Advanced Structured Materials 235, https://doi.org/10.1007/978-3-031-90979-5_5

insulation on our electrical connections (Imoisili and Jen 2024; Sisanth et al. 2017; Imoisili et al. 2023). Rubber is a unique class of polymers renowned for its elasticity and toughness. Still, the term "polymer" refers to a large range of materials with various qualities and applications. Rubber and polymers work hand in hand in a variety of applications, from electronics and medical gadgets to building and aerospace engineering (Rodgers and Waddell 2005; Song 2017; Ikeda et al. 2018; Kohjiya and Ikeda 2021).

5.1.1 Comprehending Polymers: The Basis of Contemporary Materials Science

Large molecules known as polymers, or macromolecules, are composed of repeated structural units known as monomers joined by chemical bonds. Since the kinds of monomers, the length of polymer chains, and the configuration of these chains all affect the material's properties, this structure gives polymers their distinct qualities (Ikeda et al. 2018; Kohjiya and Ikeda 2021; Kauffman 2001). Polymers are highly diverse because different molecular arrangements can result in materials with different chemical, mechanical, and thermal properties. For example, high-density polyethylene (HDPE) is stiff and durable, making it suitable for pipelines and containers. In contrast, low-density polyethylene (LDPE) is more flexible and commonly used in plastic bags and film applications (Guerra et al. 2021; Mooibroek and Cornish 2000; Venkatachalam et al. 2013).

Natural and synthetic polymers are the two primary groups into which polymers are often divided. While synthetic polymers like polystyrene, polyvinyl chloride (PVC), and polypropylene are produced by industrial chemical processes, natural polymers like cellulose, proteins, and rubber come from biological origins. Developments in polymer science have made it possible to develop ecologically friendly and functional materials. Bioplastics and biodegradable polymers, for example, provide substitutes for conventional plastics while promoting sustainability and lessening their negative effects on the environment (Sionkowska 2011; Bhatia and Bhatia 2016; Wojtowicz et al. 2009). Synthetic polymers have become especially important in industries such as packaging, healthcare, and transportation. Their versatility allows them to be designed to meet specific needs, such as being lightweight, strong, flexible, or even resistant to chemicals and UV radiation. In recent years, the development of advanced polymers, such as conductive polymers and shape-memory polymers, has opened new possibilities for smart materials and responsive technologies, paving the way for innovative applications in electronics, medicine, and wearable technology (Bhatia and Bhatia 2016; Wojtowicz et al. 2009; Boon et al. 2022).

5.1.2 *Rubber: A Unique Class of Polymers with Exceptional Elasticity*

Rubber, a subcategory of polymers, is distinguished by its extraordinary elasticity, which allows it to return to its original shape after stretching or deformation. This characteristic is essential in applications where flexibility and resilience are critical, such as automotive tires, gaskets, seals, and vibration-dampening materials. Rubber exists in natural and synthetic forms, each with distinct properties and applications (Boon et al. 2022; Wiśniewska et al. 2023; Imoisili et al. 2013).

The latex of the Hevea brasiliensis tree, which is mostly found in tropical areas, is the source of natural rubber (NR). It can stretch and bounce because of its molecular structure, which comprises long, coiled polymer chains that are weakly cross-linked. NR has long been an important resource because of its durability; it was first employed in crude ways and became a mainstay in the developing industrial sector in the nineteenth and twentieth centuries. NR has become the preferred material in many industries due to its tensile strength, abrasion resistance, and capacity to retain elasticity across a broad temperature range (Bokobza 2018; Cruz-Morales et al. 2023; Ansari et al. 2021). On the contrary, synthetic rubber is made from petrochemicals like butadiene and styrene. In response to NR shortages, synthetic rubber production got underway, and since then, different kinds of synthetic rubber have been developed, each with a unique purpose. For instance, the durability of styrene-butadiene rubber (SBR) makes it a popular material for tires, whereas neoprene is preferred for its ability to withstand weathering, oils, and chemicals. Synthetic rubber has proven invaluable when qualities beyond NR are needed, such as increased resilience to heat, cold, and environmental deterioration (Ansari et al. 2021; Shah et al. 2013; Dhanorkar et al. 2021).

5.1.3 *Polymer/Rubber Blends and Nanoblends: Enhancing Material Properties*

Developing polymer/rubber blends, especially those at the nanoscale, or nanoblends, has been a focus of significant research and innovation as the need for high-performance materials increases. One benefit of blending rubber and polymers is that the desired qualities of both materials can be combined to create new materials with better mechanical, thermal, and chemical qualities. For example, combining rubber and a rigid polymer can create a flexible and durable material, making it perfect for applications requiring resilience and impact resistance (Shah et al. 2013; Dhanorkar et al. 2021; Jourabchi and Ostad Movahed 2024).

The characteristics of polymer/rubber mixes are further improved by nanoblends, including nanoscale fillers like carbon black, silica, or graphene. Nanoscale fillers provide greater thermal stability, abrasion resistance, and reinforcement by enhancing the contact between the rubber and polymer phases. For instance, carbon black is frequently added to rubber compounds used in tires to improve durability and wear resistance. Furthermore, silica nanoparticles can increase rubber's strength and flexibility while lowering rolling resistance in tire applications, which helps cars use less fuel (Dhanorkar et al. 2021; Jourabchi and Ostad Movahed 2024; Khanra et al. 2020). The ability of polymer/rubber nanoblends to be tailored for specific uses is one of their most significant benefits. Scientists can modify these materials to satisfy exacting performance standards by carefully choosing the kind and quantity of fillers. For industries like automotive, aerospace, and healthcare, where materials must resist harsh conditions and demonstrate high levels of reliability, the capacity to fine-tune characteristics at the molecular and nanoscale levels is essential (Khanra et al. 2020; Popescu et al. 2017; Nirmal Ghosh et al. 2017).

5.1.4 Future Prospects and Emerging Applications for Polymer/Rubber Research

Research on rubber and polymer blends, especially nanoblends, keeps creating new opportunities for material development. By offering materials that reduce weight without sacrificing strength or durability, polymer/rubber composite developments, for example, help the automobile industry create lightweight, fuel-efficient vehicles. Polymer/rubber composites are utilized in aircraft for gaskets, seals, and insulating materials that withstand high pressures and temperatures. Additionally, polymer/rubber mixes are increasingly used in the medical industry for applications, including flexible medical tubing, implants, and wearable technology that need to be durable and biocompatible (Dhanorkar et al. 2021; Archibong et al. 2023). Another critical area of research for rubber and polymers is sustainability. Rubber and synthetic polymers, mostly made from non-renewable resources, harm the environment, prompting attempts to develop sustainable substitutes. The creation of recyclable polymer/rubber composites and advancements in biopolymers have the potential to save resources and cut waste. For instance, the environmental issues related to traditional rubber products may be lessened by employing NR or biodegradable polymers in particular applications (Archibong et al. 2023; Singh et al. 2023).

Rubber and polymer science have a promising future because to continuous development into more sustainable, robust, and effective materials. Scientists can push the limits of what these materials can accomplish thanks to ongoing innovations in materials engineering, nanotechnology, and polymer chemistry (Singh et al. 2023). Rubber and polymer materials will surely be crucial in forming the next generation of high-performance solutions as companies look to address contemporary issues, including energy efficiency, environmental sustainability, and technological innovation. Fundamental materials like rubber and polymers have transformed industries and are still evolving due to scientific advancements (Dhanorkar et al. 2021; Nirmal Ghosh et al. 2017). The endless possibilities of polymer/rubber mixes and nanoblends, from traditional to novel applications, are demonstrated via research and development. As we strive for a more sustainable and efficient future, innovative polymer and rubber technology will be essential in meeting the demands of a rapidly changing world (Archibong et al. 2023; Singh et al. 2023; Neethirajan et al. 2022).

5.2 Introduction to Polymer/Rubber Blends

Combining the unique qualities of synthetic polymers and elastomers to produce materials with a mix of flexibility, durability, and resilience has made polymer/rubber blends a crucial field in materials science (Ansari et al. 2021; Shah et al. 2013; Dhanorkar et al. 2021; Jourabchi and Ostad Movahed 2024; Khanra et al. 2020; Popescu et al. 2017; Nirmal Ghosh et al. 2017; Archibong et al. 2023). These blends are utilized in various industries, including consumer products, healthcare, automotive, and aerospace, where they are becoming increasingly popular due to their capacity to satisfy strict performance and sustainability requirements. By combining polymers and rubber, scientists and engineers can produce materials that have the strength, chemical resistance, and adaptability of polymers with the flexibility and elasticity of rubber, creating hybrid materials that can fulfill various functional needs (Lendvai 2021; Caldona et al. 2017).

5.2.1 The Nature of Polymers and Rubber: Building Blocks of Modern Blends

Polymers are macromolecules formed from long chains of repeating units, called monomers, which link together through chemical bonds. These chains can be manipulated in terms of structure, length, and arrangement, which allows scientists to create a diverse range of materials with specific characteristics, such as rigidity, flexibility, chemical resistance, and thermal stability. Some common synthetic polymers include polyethylene, polypropylene, polystyrene, and polyvinyl chloride (PVC), each offering unique properties suited to different applications (Caldona et al. 2017; Abbas-Abadi et al. 2021; Costa et al. 2010; Ciliz et al. 2004). On the other hand, rubber is a specific category of polymer known for its high elasticity and resilience. Its molecular structure of coiled chains with cross-linked points allows it to stretch and return to its original shape (Caldona et al. 2017; Abbas-Abadi et al. 2021).

Rubber exists in both natural and synthetic forms. NR, derived from the latex of the Hevea brasiliensis tree, has been used for centuries and is valued for its flexibility and tensile strength. Synthetic rubbers, such as SBR, nitrile rubber, and neoprene, were developed to address natural rubber's limitations and supply shortages, especially during World War II. Each type of rubber offers specific advantages, such as oil resistance, temperature tolerance, or enhanced durability. Combining these with synthetic polymers through blending creates versatile materials that capitalize on the benefits of both polymers and rubbers, making them highly adaptable for various applications (Ansari et al. 2021; Shah et al. 2013; Dhanorkar et al. 2021; Jourabchi and Ostad Movahed 2024; Khanra et al. 2020; Popescu et al. 2017; Nirmal Ghosh et al. 2017; Archibong et al. 2023; Singh et al. 2023; Neethirajan et al. 2022; Lendvai 2021; Caldona et al. 2017; Abbas-Abadi et al. 2021).

5.2.2 *The Science of Polymer/Rubber Blending*

Successful polymer/rubber mix creation is a complex procedure requiring exact material property management to ensure compatibility and functionality. When rubber and polymers are blended, their chemical structures are frequently changed to improve adhesion, heat stability, and deformation resistance. Usually, the objective is to produce a material that demonstrates a balance between rubber's elastic properties and polymers' rigid qualities (Lendvai 2021; Caldona et al. 2017; Abbas-Abadi et al. 2021; Costa et al. 2010; Ciliz et al. 2004).

The fundamental distinctions between rubber and polymers make mixing them one of the most challenging tasks. Rubbers are more amorphous and flexible than polymers, frequently stiff, crystalline structures with intermolecular solid interactions. To maintain compatibility between the two phases, researchers employ a variety of approaches to address these disparities. Compatibilizers, additives that improve the interaction between the rubber and polymer phases, are frequently used in this process. Compatibilizers increase the blend's mechanical strength, decrease phase separation, and alter the interface between the materials. The two phases can also be chemically bonded using grafting and cross-linking techniques to create a more cohesive and stable substance (Chaudhary et al. 2000; Fasihi and Mansouri 2016; Rangsantham et al. 2024).

Beyond compatibilizers, incorporating fillers, such as carbon black, silica, or nanoclay, is common in polymer/rubber blends. Fillers reinforce the material, improving properties such as tensile strength, abrasion resistance, and thermal stability. Carbon black, for example, is widely used in the tire industry for its ability to enhance the durability and performance of rubber. The selection of fillers and their distribution within the polymer/rubber matrix play crucial roles in determining the overall behavior of the blend, making the formulation process highly customizable for specific end-use requirements (Rangsantham et al. 2024; Mora-Barrantes et al. 2011; Abdul Sattar 2024).

5.2.3 *Properties of Polymer/Rubber Blends*

The types of polymers and rubbers utilized, the mix ratio, and the processing techniques can all substantially impact the properties of polymer/rubber blends. Blends of polymers and rubber are made to combine the best qualities of both materials,

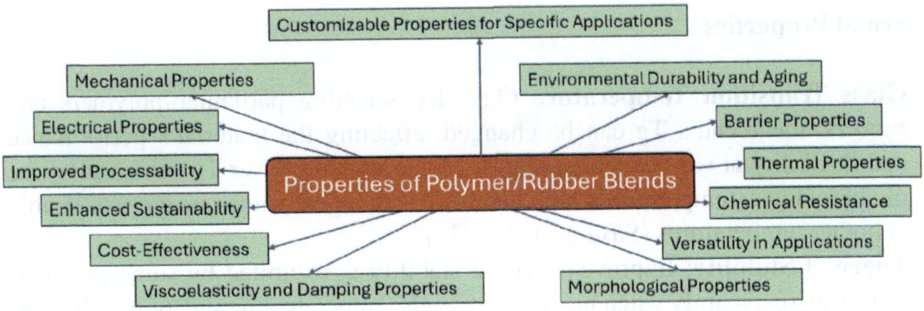

Fig. 5.1 Properties of polymer/rubber blends

frequently leading to improved processing, mechanical, thermal, and chemical properties (Abdul Sattar 2024; Paul et al. 2009; Rane et al. 2017). Blends of polymer and rubber can have highly flexible qualities that can be tailored to meet specific requirements, making them useful in various sectors. The blend design usually entails balancing mechanical, thermal, chemical, and environmental qualities for the best performance in the intended applications. Polymer/rubber mixes improve performance, longevity, and cost-effectiveness by providing the ideal strength, flexibility, chemical resistance, and processability ratio. They are extremely useful in many high-performance applications because of their adaptability and customizability (Paul et al. 2009; Rane et al. 2017; Bhattacharya et al. 2020). The schematic of polymer/rubber blends properties is shown in Fig. 5.1.

Mechanical Properties

- **Tensile Strength**: Blends of polymers and rubber can be made to have high tensile strength, which is necessary for applications like industrial seals and automobile tires that need robust, long-lasting materials. The type of rubber and polymer employed and the degree of cross-linking can all affect the tensile strength (Lendvai 2021; Hassan et al. 2005).
- **Elasticity and Flexibility**: The combination gains elasticity and flexibility from the rubber components, which enable it to stretch and regain its original shape. This characteristic is essential for shock absorbers and gaskets, among other applications (Rangsantham et al. 2024; Zhang et al. 2021).
- **Impact Resistance**: Rubber and polymer blending can improve impact resistance, which is advantageous for goods that must withstand abrupt forces or absorb shocks. This is especially useful for sporting goods and automotive applications (Lendvai 2021; Han et al. 2023).
- **Tear and Abrasion Resistance**: Blends of polymers and rubber frequently exhibit enhanced resistance to tearing and abrasion, which is essential for applications subjected to physical wear, including industrial hoses and conveyor belts (Aderikha and Marusenko 2024; Buitrago-Suescún and Britto 2020; Ezenkwa et al. 2021).

Thermal Properties

- **Glass Transition Temperature (Tg)**: By selecting particular polymers and rubbers, the blend's Tg can be changed, affecting the material's performance and flexibility at various temperatures. Materials with lower Tg values are more pliable at lower temperatures, whereas those with higher values are more inflexible at higher temperatures (Saxena et al. 2021).
- **Thermal Stability**: Improved thermal stability is exhibited by some polymer/rubber mixes, mainly when intended for high-temperature applications. They are appropriate for uses such as automobile parts since this stability can be altered by adding stabilizers or fillers (Saxena et al. 2021; Chen et al. 2021).
- **Heat Resistance**: Blends of polymers and rubber are perfect for settings with large temperature fluctuations because they can tolerate high temperatures without compromising their mechanical integrity (Saxena et al. 2021; Chen et al. 2021; Zhang et al. 2011).

Chemical Resistance

- **Oil and Solvent Resistance**: Nitrile rubber (NBR) is one of the several rubber compounds with inherent resistance to oils, fuels, and solvents. This resistance can be further increased by blending with polymers, which makes these blends perfect for use in the oil and gas, automotive, and aerospace sectors (Chen et al. 2021; Laundrie et al. 1954).
- **Weathering Resistance**: Some blends are more resistant to environmental deterioration than others, especially those that contain ozone-resistant or UV-stabilized rubbers. These mixtures are appropriate for outdoor settings where exposure to ozone and sunlight poses a risk (Saxena et al. 2021; Urtekin et al. 2023).
- **Corrosion and Chemical Attack**: Certain rubber and polymer blends are appropriate for use in hostile chemical environments because they can withstand corrosive gasses and chemicals (Zhang et al. 2011; Laundrie et al. 1954; Urtekin et al. 2023; Saleh and Farag 2020).

Viscoelasticity and Damping Properties

- **Damping Capacity**: Because polymer/rubber blends are viscoelastic, they can absorb and release energy, giving them a high damping capacity. In applications that dampen noise and vibration, such as automotive mounts and insulating materials, this is essential (Arroyo et al. 2007).
- **Creep and Stress Relaxation**: Certain creep and stress relaxation behaviors are also brought about by the viscoelastic nature of polymer/rubber mixes. For load-bearing applications, creep resistance can be improved in blends to prevent deformation over time under continuous stress (Wang et al. 2021).

Electrical Properties

- **Dielectric Properties**: Due to their high dielectric strength, some polymer/rubber blends are utilized in electrical insulation. These qualities are improved when rubber is mixed with insulating polymers like polyethylene or polyvinyl chloride (Khan and Iqbal 2023).
- **Conductivity or Antistatic Qualities**: To give electrical conductivity or antistatic qualities, particularly in electronic applications where static electricity must be reduced, certain mixes are altered with conductive fillers (such as metal particles or carbon black) (Urtekin et al. 2023; Bhavitha et al. 2021).

Barrier Properties

- **Gas and Vapor Barrier**: Considering their barrier qualities, rubber and polymer mixes are frequently utilized in packaging applications. For instance, polymers like ethylene vinyl alcohol (EVOH) improve the oxygen and moisture barrier qualities of packing materials, while butyl rubber offers superior gas impermeability (Viswanathan et al. 2024).
- **Water and Moisture Resistance**: It is possible to create blends that are resistant to moisture penetration, which is crucial for flexible packaging, automotive components, and protective coatings (Urtekin et al. 2023; Viswanathan et al. 2024).

Morphological Properties

- **Compatibility and Morphology**: The compatibility of the two components frequently affects the morphology of polymer/rubber mixtures. Compatibilizers may be necessary for incompatible mixes to enhance interfacial adhesion and guarantee a consistent distribution of characteristics (Zhao et al. 2020).
- **Processability**: Blends of polymers and rubber are frequently created to increase processability, making molding and shaping simpler. For instance, thermoplastic elastomers (TPEs) combine the processability of thermoplastics with the elasticity of rubber (Zhao et al. 2020).
- **Blend Stability**: If the components in a blend are not entirely compatible, phase separation may occur over time. To preserve blend stability and guarantee consistency in qualities, compatibilizers and other additives should be carefully chosen and designed (Thomas et al. 2023).

Environmental Durability and Aging

- **Resistance to Oxidation and UV Light**: Many polymers and rubbers can degrade when exposed to oxygen and UV light. Stabilizers and UV absorbers

are frequently added to polymer/rubber mixtures to increase resistance to these elements (Abraham et al. 2023).
- **Resistance to Fatigue and Cracking**: Rubber and polymer blending can improve resistance to fatigue and cracking, which is important for materials like tires and belts that are subjected to repeated stress (Madhavan et al. 2024).

Customizable Properties for Specific Applications

- **Customized Performance Features**: Blends of polymers and rubber can be precisely tailored to fulfill requirements, which allows them to have unique mechanical, thermal, and chemical qualities that make them suitable for a range of applications (Sidiqui and Waris Khan 2024).
- **Improved Morphology and Compatibility**: To increase the distribution of rubber throughout the polymer matrix and produce a homogeneous blend with more consistent properties—perfect for applications where performance is crucial—additives or compatibilizers are occasionally included (Urtekin et al. 2023; Zhao et al. 2020).

Improved Processability

- **Easy Molding and Shaping**: Rubber's processing qualities are enhanced by blending with specific polymers, which makes it simpler to mold, extrude, and shape. For instance, thermoplastic elastomers (TPEs), which combine the elasticity of rubber with the ease of thermoplastic manufacturing, are more straightforward to produce than conventional rubbers (Urtekin et al. 2023; Saleh and Farag 2020; Arroyo et al. 2007; Wang et al. 2021; Khan and Iqbal 2023; Bhavitha et al. 2021; Viswanathan et al. 2024; Zhao et al. 2020).
- **Reduced Processing Costs**: Since thermoplastics typically call for lower processing temperatures and shorter cycle times, the ability to combine rubber with thermoplastic polymers frequently results in reduced processing costs (Saxena et al. 2021; Chen et al. 2021; Zhang et al. 2011; Laundrie et al. 1954; Urtekin et al. 2023; Saleh and Farag 2020; Arroyo et al. 2007; Wang et al. 2021; Khan and Iqbal 2023; Bhavitha et al. 2021; Viswanathan et al. 2024; Zhao et al. 2020; Thomas et al. 2023; Abraham et al. 2023; Madhavan et al. 2024; Sidiqui and Waris Khan 2024).

Cost-Effectiveness

- **Lower Material Costs**: Manufacturers can lower the amount of costly rubber required for a finished product without sacrificing performance by mixing rubber with other polymers (Boon et al. 2022; Urtekin et al. 2023; Madhavan et al. 2024).

- **Extended Product Lifespan**: Longer-lasting products with improved durability due to polymer/rubber blends require less regular maintenance and replacements, which can ultimately save money (Dong et al. 2022).

Enhanced Sustainability

- **Recyclability Potential**: Some polymer/rubber mixes, especially thermoplastic elastomers (TPEs), can be recycled more efficiently than conventional vulcanized rubber, which supports initiatives for a circular economy (Boon et al. 2022; Urtekin et al. 2023).
- **Minimized Environmental Impact**: Manufacturers can use fewer raw materials and possibly lessen their environmental impact by optimizing their materials. Furthermore, polymer/rubber mixes' durability lowers the need for replacement and disposal (Chang et al. 2021).

Versatility in Applications

- **Numerous Applications**: Because of their versatile qualities, polymer/rubber blends find application in a wide range of industries, including consumer goods (sports gear, footwear), healthcare (medical gloves, tubing), automotive (tires, seals), and aerospace (insulation materials) (Popescu et al. 2017; Nirmal Ghosh et al. 2017; Archibong et al. 2023; Singh et al. 2023; Neethirajan et al. 2022; Lendvai 2021; Caldona et al. 2017; Abbas-Abadi et al. 2021; Costa et al. 2010; Ciliz et al. 2004; Chaudhary et al. 2000; Fasihi and Mansouri 2016; Rangsantham et al. 2024; Mora-Barrantes et al. 2011; Abdul Sattar 2024; Paul et al. 2009; Rane et al. 2017; Bhattacharya et al. 2020; Hassan et al. 2005; Zhang et al. 2021).
- **Greater Market Reach**: Mixes made for uses, such as medical-grade elastomers or flame-retardant mixes for electronics, enable producers to serve niche markets with particular needs (Frisch 2021).

5.2.4 Nanotechnology in Polymer/Rubber Blends: The Rise of Nanoblends

Incorporating nanotechnology has revolutionized the blending of polymers and rubbers by enabling scientists to work with these materials at the nanoscale. To improve the blend's mechanical, thermal, and electrical qualities, nanoblends use nanoscale fillers like graphene, silica nanoparticles, carbon nanotubes, or nanoclay (Imoisili et al. 2020a, 2020b; Okafor et al. 2021). Silica nanoparticles, for instance, improve the flexibility and abrasion resistance of tires rubber while lowering rolling resistance and boosting fuel economy (Imoisili et al. 2020b, 2022; Sethulekshmi et al. 2022; Singha et al. 2019; Imoisili and Jen 2022). Likewise, adding electrical conductivity by carbon nanotubes can make the mixture appropriate for electronic applications (Okafor et al. 2021; Imoisili et al. 2020b, 2022; Sethulekshmi et al. 2022; Singha et al. 2019; Imoisili and Jen 2022).

Applications needing highly specific performance characteristics benefit greatly from nanoblends because nanoscale fillers greatly enhance material qualities without sacrificing flexibility or adding bulk. For example, nanoclay fillers can improve the material's barrier qualities, making it more resistant to gas permeation. This is helpful in fuel systems, packaging, and other applications where containment is crucial. Because of their nanoscale size, these fillers can also be better dispersed throughout the blend, which results in more consistent reinforcement and longer durability (Chang et al. 2021; Imoisili and Jen 2022; Imoisili and Jen 2023; Kumar et al. 2021). Despite the many advantages, there are still difficulties. These include controlling the price of premium nanoparticles, guaranteeing consistent dispersion of nanomaterials, and addressing any possible adverse effects of nanomaterials on human health and the environment. To further improve the properties and uses of polymer/rubber nanoblends, research is being done to find solutions to these problems. Future developments in bio-based nanocomposites, sustainable nanomaterials, and "smart" features will probably increase the use of polymer/rubber mixes even more to satisfy the industry's increasing needs for long-lasting, high-performing, and ecologically friendly products (Boon et al. 2022; Viswanathan et al. 2024; Zhao et al. 2020; Thomas et al. 2023; Abraham et al. 2023; Madhavan et al. 2024; Sidiqui and Waris Khan 2024; Dong et al. 2022; Chang et al. 2021; Frisch 2021; Imoisili et al. 2020a, b, 2022; Okafor et al. 2021; Sethulekshmi et al. 2022; Singha et al. 2019; Imoisili and Jen 2022).

5.3 Polymer/Rubber Blends Applications

Blends of polymers and rubber are recognized for their unique qualities, such as flexibility, durability, chemical resistance, electrical insulation, and thermal stability, making them indispensable across a variety of industries. Their adaptability and resilience allow for customization to meet specific requirements in sectors like automotive, aerospace, healthcare, consumer goods, and electronics. Advances in material science have further enhanced the versatility of these blends, enabling them to address the evolving demands of modern applications (Archibong et al. 2023; Singha et al. 2019). Figure 5.2 shows the various applications of polymer/rubber blend.

5.3.1 Automotive Industry

- **Tires and Inner Tubes**: Rubber and polymer/rubber mixes are frequently used in tire production. Rubber offers flexibility and resistance to abrasion, while polymer additions improve performance and durability across a range of temperatures (Rahman et al. 2023).

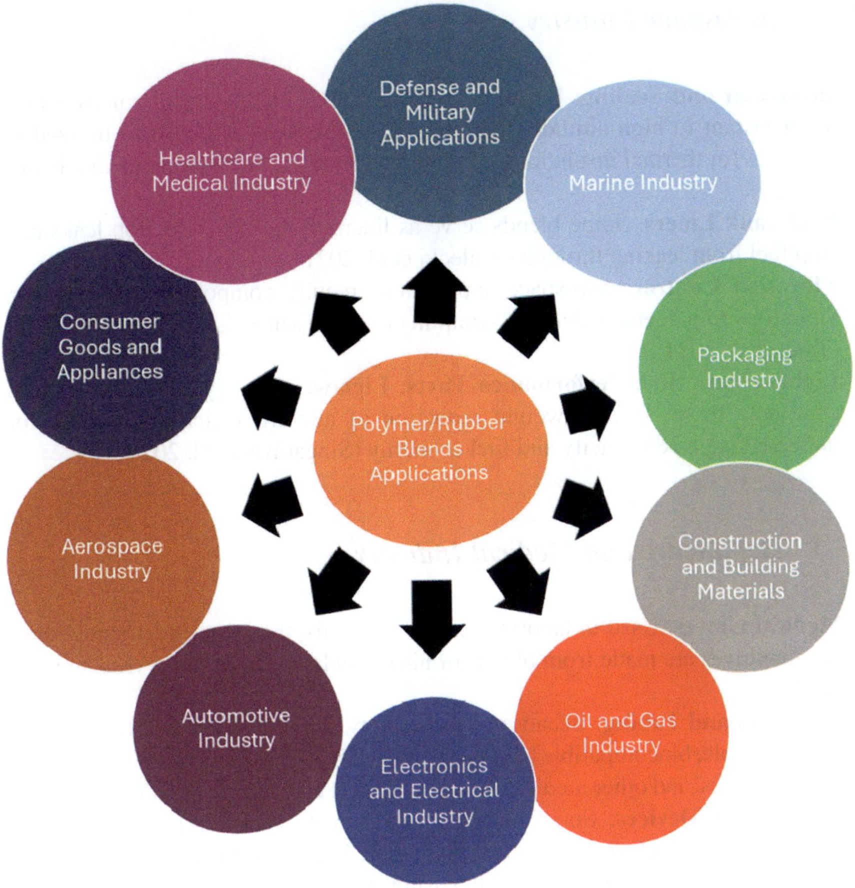

Fig. 5.2 Different applications of polymer/rubber blends

- **Seals and Gaskets**: Due to their resilience to heat, fuels, and oils, these mixes are utilized for gaskets and seals in engines, fuel systems, and braking systems (Yang et al. 2018).
- **Vibration-Dampening Components**: Motor mounts and suspension systems employ blends with a high damping capacity to lessen harshness, noise, and vibration, making rides quieter and more comfortable (Chen et al. 2022).
- **Hoses and Tubing**: Polymer/rubber blends used to make fuel and coolant hoses offer flexibility and resistance to chemical deterioration, guaranteeing the safe transportation of fluids inside automobiles (Vayyaprontavida Kaliyathan et al. 2020).

5.3.2 *Aerospace Industry*

- **Insulation and Sealing**: Because polymer/rubber blends are thermally stable and resistant to high-altitude, low-pressure circumstances, they are utilized in airplanes for thermal insulation and sealing purposes (Dahi Taleghani and Santos 2023).
- **Fuel Tank Liners**: Some blends serve as liners in fuel tanks to stop leaks and stop fuel from leaking through (Białecki et al. 2021).
- **Vibration Control**: Aerospace applications require components that dampen vibrations to safeguard delicate equipment and lessen noise in airplane cabins (Tang et al. 2024).
- **Lightweight, High-Performance Parts**: Lightweight polymer/rubber blends contribute to the overall weight reduction of aircraft, which is essential for increased payload capacity and fuel economy (Singaravel et al. 2024).

5.3.3 *Healthcare and Medical Industry*

- **Medical Gloves**: Used in laboratories and hospitals, these hypoallergenic, long-lasting gloves are made from blends of nitrile rubber and polymers (Lazim et al. 2024).
- **Catheters and Tubing**: Because of their safety, flexibility, and sterilizing resistance, flexible, biocompatible blends are utilized in medical devices like catheters, infusion tubes, and other medical tubing (Lazim et al. 2024; Pimentel et al. 2023).
- **Implantable Devices**: Some blends are designed to satisfy biocompatibility requirements, which qualifies them for use in implanted medical devices such as joint replacements and pacemaker casings (Ershad-Langroudi et al. 2024).
- **Medical Equipment Gaskets and Seals**: Polymer/rubber blends offer dependable gaskets and seals for medical equipment, which is essential for preserving sterility (Lazim et al. 2024; Pimentel et al. 2023; Ershad-Langroudi et al. 2024; Rahimi and Mashak 2013).

5.3.4 *Consumer Goods and Appliances*

- **Footwear**: Polymer/rubber blends are perfect for shoe soles because of their flexibility, resilience, and durability, which offer comfort and durability in high-impact regions (Wang et al. 2022).
- **Sports Equipment**: Polymer/rubber blends are used in many sports products, such as grips, pads, and protective gear, to improve comfort, durability, and shock absorption (Laad 2020).

- **Home Appliances**: Due to their flexibility and chemical resistance, these blends make parts like hoses and seals for washing machines, dishwashers, and refrigerators (Athawale and Joshi 2016).
- **Protective Gear**: Due to its comfort, flexibility, and resistance to impact, blends are utilized in various protective gear, such as gloves, knee pads, and elbow pads (Wang et al. 2022; Laad 2020; Athawale and Joshi 2016).

5.3.5 Electronics and Electrical Industry

- **Sheathing and Insulation for Cables**: Rubber and polymer blends offer flexible, long-lasting insulation for power cables and wires that guards against physical wear, moisture, and chemicals (Hussain et al. 2020).
- **Antistatic and Conductive Materials**: To reduce static accumulation and safeguard delicate electronics, electronic housings, and components employ polymer/rubber blends with antistatic or conductive qualities (Soto-Oviedo et al. 2006).
- **Coatings and Encapsulation**: These mixtures are applied to electronic components to prolong their lifespan by shielding them from dust, moisture, and chemicals (do Nascimento et al. 2019).
- **Flexible Circuitry**: Certain blends are utilized in flexible circuits, which are necessary for applications requiring bending or movement, such as flexible displays and wearable technology (Han et al. 2019).

5.3.6 Oil and Gas Industry

- **Seals and Gaskets for High-Pressure Environments**: Pipeline, valve, and pump seals and gaskets are made of polymer/rubber blends that are resistant to severe chemicals, oils, and high temperatures (Rangsantham et al. 2024; Zhang et al. 2021; Yang et al. 2018).
- **Oil Transport Hoses and Tubing**: These mixes are crucial for hoses that move gas and oil because they provide resistance to chemicals and the environment and stop leaks (Buitrago-Suescún and Britto 2020; Ezenkwa et al. 2021; Vayyaprontavida Kaliyathan et al. 2020).
- **Downhole Components**: Polymer/rubber mixes are utilized in downhole instruments used in drilling and extraction, which are subjected to extreme temperatures, pressures, and chemicals (Ahmed et al. 2019).
- **Pipeline Coatings**: These mixes are used to make coatings that maintain pipeline integrity by preventing leaks and preventing corrosion (Zhang et al. 2021; Yang et al. 2018; Azevedo et al. 2024)

5.3.7 *Construction and Building Materials*

- **Materials for Weatherproofing and Roofing**: Polymer/rubber mixes are frequently utilized in roofing materials and weatherproofing membranes because they offer ozone and UV resistance for extended outdoor exposure (Hancharoen et al. 2024).
- **Sealants and Expansion Joints**: These blends are perfect for expansion joints and construction sealants since they are flexible, long-lasting, and resistant to changes in humidity and temperature (Li et al. 2023).
- **Flooring and Soundproofing**: These blends provide commercial and residential areas with durability, cushioning, and noise reduction in rubberized flooring and soundproofing materials (Zeng et al. 2023).
- **Waterproofing Systems**: To stop water from seeping into buildings, tunnels, and bridges, waterproofing systems employ mixtures of polymers and rubber (Xiao et al. 2021).

5.3.8 *Packaging Industry*

- **Flexible Packaging Films**: These films are made of blends that give them strength, flexibility, and barrier qualities that keep out moisture and oxygen, preserving food and medications (Urtekin et al. 2023; Viswanathan et al. 2024).
- **Chemical and Moisture-Resistant Containers**: Blends produce containers that are extremely resistant to chemicals and moisture, making them perfect for the transportation and storage of chemicals, fertilizers, and medications (Popescu et al. 2017).
- **Liners and Seals**: Polymer/rubber mixes are frequently used in bottle caps, jar seals, and other closures to create leak-proof seals that preserve freshness and guard against contamination (Boon et al. 2022; Wiśniewska et al. 2023; Imoisili et al. 2013).

5.3.9 *Marine Industry*

- **Maritime Hoses and Seals**: Polymer/rubber mixes are appropriate for maritime hoses, gaskets, and seals because they resist abrasion, UV radiation, and seawater (Yang et al. 2018; Pimentel et al. 2023; Rahimi and Mashak 2013).
- **Boat Bumpers and Fenders**: These mixes are used in bumpers and fenders to absorb impact and keep boats safe from crashes (Saputra et al. 2020).
- **Underwater Cables and Coatings**: To guard against mechanical wear and corrosion from seawater, blends are utilized in underwater cables and coatings (do Nascimento et al. 2019; Azevedo et al. 2024).

Defense and Military Applications

- **Protective Coatings**: To give military vehicles, equipment, and structures resistance to abrasion, chemicals, and extreme environmental conditions, certain blends are utilized in coatings (Braihi et al. 2020).
- **Shock-Absorbing Materials**: To protect people and equipment, these materials are utilized in armor, protective padding, and vibration-damping components [105].
- **Sensitive Equipment Sealing and Insulation**: Weather-resistant sealing for sensitive military equipment is made possible by polymer/rubber mixes, which guarantee operational dependability even under the most adverse circumstances (Yang et al. 2018; Dahi Taleghani and Santos 2023).

Future Trends and Sustainability in Polymer/Rubber Blends

Future developments in polymer/rubber blends are concentrating on advances that address both environmental responsibility and performance enhancements as the industry moves toward sustainable practices. Some of these themes include the creation of bio-based materials, improvements in durability and recyclability, and the application of nanotechnology to produce high-performing, environmentally friendly mixes [106]. Key developments and sustainability initiatives in the realm of polymer/rubber mixes include the following:

Bio-Based and Renewable Materials

- **Biodegradable Rubbers and Polymers**: Studies are being conducted on rubbers and biodegradable polymers made from natural sources such as natural rubber, polylactic acid (PLA), and starch. These materials decompose more readily in the environment, which lowers pollution and trash in landfills [107].
- **Plant-Based Polymers**: Castor oil, maize starch, and soy-based polyols are among the materials being developed as substitutes for petroleum-based components in polymer blends. These renewable resources help to lower carbon emissions and lessen reliance on fossil fuels [108].
- **Green Synthesis Processes**: The sustainability profile of these materials is further improved by emerging innovations in the synthesis of polymer/rubber blends with low environmental impact, such as the use of bio-catalysis and non-toxic solvents [109].

Advanced Nanotechnology for High-Performance Blends

- **Nano-Enhanced Durability and Longevity**: Polymer/rubber mixes benefit from the mechanical, thermal, and chemical resistance qualities of nanoparticles like graphene, carbon nanotubes, and nanoclays. Improved durability can prolong product lifespan and lessen environmental impact by eliminating the need for frequent replacements [110].

- **Decreased Material Usage with Nanofillers**: Manufacturers can attain the required characteristics while using less raw material overall by including tiny amounts of nanoparticles. Through waste reduction and resource conservation, this optimization promotes sustainability [111].
- **Self-Healing and Smart Materials**: Nanotechnology makes it possible to create materials that can mend little damage on their own. This feature increases a product's useful life, particularly in industries like construction and automotive materials that are prone to wear and tear (Singh et al. 2023, Imoisili et al. 2020a) [106].

Eco-Friendly Additives and Fillers

- **Natural Fiber Fillers**: In polymer/rubber blends, fibers derived from agricultural waste products—such as rice, bamboo, and coconut husks—are being used as fillers. These natural fibers lessen the need for artificial fillers while improving mechanical qualities [112].
- **Non-Toxic Stabilizers and Plasticizers**: Conventional additives may be hazardous to human health and the environment. To make blends safer and more ecologically conscious, eco-friendly substitutes like non-toxic stabilizers and bio-based plasticizers are being used (Saxena et al. 2021; Wang et al. 2021; Abraham et al. 2023).
- **Recycled Carbon Black and Other Fillers**: Reclaimed carbon black from used tires is one example of a recycled filler that helps reduce production costs and environmental effects. The reinforcing qualities of these recycled fillers are comparable to those of virgin fillers.

Lightweight and High-Performance Materials

- **Lightweighting for Energy Efficiency**: Manufacturers can produce lighter materials by lowering material density without sacrificing performance. This is especially advantageous in automotive and aeronautical applications, where lighter weight results in better fuel economy and lower emissions [113].
- **Foamed Polymer/Rubber Mixes**: Foamed mixes with cellular structures provide excellent impact resistance, thermal insulation, and strength while being lightweight. Because of their energy-efficient properties, these materials are becoming increasingly popular in the packaging, automotive, and construction sectors [114].

Digital Manufacturing and Customization

- **3D Printing of Polymer/Rubber Blends**: Polymer/rubber blend additive manufacturing is becoming more popular, enabling quick prototyping and customized parts creation. By creating only what is required, 3D printing minimizes waste and makes it possible for creative designs that utilize less material without sacrificing strength or usefulness [115].
- **Customized Material Characteristics Through Simulation**: By optimizing material characteristics prior to manufacturing, cutting down on waste and resource consumption during the development process, this is possible with the use of sophisticated simulation software [116].

Applications in Renewable Energy and Sustainability-Oriented Products

- **Renewable Energy Components**: Battery casings, wind turbine seals, and solar panel mounts are examples of renewable energy systems that use polymer/rubber mixes with superior durability and chemical resistance. By offering durable, effective materials, these mixes aid in the shift to sustainable energy (Sionkowska 2011; Jourabchi and Ostad Movahed 2024; Lendvai 2021; Caldona et al. 2017).
- **Eco-Friendly Packaging Solutions**: Because sustainable blends break down more quickly than conventional packaging, they are increasingly utilized in eco-friendly packaging, such as food containers, biodegradable films, and protective wraps (Arroyo et al. 2007; Viswanathan et al. 2024).
- **Sustainable Infrastructure**: To promote ecologically friendly infrastructure projects, these mixes are utilized in construction for energy-efficient insulating materials, long-lasting waterproofing, and green roofing systems [117].

5.4 Summary

Polymer/rubber-based nanoblends, which combine the strength and versatility of polymers with the flexibility and durability of rubber while utilizing the special qualities of nanofillers, mark a substantial advancement in materials science. Materials with improved mechanical durability, thermal stability, and chemical resistance are produced via this fusion, satisfying the high-performance requirements of sectors like healthcare, electronics, automotive, and aerospace. Graphene, silica, and carbon nanotubes are nanofillers that greatly enhance these qualities, enabling nanoblends to offer enhanced conductivity, strength, and impact resistance where necessary. With compatibilizers and careful component tuning, these nanoblends may be created to exact specifications, offering amazing adaptability in solving challenging engineering problems. Creating environmentally friendly polymer/rubber nanoblends is accelerating as companies work toward sustainability. These materials are being positioned as promising solutions for lowering environmental impact through using recyclable or bio-based polymers and investigating green chemistry concepts. It is anticipated that developments in processing techniques and the creation of nanofillers will increase the use of nanoblends and make them more widely available and adaptable. To put it briefly, polymer/rubber-based nanoblends provide a route to high-performance materials that complement sustainability objectives, laying the groundwork for further advancements in durable, sustainable material solutions in various industries.

References

Abbas-Abadi MS, Van Geem KM, Fathi M, Bazgir H, Ghadiri M (2021) The pyrolysis of oak with polyethylene, polypropylene and polystyrene using fixed bed and stirred reactors and TGA instrument. Energy 232:121085

Abdul Sattar M (2024) Surface activated pyrolytic carbon black: a dual functional sustainable filler for natural rubber composites. Chemsuschem 17(2):e202301001

Abraham SM, Verma M, Kakade V (2023) Aging resistance of bitumen modifiers: a comprehensive review. J Test Eval 51(5):3533–3560

Aderikha VN, Marusenko NA (2024) Abrasive wear of rubbers based on natural rubber, carbon black, and polyoxadiazole fiber. J Friction Wear 45(2):57–65

Ahmed S, Salehi S, Ezeakacha C, Teodoriu C (2019) Experimental investigation of elastomers in downhole seal elements: implications for safety. Polym Testing 76:350–364

Ansari AH, Jakarni FM, Muniandy R, Hassim S, Elahi Z (2021) Natural rubber as a renewable and sustainable bio-modifier for pavement applications: a review. J Clean Prod 289:125727

Archibong FN, Orakwe LC, Ogah OA, Mbam SO, Ajah SA, Okechukwu ME, Igberi CO, Okafor KJ, Chima MO, Ikelle II (2023) Emerging progress in montmorillonite rubber/polymer nanocomposites: a review. J Mater Sci 58(6):2396–2429

Arroyo M, Lopez-Manchado MA, Valentin JL, Carretero J (2007) Morphology/behaviour relationship of nanocomposites based on natural rubber/epoxidized natural rubber blends. Compos Sci Technol 67(7–8):1330–1339

Athawale AA, Joshi AM (2016) Electronic applications of ethylene propylene diene monomer rubber and its composites. Flexible Stretchable Electron Comp 305–333

Azevedo FB, Gudme C, Nunes E (2024) Pipeline Coatings. Handbook of pipeline engineering. Springer International Publishing, Cham, pp 367–394

Bhatia S, Bhatia S (2016) Natural polymers vs synthetic polymer. Natural Polymer Drug Delivery Systems: Nanoparticles, Plants, and Algae, 95–118

Bhattacharya AB, Raju AT, Chatterjee T, Naskar K (2020) Development and characterizations of ultra-high molecular weight EPDM/PP based TPV nanocomposites for automotive applications. Polym Compos 41(12):4950–4962

Bhavitha KB, Yaragalla S, China Satyanarayana CH, Kalarikkal N, Thomas S (2021) Natural rubber/graphene nanocomposites and their applications. Graphene Based Biopolymer Nanocomposites, 203–220

Białecki T, Sitkiewicz A, Giemza B, Sarnecki J, Skolniak M, Gawron B (2021) Compatibility of different automotive elastomers in paraffinic diesel fuel. Appl Sci 11(23):11312

Bokobza L (2018) Natural rubber nanocomposites: a review. Nanomaterials 9(1):12

Boon ZH, Teo YY, Ang DTC (2022) Recent development of biodegradable synthetic rubbers and bio-based rubbers using sustainable materials from biological sources. RSC Adv 12(52):34028–34052

Braihi AJ, Jawad AJ, Kadhum AAH, Aljibori HSS, Al-Amiery AA (2020) Chemical resistance of NR/SBR rubber blends for surfaces corrosion protection of metallic tanks in petrochemical industries. KOM–Corrosion Mater Protect J 64(2):65–71

Buitrago-Suescún O, Britto R (2020) Devulcanization of ground tire rubber: thermo-oxidation followed by microwave exposure in the presence of devulcanizing agent. Iran Polym J 29(7):553–567

Caldona EB, De Leon ACC, Pajarito BB, Advincula RC (2017) A review on rubber-enhanced polymeric materials. Polym Rev 57(2):311–338

Chang BP, Gupta A, Muthuraj R, Mekonnen TH (2021) Bioresourced fillers for rubber composite sustainability: current development and future opportunities. Green Chem 23(15):5337–5378

Chaudhary BI, Barry RP, Tusim MH (2000) Foams made from blends of ethylene styrene interpolymers with polyethylene, polypropylene and polystyrene. J Cell Plast 36(5):397–421

Chen W, Qi C, Li Y, Tao H (2021) The degradation investigation of biodegradable PLA/PBAT blend: thermal stability, mechanical properties and PALS analysis. Radiat Phys Chem 180:109239

Chen B, Dai J, Song T, Guan Q (2022) Research and development of high-performance high-damping rubber materials for high-damping rubber isolation bearings: a review. Polymers 14(12):2427

Ciliz NK, Ekinci E, Snape CE (2004) Pyrolysis of virgin and waste polypropylene and its mixtures with waste polyethylene and polystyrene. Waste Manage 24(2):173–181

Costa P, Pinto F, Ramos AM, Gulyurtlu I, Cabrita I, Bernardo MS (2010) Study of the pyrolysis kinetics of a mixture of polyethylene, polypropylene, and polystyrene. Energy Fuels 24(12):6239–6247

Cruz-Morales JA, Gutiérrez-Flores C, Zárate-Saldaña D, Burelo M, García-Ortega H, Gutiérrez S (2023) Synthetic polyisoprene rubber as a mimic of natural rubber: recent advances on synthesis, nanocomposites, and applications. Polymers 15(20):4074

Dahi Taleghani A, Santos L (2023) Elastomeric sealing assemblies. Wellbore Integrity: From Theory to Practice, 165–193

Dhanorkar RJ, Mohanty S, Gupta VK (2021) Synthesis of functionalized styrene butadiene rubber and its applications in SBR–silica composites for high performance tire applications. Ind Eng Chem Res 60(12):4517–4535

do Nascimento RM, de Paula AJ, Oliveira NC, Alves AC, de Oliveira Aquino YML, Souza Filho AG, Hernandes AC et al (2019) Towards the production of natural rubber-calcium phosphate hybrid for applications as bioactive coatings. Mater Sci Eng C 94:417–425

Dong Y, Zhao Y, Wang H, Liu P, He Y, Lin G (2022) Integration of life cycle assessment and life cycle costing for the eco-design of rubber products. Sci Rep 12(1):595

Ershad-Langroudi A, Babazadeh N, Alizadegan F, Mousaei SM, Moradi G (2024) Polymers for implantable devices. J Ind Eng Chem

Ezenkwa OE, Hassan A, Samsudin SA (2021) Influence of different surface treatment techniques on properties of rice husk incorporated polymer composites. Rev Chem Eng 37(8):907–930

Fasihi M, Mansouri H (2016) Effect of rubber interparticle distance distribution on toughening behavior of thermoplastic polyolefin elastomer toughened polypropylene. J Appl Polym Sci 133(40)

Frisch EE (2021) High performance medical grade silicone elastomer. In: Advances in biomaterials. CRC Press, pp 143–156

Guerra NB, Pegorin GSA, Boratto MH, de Barros NR, de Oliveira Graeff CF, Herculano RD (2021) Biomedical applications of natural rubber latex from the rubber tree Hevea brasiliensis. Mater Sci Eng, C 126:112126

Han J, Lu K, Yue Y, Mei C, Huang C, Wu Q, Xu X (2019) Nanocellulose-templated assembly of polyaniline in natural rubber-based hybrid elastomers toward flexible electronic conductors. Ind Crops Prod 128:94–107

Han IS, Park C, Jeon H, Ahn S, Lee HG, Yoo J, Yoo Y (2023) Polypropylene/polyolefin elastomer composites with enhanced impact strength: the effect of rubber domain size on toughness. J Polym Res 30(6):224

Hancharoen K, Kamhangrittirong P, Suwanna P (2024) Improvement of natural fiber cement composite for roofing applications through addition of waste tire rubber: an investigation of the physical, mechanical, thermal, and acoustic properties. Cleaner Materials 13:100265

Hassan A, Wahit MU, Chee CY (2005) Mechanical and morphological properties of PP/LLPDE/NR blends—effects of polyoctenamer. Polym-Plast Technol Eng 44(7):1245–1256

Hussain Z, Lizhen G, Moeen M (2020) Treatment of coal fly ash and environmentally friendly use with rubber in cable wires as insulation material. Sustainability 12(12):5218

Ikeda Y, Kato A, Kohjiya S, Nakajima Y (2018) Rubber science. Springer, Singapore, 10, 978–981

Imoisili PK, Jen TC (2022) Microwave-assisted sol-gel template-free synthesis and characterization of silica nanoparticles obtained from South Africa coal fly. Nanotechnol Rev 11:1–11

Imoisili PE, Jen TC (2023) Effect of physical modification on the tensile and thermal properties of plantain fibre polymer composite. In: ASME international mechanical engineering congress and exposition, vol 87615. American Society of Mechanical Engineers, p V004T04A004

Imoisili PE, Jen TC (2024) Mechanism and kinetics of water absorption of plantain fibre reinforced bio-composites. In: Materials science forum, vol 1115. Trans Tech Publications Ltd, pp 55–62

Imoisili PE, Ukoba KO, Adejugbe IT, Adgidzi D, Olusunle SOO (2013) Mechanical properties of rice husk/carbon black hybrid natural rubber composite. Chem Mater Res 3(8):12–16

Imoisili PE, Ukoba KO, Jen T-C (2020a) Synthesis and characterization of amorphous mesoporous silica from palm kernel shell ash. Boletín De La Sociedad Española De Cerámica y Vidrio. 59:159–164

Imoisili PE, Ukoba KO, Jen T-C (2020b) Green technology extraction and characterisation of silica nanoparticles from palm kernel shell ash via sol–gel. J Mater Res Technol 9(1):307–313

Imoisili PE, Nwanna EC, Jen T-C (2022) Facile Preparation and characterization of silica nanoparticles from south Africa fly ash using a sol-gel hydrothermal method. Process 10(11):2440

Imoisili PE, Adeleke O, Makhatha ME, Jen TC (2023) Response surface methodology (RSM)-artificial neural networks (ANN) aided prediction of the impact strength of natural fibre/carbon nanotubes hybrid reinforced polymer nanocomposite. Engineered Science 23:852

Jourabchi S, Ostad Movahed S (2024) The effect of microwave irradiation on the molecular diffusion of the ethyl acetate (EA) into the blend of styrene-butadiene rubber (SBR)/ethylene-propylene-diene monomer rubber (EPDM). Progress in Rubber, Plastics and Recycling Technology, 14777606241262875

Kauffman GB (2001) Charles Goodyear (1800–1860), American Inventor, on the Bicentennial of His Birth. Chem Educ 6(1):50–54

Khan S, Iqbal A (2023) Organic polymers revolution: applications and formation strategies, and future perspectives. J Polym Sci Eng 6(1):3125

Khanra S, Ganguly D, Ghorai SK, Goswami D, Chattopadhyay S (2020) The synergistic effect of fluorosilicone and silica towards the compatibilization of silicone rubber and fluoroelastomer based high performance blend. J Polym Res 27:1–17

Kohjiya S, Ikeda Y (2021) A short history of natural rubber research. In: Chemistry, manufacture, and applications of natural rubber. Woodhead Publishing, pp 407–427

Kumar V, Alam MN, Manikkavel A, Song M, Lee DJ, Park SS (2021) Silicone rubber composites reinforced by carbon nanofillers and their hybrids for various applications: a review. Polymers 13(14):2322

Laad MS (2020) Polymers in sports. In: Polymer science and innovative applications. Elsevier, pp 485–523

Laundrie RW, Feldon M, Rodde AL (1954) Evaluation of diene-type elastomers for oil and solvent resistance and low temperature flexibility. Ind Eng Chem 46(4):794–803

Lazim NH, Shamsudin SA, Hidzir NM, Keong CC, Ichinose I (2024) Optimizing natural rubber blends for examination gloves: Vacuum radiation for mechanical strength and solvent barrier. J Appl Polym Sci 141(1):e54741

Lendvai L (2021) A novel preparation method of polypropylene/natural rubber blends with improved toughness. Polym Int 70(3):298–307

Li K, Yan X, Ai T, Niu Y, Jiang S (2023) Study on properties and application of chloroprene rubber/polyurethane modified asphalt sealant. Constr Build Mater 406:133177

Madhavan MK, Nambaithodi V, Krishnamoorthy A, Palanisamy S (2024) Hemicellulose rubber composites and rubber bionanocomposites. In: Rubber based bionanocomposites: characterisation. Springer Nature Singapore, Singapore, pp 105–127

Mooibroek H, Cornish K (2000) Alternative sources of natural rubber. Appl Microbiol Biotechnol 53:355–365

Mora-Barrantes I, Ibarra L, Rodriguez A, Gonzalez L, Valentín JL (2011) Elastomer composites based on improved fumed silica and carbon black. advantages of mixed reinforcing systems. J Mater Chem 21(43):17526–17533

Neethirajan J, Parathodika AR, Hu GH, Naskar K (2022) Functional rubber composites based on silica-silane reinforcement for green tire application: the state of the art. Functional Composite Materials 3(1):7

Nirmal Ghosh OS, Gayathri S, Sudhakara P, Misra SK, Jayaramudu J (2017) Natural rubber nanoblends: preparation, characterization and applications. Rubber Nano Blends: Preparation, Characterization and Applications, 15–65

Okafor CE, Onovo AC, Imoisili PE, Kulkarni KM, Ihueze CC (2021) Optimal route to robust hybridization of banana-coir fibre particulate in polymer matrix for automotive applications. Materialia, 101098

Paul KT, Pabi SK, Chakraborty KK, Nando GB (2009) Nanostructured fly ash–styrene butadiene rubber hybrid nanocomposites. Polym Compos 30(11):1647–1656

Pimentel E, Costa P, Tubio CR, Vilaça JL, Costa CM, Lanceros-Méndez S, Miranda D (2023) Printable piezoresistive polymer composites for self-sensing medical catheter device applications. Compos Sci Technol 239:110071

Popescu RC, Popescu D, Grumezescu AM (2017) Applications of rubber-based blends. In: Recent developments in polymer macro, micro and nano blends. Woodhead Publishing, pp 75–109

Rahimi A, Mashak A (2013) Review on rubbers in medicine: natural, silicone and polyurethane rubbers. Plast, Rubber Compos 42(6):223–230

Rahman MZ, Rahman M, Mahbub T, Ashiquzzaman M, Sagadevan S, Hoque ME (2023) Advanced biopolymers for automobile and aviation engineering applications. J Polym Res 30(3):106

Rane AV, Kanny K, Abitha VK, Patil SS, Thomas S (2017) Clay–polymer composites: design of clay polymer nanocomposite by mixing. In: Clay-Polymer nanocomposites. Elsevier, pp 113–144

Rangsantham P, Nonthiphalang T, Wongwitthayakool P, Sirisinha C, Krajangta N, Phumpatrakom P (2024) The flexural strength and the effect of the autoclave sterilization of polypropylene/natural rubber blended materials. Dentistry J 12(11):361

Rodgers B, Waddell W (2005) The science of rubber compounding. In Science and technology of rubber. Academic Press, pp 401–454

Saleh AMM, Farag RK (2020) Recycling of reactive and functional polymers. reactive and functional polymers volume four: surface, interface, biodegradability, compostability and recycling, 129–157

Saputra DA, Husin S, Gumelar MD, Aisah N, Susanto H, Admi RI, Anindita GL (2020) Preparation and characterization of hard rubber and soft rubber for marine rubber fender. In: Macromolecular symposia, vol 391, no 1, p 1900189

Saxena P, Shukla P, Gaur MS (2021) Thermal analysis of polymer blends and double layer by DSC. Polym Polym Comp 29(9_suppl):S11-S18

Sethulekshmi AS, Saritha A, Joseph K (2022) A comprehensive review on the recent advancements in natural rubber nanocomposites. Int J Biol Macromol 194:819–842

Shah AA, Hasan F, Shah Z, Kanwal N, Zeb S (2013) Biodegradation of natural and synthetic rubbers: a review. Int Biodeterior Biodegradation 83:145–157

Sidiqui AJ, Waris Khan MG (2024) Emerging trends of polymer materials for footwear applications in India. Engineering Headway 11:17–26

Singaravel DA, Veerapandian P, Rajendran S, Dhairiyasamy R (2024) Enhancing high-performance concrete sustainability: integration of waste tire rubber for innovation. Sci Rep 14(1):4635

Singh P, Goli VSNS, Singh DN (2023) Current practices and the way forward in waste rubber utilization for infrastructural and geoenvironmental engineering applications. Environ Technol Rev 12(1):2252164
Singha NR, Mahapatra M, Karmakar M, Chattopadhyay PK (2019) Processing, characterization, and application of natural rubber-based environmentally friendly polymer composites. Sustainable polymer composites and nanocomposites, pp 855–897
Sionkowska A (2011) Current research on the blends of natural and synthetic polymers as new biomaterials. Prog Polym Sci 36(9):1254–1276
Sisanth KS, Thomas MG, Abraham J, Thomas S (2017) General introduction to rubber compounding. In Progress in rubber nanocomposites. Woodhead Publishing, pp 1–39
Song K (2017) Micro-and nano-fillers used in the rubber industry. In: Progress in rubber nanocomposites. Woodhead Publishing, pp 41–80
Soto-Oviedo MA, Araújo OA, Faez R, Rezende MC, De Paoli MA (2006) Antistatic coating and electromagnetic shielding properties of a hybrid material based on polyaniline/organoclay nanocomposite and EPDM rubber. Synth Met 156(18–20):1249–1255
Tang S, Li Z, Sun W, Liu Y, Wang J, Wang X, Lin J (2024) Natural rubber/styrene–butadiene rubber blend composites potentially applied in damping bearings. Polymers 16(13):1945
Thomas RM, Sambhudevan S, Hema S, Reghunadhan A (2023) PTT/Rubber, thermoplastic and thermosetting polymer blends and IPNs. In: Poly trimethylene terephthalate: based blends, IPNs, composites and nanocomposites. Springer Nature Singapore, Singapore, pp 49–77
Urtekin G, Ullah MS, Yildirim R, Ozkoc G, Kodal M (2023) A comprehensive review of the recent developments in thermoplastics and rubber blends-based composites and nanocomposites. Polym Compos 44(12):8303–8329
Vayyaprontavida Kaliyathan A, Varghese KM, Nair AS, Thomas S (2020) Rubber–rubber blends: a critical review. Progress Rubber, Plast Recycling Technol 36(3):196–242
Venkatachalam P, Geetha N, Sangeetha P, Thulaseedharan A (2013) Natural rubber producing plants: an overview. Afr J Biotechnol 12(12)
Viswanathan VP, Kulandhaivelu SV, Manivasakan K, Ramakrishnan R (2024) Development of biodegradable packaging films from carboxymethyl cellulose and oxidised natural rubber latex. Int J Biol Macromol 262:129980
Wang ZN, Shen SL, Zhou A, Lyu HM (2021) Investigation of time-dependent characteristics of EPDM rubber gasket used for shield tunnels. J Mater Civ Eng 33(9):04021251
Wang Y, Yu Z, Phule AD, Zhao Y, Wen S, Zhang ZX (2022) A lightweight, abrasion-resistant polybutadiene rubber/styrene butadiene rubber foam prepared by three-step process for footwear outsole applications. Macromol Mater Eng 307(2):2100702
Wiśniewska P, Haponiuk JT, Colom X, Saeb MR (2023) Green approaches in rubber recycling technologies: present status and future perspective. ACS Sustain Chem Eng 11(24):8706–8726
Wojtowicz A, Janssen LP, Moscicki L (2009) Blends of natural and synthetic polymers. Thermoplastic Starch: A Green Material for Various Industries, 35–53
Xiao X, Wang J, Cai D, Lou L, Xiao F (2021) A novel application of thermoplastic polyurethane/waste rubber powder blend for waterproof seal layer in high-speed railway. Transportation Geotechnics 27:100503
Yang C, Shen SL, Hou DW, Liao SM, Yuan DJ (2018) Material properties of the seal gasket for shield tunnels: a review. Constr Build Mater 191:877–890
Zeng X, Zhu J, Li G, Miao Q, Sain M, Jian R (2023) Formulation and performance of NBR/CR-based high-damping rubber composites for soundproof using orthogonal test. Polymers 15(9):2208
Zhang Z, Chen S, Zhang J (2011) Improvement in the heat resistance of poly (vinyl chloride) profile with styrenic polymers. J Vinyl Add Tech 17(2):85–91
Zhang X, Han C, Yang J, Xu X, Zhang F (2021) Evaluating the rheological properties of high-modulus asphalt binders modified with rubber polymer composite modifier. Materials 14(24):7727
Zhao X, Hu H, Wang X, Yu X, Zhou W, Peng S (2020) Super tough poly (lactic acid) blends: a comprehensive review. RSC Adv 10(22):13316–13368

Chapter 6
Eco-Friendly Fabrication of Nanoparticles (NP) Particles and Fibrils Using Polymer Blends as Templates

Patrick Ehi Imoisili

Abstract A sustainable and adaptable method for producing high-performance nanomaterials is the environmentally friendly production of NP particles and fibrils utilizing polymer blends as templates. This technique allows for exact control over the shape and functionality of nanostructures by taking advantage of the phase separation, compatibility, and selective removal characteristics of polymer blends. This approach's scalability and versatility are increased by methods including solvent evaporation, electrospinning, and template-assisted synthesis, which enable the creation of nanostructures with a variety of uses in energy storage, biomedicine, environmental remediation, and agriculture. Sustainable development goals are supported, and the worldwide problem of plastic waste is addressed by minimizing environmental impact through the use of biodegradable polymers, renewable resources, and green solvents. This approach reduces waste production and encourages resource efficiency, which is consistent with the ideas of the circular economy. Despite obstacles like as maximizing polymer compatibility, customizing degradation profiles, and guaranteeing industrial scalability, developments in green chemistry and material science are improving and extending the possibilities of polymer blend templating. This creative approach not only produces environmentally friendly nanostructures with exceptional performance but also provides a roadmap for green nanotechnology, striking a balance between ecological responsibility and technological growth to satisfy the expanding need for sustainable materials in various industries.

Keywords Nanoparticles · Nanomaterials · Polymer blend templating · Biodegradable polymers · Template-assisted synthesis · Nanostructure

P. E. Imoisili (✉)
Mechanical Engineering Science Department, University of Johannesburg, Johannesburg, South Africa
e-mail: pimoisili@uj.ac.za

P. M. Visakh (ed.), *Engineering Applications of Polymer based Nano Blends*, Advanced Structured Materials 235, https://doi.org/10.1007/978-3-031-90979-5_6

6.1 Introduction

Plastic trash is widely present in the environment due to the quick development and wide-ranging use of plastic-based materials in recent decades. Because of their longevity and resistance to deterioration, traditional plastics lead to the build-up of persistent debris, which presents serious health and environmental risks (Walker and Fequet 2023). This has spurred a global drive to manage plastic garbage sustainably and, more importantly, to develop eco-friendly substitutes through materials science innovation (Walker and Fequet 2023; Imoisili and Jen 2023a). Because of its importance in various industries, including electronics, medicine, packaging, and environmental engineering, the creation of NP particles and fibrils—tiny plastic pieces on a nanoscale have attracted much attention. However, the traditional methods of producing these nano-plastics (NP) frequently use materials and procedures that are not sustainable, which emphasizes the need for ecologically friendly fabrication methods (Walker and Fequet 2023; Imoisili and Jen 2023a; Rai et al. 2021a).

Larger plastic materials can degrade accidentally or intentionally, producing NP, commonly described as particles smaller than one millimetre. These particles' capacity to enter food chains, penetrate natural ecosystems, and even build up inside living things has sparked worries. In controlled applications, such as focused medicine administration, where their small size and adaptable qualities might be useful, NP may also be beneficial despite these hazards. Therefore, the goal is to eradicate NP and develop methods that enable their environmentally beneficial, sustainable, and safe manufacture (Walker and Fequet 2023; Imoisili and Jen 2023a; Rai et al. 2021a). Blends of polymers have drawn a lot of interest as a potential environmentally friendly method of producing NP. By physically combining various polymer types, these blends can produce special material qualities that are not possible with just one polymer. Crucially, by utilizing biodegradable or bio-based polymers, polymer blends provide a sustainable solution while lowering dependency on non-renewable resources (Imoisili and Jen 2024a; Sisanth et al. 2017; Geyer 2020). Green chemistry, which emphasizes less chemical waste, less energy use, and less toxicity, is in line with the use of polymer blends as templates for creating NP (Zimmerman et al. 2020).

The utilization of polymer blends as templates in the manufacturing of NP presents a method in which bio-based polymers, polylactic acid (PLA), polyhydroxyalkanoates (PHAs), and other eco-friendly polymers act as matrices that allow the formation of nano-plastic fibrils and particles (Li et al. 2023; Tan et al. 2021). This template-based method can precisely control the size, form, and functional characteristics of NP. Additionally, scientists can balance the biodegradability needed for environmental friendliness and the durability required for certain applications by employing mixes containing natural and synthetic polymers. These templates lessen the negative effects that traditional plastics have on the environment while simultaneously making it easier to create NP with the appropriate properties (Geyer 2020; Zimmerman et al. 2020; Li et al. 2023; Tan et al. 2021). A variety of Sustainable Development Goals (SDGs) of the UN, including SDG 12 (Responsible Consumption and Production), SDG 14 (Life Below Water), and SDG 15 (Life on Land), are also in line with the environmentally friendly production of NP employing polymer blends. This strategy offers a promising contribution to international efforts to reduce plastic pollution by manufacturing materials that decompose more easily in natural environments and developing environmentally friendly methods. Additionally, environmentally friendly techniques for fabricating NP foster creativity and interdisciplinary cooperation, improving the sustainability of sectors that depend on plastic materials (Pramudianto 2024; Imoisili et al. 2023; Rai et al. 2021b).

This chapter will give an extensive overview of the processes and principles underlying the environmentally friendly production of NP employing mixes of polymers as templates. The technological advancements enabling the production of sustainable NP, an examination of the structural and functional properties possible with template-based techniques, and a thorough investigation of different polymer blends and their effects on the environment will all be important topics. The potential of eco-friendly NP to promote sustainability across industries will also be demonstrated through case studies that showcase successful applications of these materials (Rai et al. 2021b).

6.2 Method for Fabricating NP Materials

Various methods are used to fabricate NP fibrils and particles to attain the appropriate morphologies, sizes, and environmental compatibility. The following strategies, which highlight the usage of polymer blends and low-impact processing technologies, are examples of sustainable practices (Imoisili et al. 2020; Rane et al. 2018; Imoisili and Jen 2021). Figure 6.1 shows the Schematics of the major fabrication methods of nano-plastics.

6.2.1 Solvent Evaporation Method

Solvent evaporation is a popular approach for producing NP particles, particularly when making eco-friendly or biodegradable nanoparticles. Using this technique, a polymer is dissolved in a volatile organic solvent and emulsified into an aqueous phase. The polymer precipitates out as particles or fibrils of NP as the solvent evaporates (Fig. 6.2). This technique is adaptable and frequently used due to its ease of use and controllable particle size (Deshmukh et al. 2016).

Overview of the Solvent Evaporation Process

- **Step 1: Polymer Dissolution**
  - To create a homogeneous polymer solution, a polymer (or polymer blend) is dissolved in a volatile organic solvent, like acetone or dichloromethane (Deshmukh et al. 2016; Yabu et al. 2005; Zheng et al. 2010).

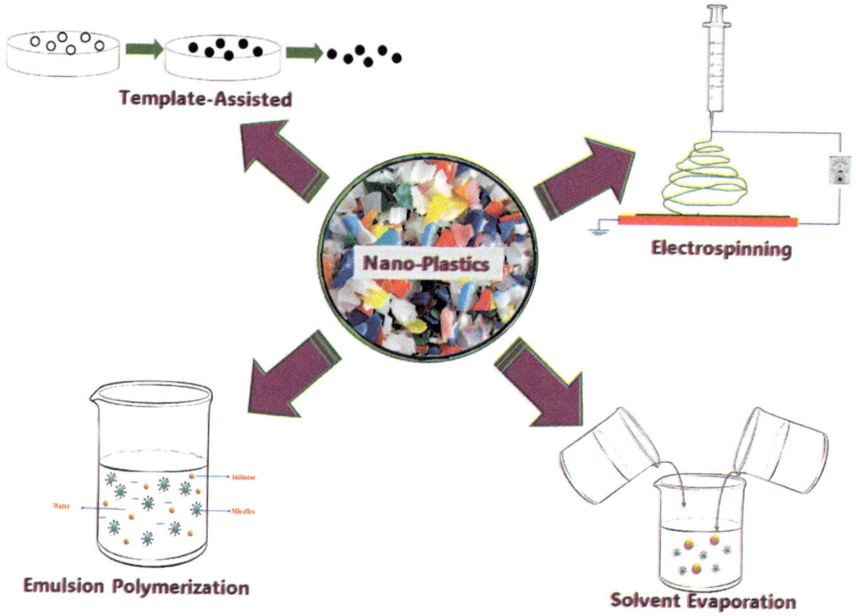

Fig. 6.1 Schematics of the different methods of nano-plastic fabrication

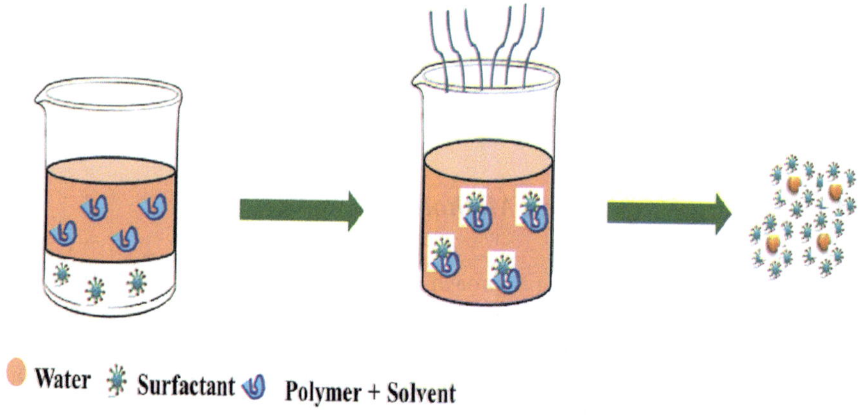

Fig. 6.2 Schematics of the solvent evaporation method of nano-plastic fabrication

- **Step 2: Emulsification**
  - Tiny droplets of the polymer–solvent mixture suspended in water create an emulsion when the polymer solution is introduced to an aqueous phase that contains a stabilizer or surfactant (Deshmukh et al. 2016; Yabu et al. 2005; Zheng et al. 2010).

- **Step 3: Solvent Evaporation**
  - When the emulsion is agitated, the solvent separates from the droplets, and the polymer precipitates, leaving behind solid NP particles (Deshmukh et al. 2016; Yabu et al. 2005; Zheng et al. 2010).
- **Step 4: Particle Collection and Purification**
  - Centrifugation, filtering, or washing are the usual methods used to separate the nanoparticles from the aqueous phase (Deshmukh et al. 2016; Yabu et al. 2005; Zheng et al. 2010).

Eco-Friendly Aspects of the Solvent Evaporation Method

- **Green Solvent Alternatives**
  - Eco-friendly substitutes like acetone or ethyl acetate are being utilized more frequently because of their reduced toxicity and biodegradability, even if conventional solvents might still be hazardous (Munir et al. 2022).
- **Reduction in Solvent Volume**
  - Optimizing solvent concentrations can lessen the process's overall toxicity and negative environmental effects (Yabu et al. 2005; Zheng et al. 2010; Munir et al. 2022).
- **Biodegradable Polymers**
  - The resulting NP is more environmentally friendly since biodegradable polymers like PLA (polylactic acid), PCL (polycaprolactone), or PHA (polyhydroxyalkanoate) are frequently used (Zheng et al. 2010; Munir et al. 2022; Li et al. 2021a).
- **Water-based Emulsification**
  - Utilizing water as the continuous phase lowers the waste from organic solvents and improves environmental safety (Abdenour et al. 2024).

Factors Affecting NP Particle Size and Morphology

- **Concentration of Polymers**
  - Higher polymer concentrations usually produce larger particles, while lower concentrations produce finer particles (Caputo et al. 2021).
- **Choice of Solvent and Rate of Evaporation**
  - The evaporation rate is influenced by solvent selection; faster rates encourage smaller particle sizes because of their quick solidification (Caputo et al. 2021).

- **Type and Concentration of Surfactants**
  - The kind and concentration of surfactants affect the stability and homogeneity of the particles; they stabilize the emulsion by preventing droplets from coalescing (Li et al. 2021a; Abdenour et al. 2024; Caputo et al. 2021).
- **Speed of Agitation**
  - Smaller droplets produced by faster emulsification stirring can result in smaller NP particles (Caputo et al. 2021).

Applications of Solvent Evaporation Fabricated from NP Particles

- **Biomedical Field**
  - Drug delivery systems: Drugs can be encapsulated in NP particles for controlled and targeted release (Carreón et al. 2024).
- **Agriculture**
  - Controlled-release pesticides: Biodegradable polymer-based nanoparticles can offer ecologically beneficial pesticide delivery methods (Carreón et al. 2024).
- **Packaging and Coatings**
  - Eco-friendly coatings: NP particles can improve packaging materials' durability and biodegradability (Carreón et al. 2024; Ivleva 2021).

Advantages of the Solvent Evaporation Method

For creating NP particles, the solvent evaporation approach is popular because it offers several significant benefits, most notably ease of use, control, and scalability (Pulingam et al. 2022). The main benefits of this approach are as follows:

- **Simplicity and Implementation Ease**
  - **Simple Procedure**: The solvent evaporation method consists of a few comparatively easy procedures, such as solvent removal, emulsification, and polymer dissolution. Because of this, it is simple to use in both industrial and laboratory settings (Pulingam et al. 2022).
  - **Minimal Equipment Needs**: The procedure only requires rudimentary laboratory equipment, such as centrifuges, heating elements, and stirrers, which makes it economical, particularly for small-scale operations (Pulingam et al. 2022).
- **High Control over Morphology and Particle Size**
  - **Fine-tuning Particle Size**: By modifying variables such as solvent selection, polymer concentration, and evaporation rate, the size of the NP particles can

be managed. This makes achieving the required particle size and dispersion possible with a high degree of precision (Pulingam et al. 2022; Szczęch and Szczepanowicz 2020).
- **Morphological Control**: By adjusting surfactants, polymer mixes, and processing parameters, the method can create various particle forms, such as fibrils, spherical particles, and other nanostructures (Pulingam et al. 2022; Szczęch and Szczepanowicz 2020).

- **Scalability**
  - **Large-Scale Production**: The solvent evaporation process can be readily scaled from laboratory to industrial production with few adjustments. Because of its scalability may be used in various sectors to produce NP in large quantities (Ivleva 2021; Pulingam et al. 2022; Szczęch and Szczepanowicz 2020).
  - **Continuous Processing**: By modifying the procedure, production time can decrease and throughput can increase for continuous manufacturing (Pulingam et al. 2022; Szczęch and Szczepanowicz 2020).
- **Flexibility in the Selection of Polymers**
  - **Numerous Polymers**: Depending on the desired characteristics of the finished NP particles, various natural and synthetic polymers can be utilized, providing flexibility. This comprises environmentally friendly biodegradable polymers, including PLA, PCL, and PHA (Ivleva 2021; Pulingam et al. 2022; Szczęch and Szczepanowicz 2020).
  - **Polymer Blending**: By allowing for the combination of several polymers, the technique makes it possible to create composite NP with specific qualities like increased strength, flexibility, or rates of degradation (Li et al. 2021a; Caputo et al. 2021; Pulingam et al. 2022).
- **Potential for Eco-Friendliness**
  - **Green Solvents**: By selecting environmentally friendly solvents (such as water, ethanol, or ethyl acetate), the process's negative environmental effects can be reduced, as can the toxicity and damage connected to conventional solvent-based techniques (Alam et al. 2020).
  - **Biodegradable NP**: The solvent evaporation technique produces NP particles that break down more quickly in the environment when biodegradable polymers are used, which promotes sustainability (Zimmerman et al. 2020; Zheng et al. 2010; Munir et al. 2022; Li et al. 2021a).
- **Control Over Surface Chemistry and Functionalization**
  - **Surface Modifications**: NP particles fabricated via solvent evaporation can be functionalized with various chemical groups, making them suitable for various applications, including drug delivery, catalysis, and biosensing (Li et al. 2021a; Abdenour et al. 2024; Caputo et al. 2021; Carreón et al. 2024; Ivleva 2021; Pulingam et al. 2022; Szczęch and Szczepanowicz 2020; Alam et al. 2020).

- **Encapsulation**: The method allows for encapsulating bioactive substances, such as drugs, enzymes, or nutrients, within the NP particles, enabling controlled release and targeted delivery (Caputo et al. 2021; Carreón et al. 2024; Ivleva 2021; Pulingam et al. 2022; Szczęch and Szczepanowicz 2020; Alam et al. 2020).

- **Minimal Cost**
  - **Cost-effective**: Since the solvent evaporation approach calls for fewer complex tools and procedures than other nanomaterial manufacturing techniques, it is comparatively less expensive. This is particularly beneficial for increasing production without requiring a large financial commitment (Caputo et al. 2021; Carreón et al. 2024; Ivleva 2021; Pulingam et al. 2022; Szczęch and Szczepanowicz 2020; Alam et al. 2020).
- **High Efficiency and Yield**
  - **High Production Efficiency**: The process usually delivers large quantities of uniformly sized NP particles. This makes the procedure very effective for creating vast amounts of NP and guarantees uniformity in quality (Carreón et al. 2024; Ivleva 2021; Pulingam et al. 2022).

Disadvantages of the Solvent Evaporation Method

- **Utilizing Organic Solvents**
  - **Toxicity of traditional Solvents**: If not handled appropriately, several traditional solvents used in the solvent evaporation process, including dichloromethane, chloroform, and acetone, are toxic, harmful to human health, and potentially harmful to the environment (Bae et al. 2022).
  - **Environmental Concerns**: Even when green solvents are used, there is a chance that solvent evaporation will pollute the air, particularly if the solvents are not properly recovered or handled (Bae et al. 2022).
- **Restricted Authority Over the Distribution of Polymers**
  - **Heterogeneity and Phase Separation**: In certain polymer blends, phase separation may occur during evaporation, resulting in unequal polymer distribution or heterogeneous particle sizes. In certain situations, achieving homogeneity can be difficult and may require additional solvent combinations and temperature adjustments (Xu et al. 2023).
  - **Inconsistent Morphology**: Although the size and form of the particles can be controlled, changes in processing parameters (such as the pace at which the solvent evaporates or temperature swings) can cause irregularities in the morphology of the NP particles (Alam et al. 2020; Bae et al. 2022; Xu et al. 2023).

- **Time and Energy Used in the Process**
  - **Long Evaporation Times**: The evaporation process might take a long time, depending on the solvent and the system's architecture. Particularly in large-scale production, this may cause the total production rate to slow down (Pulingam et al. 2022; Xu et al. 2023).
  - **Energy Consumption**: Some environmental advantages of utilizing eco-friendly solvents may be outweighed by the increased energy consumption required for heating or vacuum systems to aid solvent evaporation (Szczęch and Szczepanowicz 2020; Alam et al. 2020; Bae et al. 2022; Xu et al. 2023).
- **Limited Solubility of Some Polymers**
  - **Polymer Compatibility**: Creating NP particles with qualities can be challenging since some polymers may not dissolve well in particular green solvents. The solvent selection becomes important, as this approach might not work for some polymers without using more hazardous solvents (Bae et al. 2022; Xu et al. 2023).
- **Risk of Polymer Degradation**
  - **Thermal degradation**: Polymer deterioration is possible if heating is used in solvent evaporation, particularly for heat-sensitive polymers. Desired qualities like mechanical strength, biodegradability, or biocompatibility may be lost (Veskova et al. 2022).
  - **Polymer Chain Scission**: The characteristics of the resulting NP particles may be adversely affected by polymer chain scission, which is the breaking of polymer chains caused by prolonged exposure to the solvent or high temperatures during the evaporation process (Xu et al. 2023; Veskova et al. 2022).
- **The Challenge of Producing Uniform NP**
  - **Size and dispersion Variability**: Despite its adaptability, variables including emulsion stability, stirring speed, and solvent evaporation rate can make it challenging to achieve uniform nanoparticle size and size dispersion. This may result in a wide range of particle sizes, which would not be appropriate for some uses where consistency is essential (Banerjee et al. 2024).
  - **Need for Fine-Tuning**: The process conditions need to be meticulously tuned and often monitored to provide highly consistent and repeatable results, which can increase the method's labour and time requirements (Veskova et al. 2022; Banerjee et al. 2024).
- **Scaling Challenges**
  - **Difficulties with Scaling Up**: Despite the solvent evaporation method's scalability, problems with temperature control and solvent evaporation rate might

occasionally arise as manufacturing moves from the lab to the industrial scale. These difficulties may impact the final NP's uniformity, requiring modifications to process and equipment parameters (Alam et al. 2020; Bae et al. 2022; Xu et al. 2023; Veskova et al. 2022; Banerjee et al. 2024).

- **Possibilities of Inconsistent Results**: Differences in size could result in different NP particle ultimate qualities, necessitating further refining in larger-scale manufacturing procedures (Bae et al. 2022; Xu et al. 2023; Veskova et al. 2022).

6.2.2 Emulsion Polymerization

Emulsion polymerization is a sustainable technique for creating NP particles since it uses water as the primary medium, allows particle size control, and works well with bio-based polymers. Utilizing renewable polymers, biodegradable surfactants and initiators, and reaction condition optimization, this approach fits nicely with environmentally conscious objectives. It is, therefore, perfect for applications that place a high value on sustainability. Because of its scalability, emulsion polymerization is also a viable option for satisfying the expanding need for environmentally friendly materials in nanotechnology (Lovell and Schork 2020; Chern 2006). Compared to solvent-based alternatives, emulsion polymerization is a cleaner and frequently more environmentally friendly technique since it creates polymer particles in an aqueous environment. Using water as the primary medium offers benefits in managing particle size and dispersion while lowering the demand for organic solvents (Lovell and Schork 2020). The schematic for emulsion polymerization is shown in Fig. 6.3

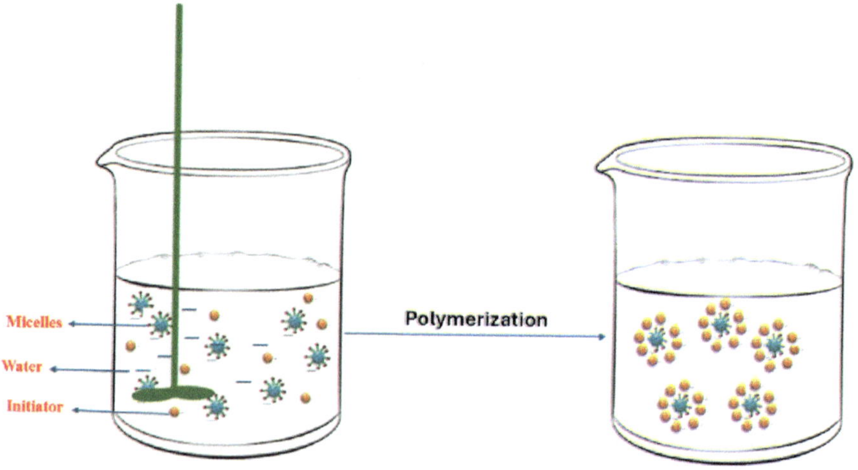

Fig. 6.3 Schematics of emulsion polymerization of nano-plastic

Overview of the Emulsion Polymerization Process

The following steps are involved in the emulsion polymerization process:

- **Preparation of the Emulsion**: A surfactant or emulsifier is used to distribute a polymer blend (or monomers that will polymerize) in water. Micelles, tiny droplets where the polymerization reaction will occur, are created as a result. These droplets are stabilized by surfactants, which are frequently naturally occurring or biodegradable, to keep them from combining during polymerization (Lovell and Schork 2020; Chern 2006; Asua 2004).
- **Initiation**: An initiator that dissolves in water is usually used to start the polymerization process inside the micelles. Free radicals are created when the initiator degrades and reacts with the micelle monomers to produce polymerization. This restricted reaction inside the micelles guarantees uniform particle development (Asua 2004).
- **Polymerization within Micelles**: More monomers are taken up by the micelles from the surrounding water as the reaction goes on, which enables the particles to enlarge. This stage is essential for managing particle size because the original micelle size determines the final nanoparticles' dimensions (Chern 2006; Asua 2004).
- **Collection of Nanoparticles**: Centrifugation, filtration, or drying are methods used to purify the nanoparticles after polymerization. After the particles are cleaned to get rid of any remaining surfactants or unreacted components, stable, distinct NP particles are produced (Lovell and Schork 2020; Chern 2006; Asua 2004).

Factors Affecting Emulsion Polymerization

During emulsion polymerization, several factors affect the size and shape of the particles:

- **Choosing a Surfactant**: Particle size and stability are greatly influenced by the kind and concentration of the surfactant. Eco-friendly options that guarantee minimal environmental impact while stabilizing the emulsion include biodegradable surfactants such as lecithin, polysaccharides (chitosan or alginate), or other bio-derived compounds (Lovell and Schork 2020).
- **Initiator Type**: Water-soluble, non-toxic initiators are favoured for ecologically friendly emulsion polymerization. For instance, potassium persulfate is a frequently used initiator that breaks down into innocuous metabolites (Faust et al. 2021).
- **Polymer Blend Composition**: Eco-friendliness is promoted by combining bio-based or biodegradable polymers (such as polylactic acid, polyhydroxyalkanoates, or starch-based compounds). The mix composition also influences particle characteristics such as hydrophobicity, mechanical strength, and degradation rate (Asua 2004; Faust et al. 2021).

- **Temperature and Reaction Time**: Although the polymerization process can be accelerated by higher temperatures, maintaining particle homogeneity requires careful temperature management. While longer reaction durations can encourage uniform particle formation, lower temperatures usually result in smaller particles (Zhao et al. 2020).

Advantages of Emulsion Polymerization

Emulsion polymerization provides several advantages for the production of environmentally friendly NP:

- **Water-Based Process**: This method is safer and more ecologically friendly by using water as the main medium instead of hazardous solvents (Asua 2004; Faust et al. 2021; Zhao et al. 2020).
- **Fine Particle Control**: This technique is appropriate for applications needing certain nanoparticle dimensions because it provides exact control over particle size by modifying variables such as surfactant content and reaction conditions (Asua 2004; Faust et al. 2021; Zhao et al. 2020).
- **High Polymerization Rates**: This technique is effective and scalable for greater production because of the accelerated polymerization caused by the limited reaction environment inside micelles (Faust et al. 2021; Zhao et al. 2020).
- **Lower Energy and Solvent Use**: Emulsion polymerization typically utilizes less energy and resources than conventional solvent-based techniques, which helps to lessen its environmental impact (Asua 2004; Faust et al. 2021; Zhao et al. 2020).

Eco-Friendly Considerations in Emulsion Polymerization

Emulsion polymerization can be made as sustainable as feasible by implementing the following green practices:

- **Eco-Friendly Surfactants and Initiators**: Environmental risks are reduced when surfactants and initiators with low toxicity and biodegradability are used. Without adding long-lasting chemicals, the emulsion can be stabilized by using natural surfactants like lecithin or chitosan (Zhang et al. 2023).
- **Blends of Biodegradable Polymers**: By combining biodegradable polymers like PLA, PHAs, or other renewable resources, the resultant nanoparticles are made to break down more readily in natural settings, lowering the danger of long-term contamination (Vayshbeyn et al. 2023).
- **Energy-Efficient Reaction Conditions**: This technique is more environmentally friendly on an industrial level when the polymerization is carried out at moderate temperatures and with optimal reaction periods (Zhao et al. 2020; Zhang et al. 2023; Vayshbeyn et al. 2023).

Applications of Emulsion Polymerization in Eco-Friendly Fabrication

Emulsion polymerization's adaptability makes it useful for creating environmentally acceptable NP in several industries:

- **Applications in Biomedicine**: Drug delivery systems may employ biodegradable nanoparticles made by emulsion polymerization when exact size control and biocompatibility are crucial (Zhang et al. 2023).
- **Applications in Agriculture**: NP made from biodegradable polymers are perfect for encapsulating controlled-release insecticides or fertilizers in particles that will break down after delivery, reducing their negative environmental effects (Saad et al. 2020).
- **Environmental Remediation**: By functionalizing nanoparticles produced using this technique, cleaned water bodies can be made less susceptible to secondary contamination thanks to environmentally benign elements (Zhang et al. 2023; Vayshbeyn et al. 2023; Saad et al. 2020).

6.2.3 Electrospinning Technique

The flexible electrospinning process is mainly employed to create nanofibers, but under certain circumstances, it can also be altered to create NP particles. Subjecting a polymer solution or melt to a high-voltage electric field produces thin fibers or particles that solidify upon reaching a grounded collector (Fig. 6.4). Because of its versatility, electrospinning can create nanostructures in various shapes, such as fibers, beads, and particles (Bavatharani et al. 2021).

Electrospinning is a potent method that allows for fine control over morphology, scalability, and compatibility with biodegradable materials when creating environmentally friendly NP fibers and particles. Using biodegradable polymer blends, green solvents, and energy efficiency optimization, this method fits nicely with sustainable objectives and is a viable way to develop environmentally friendly nanomaterials. The versatility of electrospinning makes it appropriate for a wide range of industries where controlled release, tenable degradation, and large surface area are desired (Bavatharani et al. 2021; Li et al. 2021b).

An Overview of the Process of Electrospinning

- **Preparing the Polymer Solution**: To make a solution with the required viscosity and conductivity, a blend of polymers is dissolved in an appropriate solvent. The electrospinning process depends on this solution, whose conductivity, surface tension, and concentration determine the final structure.
- **Application of Electric Field**: A syringe that is attached to a needle is filled with the polymer solution. The needle is placed between a grounded collector and a

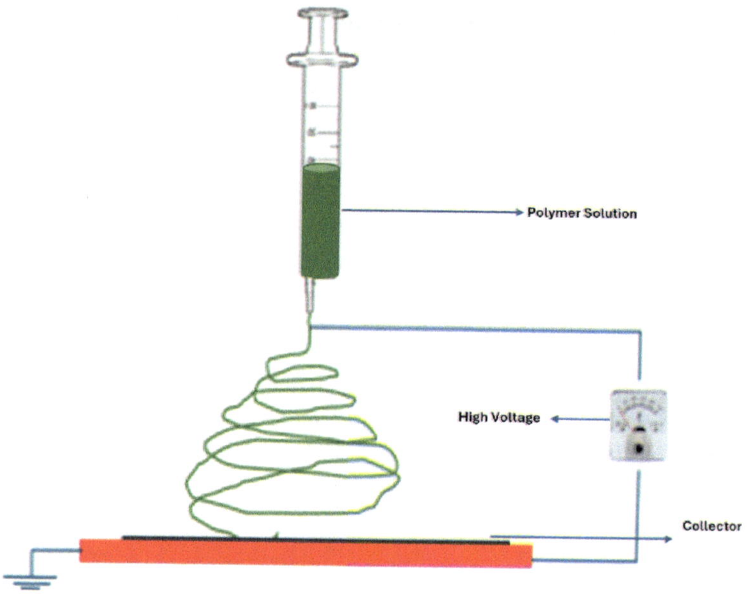

Fig. 6.4 Schematics of electrospinning method

high-voltage electric field. Electrostatic forces cause the solution droplet at the needle's tip to expand into a jet as the electric field strength increases.
- **Creation of Nanostructures**: A fine jet of the polymer solution is released when surface tension is overcome by electrostatic forces. Depending on the process settings, this jet undergoes rapid stretching and solvent evaporation as it moves toward the collector, generating either fibers or particles.
- **Collection**: On a grounded surface, the resultant nanostructures are gathered, frequently forming a fibrous mat. Instead of continuous fibers, NP particles can be produced by adjusting the solution and operating parameters.

Factors Influencing Nanostructure Formation

NP particles and fibrils can be produced by varying a number of electrospinning process parameters, such as:

- **Polymer Solution Properties**: The development of fibers or particles is greatly influenced by the polymer solution's concentration, viscosity, and conductivity. For example, a higher viscosity promotes continuous fiber creation, while a lower viscosity promotes particle formation (Bavatharani et al. 2021).
- **Voltage and Distance**: The polymer jet's stretching and the nanostructures' shape are influenced by the applied voltage and the distance between the needle and collector. Particle production may result from the finer, bead-like formations

encouraged by higher voltages and shorter distances (Li et al. 2021b; Yang et al. 2022; Coelho et al. 2021).

- **Flow Rate**: Instead of producing continuous fibers, lower flow rates might create discrete particles or bead-on-string structures by controlled jet creation and solvent evaporation (Coelho et al. 2021).
- **Solvent Selection**: Electrospinning has a less negative environmental impact when environmentally friendly solvents like ethanol, acetone, or water are used, whenever feasible. In keeping with ecological methods, green solvents can promote particle production (Saad et al. 2020; Bavatharani et al. 2021; Li et al. 2021b; Yang et al. 2022; Coelho et al. 2021).

Advantages of the Electrospinning Technique

For the production of environmentally friendly NP, electrospinning has the following benefits:

- **Control over Morphology**: Depending on the application's requirements, this technology offers exact control over the size and shape of nanostructures, enabling the production of particles, fibers, or hybrid structures (Li et al. 2021c).
- **Scalability and Simplicity**: Electrospinning can be used to fabricate environmentally friendly nanostructures on a wide scale because it is easy to set up and can be scaled up for industrial output (Coelho et al. 2021; Li et al. 2021c).
- **Compatibility with Biodegradable Polymers**: Bio-based polymers like polylactic acid (PLA) or polyhydroxyalkanoates (PHAs) can be included with electrospinning to produce environmentally friendly nanostructures (Li et al. 2021c; Keirouz et al. 2023).
- **Low Solvent Requirement**: Using greener solvents reduces solvent-related environmental concerns, and the procedure is comparatively efficient in terms of solvent usage (Coelho et al. 2021; Li et al. 2021c; Keirouz et al. 2023).

Eco-Friendly Considerations in Electrospinning

Certain procedures can be used to make the electrospinning method more environmentally friendly:

- **Green Solvents**: The process's environmental impact is reduced by using biodegradable substitutes (such as ethanol and acetone) in place of conventional, harmful solvents (Keirouz et al. 2023).
- **Renewable Polymer Blends**: Eco-friendly nanostructures that dissolve spontaneously can be produced by combining biodegradable polymers, like PLA, with starch or chitosan (Li et al. 2021c; Keirouz et al. 2023).
- **Energy Efficiency**: To improve sustainability, operate at moderate voltages and modify flow rates to prevent excessive energy usage (Coelho et al. 2021; Li et al. 2021c; Keirouz et al. 2023).

Applications of Electrospun Nanostructures

Applications for electrospun nanostructures made of environmentally acceptable materials are numerous.

- **Biomedical Applications**: Because of their large surface area and regulated rates of disintegration, electrospun biodegradable nanofibers are frequently utilized in tissue scaffolds, drug delivery systems, and wound dressings (Keirouz et al. 2023).
- **Environmental filter**: Electrospun nanofibers work well as filter media because they can absorb contaminants and microplastics. Using environmentally friendly polymers ensures that the filtration media won't contribute to persistent plastic waste (Coelho et al. 2021; Li et al. 2021c; Keirouz et al. 2023).
- **Agricultural Use**: By encasing fertilizers, insecticides, or nutrients, NP fibers and particles enable regulated release and less environmental contamination. Following application, these environmentally favourable nanostructures break down, promoting more sustainable farming practices (Bavatharani et al. 2021; Li et al. 2021b, c; Yang et al. 2022; Coelho et al. 2021; Keirouz et al. 2023).

6.2.4 *Template-Assisted Synthesis*

Using pre-made templates, a production technique called "template-assisted synthesis" guides the size, shape, and structure of nanoparticles or nanofibers (Fig. 6.5). Employing templates with predetermined dimensions provides high precision and may produce homogenous nanostructures. In environmentally friendly settings, non-toxic reagents, biodegradable polymers, and sustainable templates are prioritized to lessen environmental influence (Kim et al. 2023). Template-assisted synthesis is an effective and adaptable technique for creating environmentally acceptable NP fibrils and particles. This method offers fine control over nanostructure morphology and is in line with sustainability goals by using renewable polymers, low-toxicity solvents, and biodegradable or reusable templates. Applications for template-assisted synthesis can be found in a variety of fields, such as environmental remediation and biomedicine, where environmentally friendly nanostructures help create a more sustainable future (Kim et al. 2023; Shafqat et al. 2023).

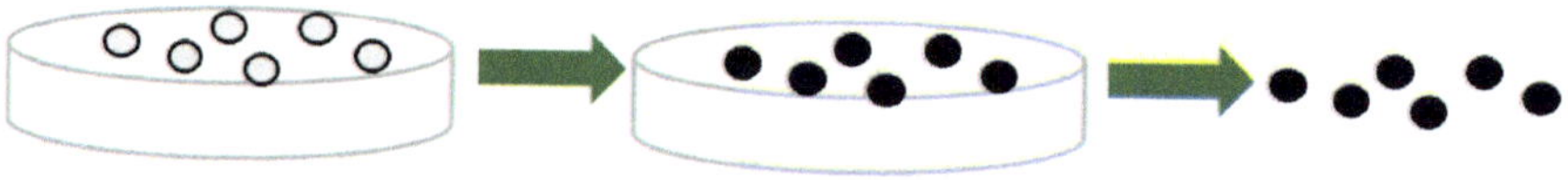

Fig. 6.5 Schematics of template-assisted synthesis method

Overview of the Template-Assisted Synthesis Process

In general, the following steps are involved in the template-assisted synthesis process:

- **Selection of the Template**: The template is selected by selecting an appropriate material, such as biodegradable templates (e.g., cellulose generated from plants), porous membranes, or nanoporous alumina. The template defines the nanostructures' dimensions, which act as a mould. Waste can be reduced for environmentally responsible synthesis by using reusable or biodegradable templates (Kim et al. 2023; Shafqat et al. 2023; Su et al. 2021).
- **Infusion of Polymer Solution**: The polymer mix solution, which frequently includes a biodegradable polymer, is infused into the channels or pores of the template. Capillary action, dip coating, or pressure application can all be used to accomplish this operation. The procedure will be environmentally benign if a low-toxicity solvent (such as acetone, ethanol, or water) is used (Kim et al. 2023; Shafqat et al. 2023; Su et al. 2021).
- **Solidification or polymerization**: The infused polymer becomes firmly embedded in the template's structure. Depending on the polymers employed, this stage may entail chemical cross-linking, solvent evaporation, or thermal curing. Heating, for example, may aid in the evaporation of water in aqueous-based procedures, leaving the solid polymer inside the template holes (Shafqat et al. 2023).
- **Removal of the Template**: The template is either dissolved or physically removed once it has solidified, leaving behind nanostructures that have the same shape as the template. Reusable, biodegradable, or readily removed templates that don't leave harmful residues are frequently utilized in environmentally friendly synthesis (Kim et al. 2023; Su et al. 2021).

Factors in Template-Assisted Synthesis

Several aspects influence the resultant nanostructures' properties in the template-assisted synthesis process:

- **Template Material and Geometry**: The size and shape of the nanostructures are directly influenced by the selection of the template material, pore size, and geometry. Commonly utilized templates include porous alumina, silica, or biodegradable plant-based materials; each has special benefits for uses (Shafqat et al. 2023; Su et al. 2021; Pavlenko et al. 2022).
- **Polymer Blend Composition**: For applications that care about the environment, biodegradable or renewable polymers such as polylactic acid (PLA), polyhydroxyalkanoates (PHA), or naturally occurring polymers like cellulose are preferred. The mechanical strength, rate of deterioration, and environmental suitability of the resultant nanostructures are all influenced by the characteristics of the polymer (Shafqat et al. 2023; Su et al. 2021; Pavlenko et al. 2022; Nwanna et al. 2020).

- **Infusion Technique**: The consistency and completeness of the filling are influenced by the technique used to infuse the polymer solution into the template (such as pressure-driven or dip coating). This, in turn, affects the nanoproduct's ultimate quality and structure (Li et al. 2021c; Shafqat et al. 2023).
- **Method of Template Removal**: To remove templates without hazardous chemicals, eco-friendly synthesis usually involves dissolving the template in ethanol or water, or mechanically detaching it. Reusable templates are beneficial since they use fewer resources and produce less waste (Lamaoui et al. 2024).

Advantages of Template-Assisted Synthesis

There are several advantages of template-assisted synthesis for environmentally friendly nanofabrication:

- **Precise Control over Nanostructure Morphology**: The template produces extremely homogeneous particles and fibrils by carefully dictating the nanostructure's size, shape, and structure. This is beneficial for applications that demand consistency (Lamaoui et al. 2024).
- **Polymer Type Versatility**: This technique works with various polymers, including biodegradable ones. Because bio-based polymers are made to break down without remaining in the environment, using them is a good way to achieve sustainability goals (Kaur et al. 2022).
- **Scalability**: Because templates can be reused and the process can be automated for mass manufacturing, template-assisted synthesis can be scaled up for industrial production rather readily (Lamaoui et al. 2024; Kaur et al. 2022).
- **Decreased Environmental Impact**: This technique reduces waste and harmful residues by utilizing environmentally acceptable templates, solvents, and polymers. Sustainability is further enhanced with biodegradable materials and reusable templates (Shafqat et al. 2023; Su et al. 2021; Pavlenko et al. 2022; Nwanna et al. 2020; Lamaoui et al. 2024).

Eco-Friendly Considerations in Template-Assisted Synthesis

Template-assisted synthesis can use several environmentally friendly techniques for applications that are more sustainable:

- **Biodegradable or Reusable Templates**: Utilizing either biodegradable or reusable (such as templates made of cellulose or starch) lowers the amount of waste produced. For example, porous alumina can be cleaned and reused (Zimmerman et al. 2020; Deshmukh et al. 2016).
- **Low-Toxicity Solvents**: Selecting renewable or low-toxicity solvents, such as water or ethanol, reduces the hazards that solvent use poses to human health and the environment (Munir et al. 2022; Li et al. 2021a).

- **Blends of Renewable Polymers**: Using a combination of natural or biodegradable polymers, such as PLA, guarantees that the final nanostructures are environmentally benign and in line with green manufacturing practices (Rajeshkumar 2022).

Applications of Template-Assisted Nanostructures

Template-assisted synthesis can create nanostructures for several environmentally beneficial uses, including:

- **Biomedicine**: Where exact control over particle morphology is crucial, biodegradable nanofibers or particles can be employed in tissue engineering, drug delivery systems, and wound dressings (Pavlenko et al. 2022).
- **Water Filtration and Environmental Remediation**: With the benefit of being biodegradable, nanostructured materials made of environmentally benign polymers can be used in soil remediation, pollutant adsorption, and water filtration systems (Kaur et al. 2022).
- **Energy storage and catalysis**: High surface area and regulated shape nanostructures are advantageous for catalytic applications, such as green energy storage, where performance and sustainability are essential (Xu et al. 2021).
- **Applications in Agriculture**: Through encapsulating and releasing fertilizers or pesticides in a regulated manner, biodegradable nanofibers or particles can improve agricultural sustainability and lessen environmental contamination (Kaur et al. 2022; Rajeshkumar 2022; Xu et al. 2021).

6.3 Characterizations

Characterization is essential for comprehending and confirming the structural, physical, chemical, and functional characteristics of the produced nanostructures in the environmentally friendly production of NP particles and fibrils. Here is a thorough examination of the several characterization methods frequently applied in this situation (Zielińska et al. 2020). Combined, each characterization methodology's distinct and complementary information provides a comprehensive picture of nanostructures made using environmentally benign processes. These evaluations guarantee that the materials fulfill application-specific requirements and validate the efficacy of the selected production technique. Since accurate characterization confirms the material's performance, sustainability, and potential environmental impact, it is crucial to harmonize with eco-friendly aims (Zielińska et al. 2020; Danielsen et al. 2021).

6.3.1 Morphological Characterization

- **Scanning Electron Microscopy (SEM)**: SEM offers high-resolution pictures of a nanostructure's surface morphology, allowing for a thorough analysis of the size, shape, and distribution of the particles. It is particularly helpful for examining nanofibers, verifying homogeneity, and identifying any flaws on the surface. SEM works well with single particles and fibril networks (Imoisili and Jen 2023b; Makhatha et al. 2024).
- **Transmission Electron Microscopy (TEM)**: The interior structure of fibrils and NP can be seen using TEM, which has an even better resolution than SEM. TEM can disclose internal crystallinity, particle stacking, or porosity for applications needing particular material qualities (Ntobeng et al. 2020; Alqaheem and Alomair 2020)
- **Atomic Force Microscopy (AFM)**: AFM is useful for researching the surface roughness and nanoscale characteristics of polymer-based nanostructures since it offers 3D surface topography. Furthermore, AFM enables the characterization of mechanical properties at the nanoscale, which is useful for determining the strength and elasticity of fibers or particles (Collinson et al. 2021).

6.3.2 Structural Characterization

- **X-Ray Diffraction (XRD)**: The crystallinity and phase composition of polymer-based nanostructures are examined using XRD. For instance, XRD can be used to differentiate between amorphous and crystalline areas in nanofibers or to identify the existence of semi-crystalline phases in polymer blends. Increased mechanical strength and thermal stability are frequently associated with high crystallinity (Imoisili and Jen 2024b).
- **Fourier-Transform Infrared Spectroscopy (FTIR)**: FTIR analyzes infrared absorption spectra to find chemical bonds and functional groups in nanostructures. To confirm polymer composition and make sure the mix contains the right polymers and additives, FTIR is crucial. Additionally, FTIR can show any chemical interactions between blends' polymers, revealing information on stability and compatibility (Imoisili and Jen 2024b; Gulmine et al. 2002).
- **Raman Spectroscopy**: This method may identify certain chemical bonds and provide information on molecular vibrations. By offering comprehensive vibrational information on polymer chains and functional groups, Raman spectroscopy enhances FTIR in the characterization of chemical composition, particularly in complex polymer blends or modified nanostructures (Vélez-Escamilla and Contreras-Torres 2022).

6.3.3 Thermal Characterization

- **Differential Scanning Calorimetry (DSC)**: DSC measures the thermal transitions of nanostructures, such as the melting and glass transition temperatures. This knowledge is essential for forecasting how nanostructures behave in various settings and comprehending the thermal stability of biodegradable polymers. The compatibility and miscibility of polymers in a blend can also be evaluated using DSC (Drzeżdżon et al. 2019).
- **Thermogravimetric Analysis (TGA)**: Nanostructures' thermal stability and breakdown behaviour are evaluated using TGA, which tracks weight loss as the temperature rises. Because it shows the temperature at which eco-friendly polymers begin to degrade, this research is especially crucial for assessing the degradation behaviour of these materials and forecasting how long they will last in natural environments (Nurazzi et al. 2021).

6.3.4 Mechanical Characterization

- **Dynamic Mechanical Analysis (DMA)**: As a function of temperature, DMA quantifies the mechanical characteristics of nanostructures, such as stiffness, elasticity, and damping behaviour. This method is essential for describing the viscoelastic characteristics of nanofibers, especially in applications requiring mechanical stability (Bashir 2021).
- **Tensile Testing**: Tensile testing evaluates characteristics, including Young's modulus, elongation, and tensile strength for nanofibers or bigger particles. These properties are crucial for applications that call for durable materials, like scaffolds for tissue engineering or filtering media (Imoisili and Jen 2023a, 2024a; Okafor et al. 2021).

6.3.5 Surface Characterization

- **Contact Angle Measurement**: This technique assesses the wettability of nanostructure surfaces by measuring the contact angle of a droplet on the material. For applications like medicine administration, water filtration, or packaging materials where surface interactions with liquids are essential, hydrophobicity or hydrophilicity is essential (Ismail et al. 2022).
- **X-Ray Photoelectron Spectroscopy (XPS)**: Through the detection of elemental composition and oxidation states on the surface of nanostructures, XPS is used to evaluate surface chemistry. When describing surface-treated or functionalized nanostructures, this method is helpful since it offers important details on surface functionality and any chemical alterations (Easton et al. 2020).

6.3.6 Particle Size and Distribution

- **Dynamic Light Scattering (DLS)**: The size distribution of nanoparticles in a fluid can be measured using the DLS approach. This technique is vital for confirming that the particle sizes in NP dispersions are homogeneous, particularly in applications such as medication delivery, where consistent particle sizes are necessary for predictable behaviour (Babick 2020).
- **Zeta Potential Analysis**: The surface charge of suspended particles, which influences colloidal stability, is measured using this method. Since a high zeta potential denotes resistance to aggregation and settling, zeta potential analysis is crucial in applications that call for stable dispersions. This is especially true for NP in liquid formulations (Gomide et al. 2020).

6.3.7 Environmental Stability and Degradation Testing

- **Biodegradability Testing**: For eco-friendly nanostructures, biodegradability testing in simulated environmental conditions (e.g., soil, water, compost) provide information on the degradation rate and behaviour. This is critical for understanding biodegradable NP's environmental impact and lifespan and ensuring they align with sustainability goals (Andréia da Silva et al. 2024).
- **UV–Visible Spectroscopy (UV–Vis)**: UV–Vis can monitor degradation by measuring absorbance changes over time, especially in polymer-based nanostructures exposed to UV light. This method is useful for studying photodegradation and the material's stability under sunlight, which is valuable for outdoor applications (Nwanna et al. 2021).

6.4 Polymer Blends as Templates

Using polymer mixes as templates is an inventive method for creating nanostructures, such as environmentally acceptable NP particles and fibrils. Using a polymer blend as a mould or framework, this technique directs the creation of nanoscale particles and fibrils with certain sizes, shapes, and characteristics. The distinct microphase separation, compatibility, and dissolving properties of the various polymers in the blend are frequently utilized in the templating process (Rajeshkumar 2022; Thomas et al. 2023).

6.4.1 *Key Concepts in Polymer Blend Templating*

- **Phase Separation**: Polymer blends can exhibit microphase separation, where different polymer components form distinct domains or phases due to differences in compatibility, polarity, or crystallinity. By controlling these properties, phase-separated domains act as natural templates for creating nanostructures. This is particularly useful for designing complex morphologies like fibrils, core–shell particles, or porous structures (Xu et al. 2023).
- **Compatibility and Miscibility**: The miscibility between polymers in the blend affects the template's stability and functionality. Blends can be chosen to be partially compatible, where selective phase separation occurs, or to have complementary properties, like one degradable polymer (e.g., polylactic acid) and one stable polymer (e.g., polyethylene oxide). This compatibility can control how one polymer serves as a framework while the other creates the final structure after selective removal or processing (Habib et al. 2024).
- **Selective Removal**: A nanostructured substance is frequently left behind after one component of the polymer blend template is eliminated (either by chemical dissolution, thermal degradation, or biodegradation). For example, to create nano-porosity or fibrous networks, a hydrophobic polymer can be mixed with a water-soluble polymer, such as polyvinyl alcohol (PVA), and selectively dissolved (Liu et al. 2021).

6.4.2 *Fabrication Process Using Polymer Blend Templates*

- **Preparation of the Polymer Blend**: A homogeneous or phase-separated substance is produced by blending two or more polymers, usually via solution mixing or melt blending. The choice of solvent significantly impacts the final blend's polymer compatibility and distribution. Green solvents like water or ethanol are better for environmentally friendly applications (Garg and Pamme 2024; Saxena et al. 2021).
- **Formation of the Template Structure**: Through casting, extrusion, electrospinning, or 3D printing, the polymer blend is transformed into a particular shape, such as fibers, particles, or thin films. Since every processing method produces a distinct templating geometry, the final nanostructure that is required determines the processing method. For instance, a blend can be electrospun to create phase-separated nanofibers with particular surface morphologies (Laoutid et al. 2021; Arefin et al. 2021).
- **Selective Removal or Phase Manipulation**: After the polymer blend is constructed, one polymer component is removed using selective removal procedures, resulting in nanostructures (Liu et al. 2021). Standard methods include:
  - **Dissolution**: using a solvent that dissolves one polymer alone, leaving the other unaltered (Banerji et al. 2021).

  - **Thermal Degradation**: A thermally stable structure is left behind when one polymer is broken down at a lower temperature by heating the mixture (Veskova et al. 2022).
  - **Biodegradation**: If the blend contains biodegradable polymers, the template may be subjected to environmental factors (such as compost or soil) that permit one polymer to degrade naturally (Liu et al. 2021).

- **Finalization and Characterization**: The resultant nanostructures are usually purified and analyzed to verify their form, content, and characteristics once the template is removed. The particle size, distribution, and chemical stability analysis are done using methods such as SEM, TEM, and FTIR (Imoisili and Jen 2023b, 2024b; Makhatha et al. 2024; Ntobeng et al. 2020; Alqaheem and Alomair 2020; Gulmine et al. 2002).

6.4.3 Advantages of Using Polymer Blends as Templates

- **Customizable Nanostructures**: By adjusting blend ratios, processing strategies, and removal processes, polymer blends give exact control over the size, shape, and morphology of nanostructures (Kelly and Wolf 2010).
- **Eco-Friendly Potential**: Polymer blends as templates align with sustainable production practices using biodegradable or renewable polymers, environmentally friendly solvent systems, and low-energy processing techniques (Kelly and Wolf 2010; Ethirajan and Landfester 2010).
- **Versatility in Applications**: The templating technique applies to various morphologies, such as porous particles, core–shell structures, and nanofibers, increasing its applicability in industrial, biomedical, and environmental settings.
- **High Yield and Scalability**: Extrusion, casting, and electrospinning are scalable processes that may produce polymer blends, making the approach appropriate for the industrial production of nanomaterials.

6.4.4 Applications of Polymer Blends as Templates

- **Biomedicine**: Polymer mixtures can be used to create nanostructures for biosensing, tissue engineering, and drug delivery. For instance, nanofibrous scaffolds that promote cell development and break down over time in tandem with tissue healing processes can be produced using biodegradable templates (Lin et al. 2017).
- **Environmental Remediation**: Pollutant absorbents or filtration membranes can be made from templated nanostructures. The increased surface area of porous particles made from polymer blends makes them efficient at absorbing pollutants (Yu et al. 2022).

- **Agricultural Use**: Fertilizers and insecticides can be encapsulated in nanostructures templated by polymer mixes, enabling controlled release and reducing environmental effects (Taware et al. 2024).
- **Energy Storage and Catalysis**: High surface area and adjustable porosity are advantageous in applications such as battery electrodes or catalytic materials and are provided by nanoporous or fibrous structures made from polymer blend templates (Liu et al. 2019).

6.5 Conclusion

The environmentally friendly production of NP particles and fibrils using polymer blends as templates is a significant development in sustainable nanotechnology. This method creates nanostructures with regulated morphology, functionality, and biodegradability by fusing flexible and adjustable polymer blends with eco-friendly manufacturing processes. The precise production of nanostructures is made possible by polymer blends' phase separation, selective removal, compatibility with biodegradable polymers and green solvents, and other unique benefits. Using blends of polymers improves processes like solvent evaporation, electrospinning, and template-assisted synthesis, opening up possibilities for producing porous particles, nanofibers, and other specialized geometries. Biomedicine, energy storage, environmental cleanup, and sustainable agriculture are just a few of the many industries that use these nanostructures.

Furthermore, the foundation of this strategy is eco-friendly behaviours. Utilizing renewable resources like cellulose and biodegradable polymers like polylactic acid (PLA) reduces the fabrication process's negative environmental effects. This approach also aligns with sustainability objectives because it uses recyclable templates and low-toxicity solvents. Despite obstacles like maximizing polymer compatibility and guaranteeing controlled degradation, green chemistry, and material science developments are increasing the possibilities of polymer blend templating. This technique helps create high-performance, sustainable nanostructures for a more environmentally friendly future while addressing environmental issues related to traditional nanomaterial manufacturing.

References

Abdenour C, Nguyen-Tri P, Chabot B, Barnabé S, Bley J, Tolnai B, Guy N, Eesaee M (2024) New stable waterborne amorphous polylactic acid/organoclay nanocomposites prepared using emulsification solvent evaporation method. Polym Compos 45(14):12975–12994

Alam P, Ezzeldin E, Iqbal M, Anwer MK, Mostafa GA, Alqarni MH, Foudah AI, Shakeel F (2020) Ecofriendly densitometric RP-HPTLC method for determination of rivaroxaban in nanoparticle formulations using green solvents. RSC Adv 10(4):2133–2140

Alqaheem Y, Alomair AA (2020) Microscopy and spectroscopy techniques for characterization of polymeric membranes. Membranes 10(2):33

Andréia da Silva S, Faccin DJL, Cardozo NSM (2024) A kinetic-based criterion for polymer biodegradability applicable to both accelerated and standard long-term composting biodegradation tests. ACS Sustain Chem Eng 12(32):11856–11865

Arefin AM, Khatri NR, Kulkarni N, Egan PF (2021) Polymer 3D printing review: materials, process, and design strategies for medical applications. Polymers 13(9):1499

Asua JM (2004) Emulsion polymerization: from fundamental mechanisms to process developments. J Polym Sci, Part A: Polym Chem 42(5):1025–1041

Babick F (2020) Dynamic light scattering (DLS). In: Characterization of nanoparticles. Elsevier, pp 137–172

Bae SR, Heo DY, Kim SY (2022) Recent progress of perovskite devices fabricated using thermal evaporation method: Perspective and outlook. Mater Today Adv 14:100232

Banerjee A, Shwetabh K, Kumar K, Poddar R (2024) Exploring the effect of size variability on efficiency of upconversion nanoparticles as optical contrast agents. J Fluoresc 34(4):1589–1602

Banerji A, Jin K, Mahanthappa MK, Bates FS, Ellison CJ (2021) Porous fibers templated by melt blowing cocontinuous immiscible polymer blends. ACS Macro Lett 10(10):1196–1203

Bashir MA (2021) Use of dynamic mechanical analysis (DMA) for characterizing interfacial interactions in filled polymers. Solids 2(1):108–120

Bavatharani C, Muthusankar E, Wabaidur SM, Alothman ZA, Alsheetan KM, Mana AL-Anazy M, Ragupathy D (2021) Electrospinning technique for production of polyaniline nanocomposites/nanofibres for multi-functional applications: a review. Synthetic Metals 271, 116609

Caputo F, Vogel R, Savage J, Vella G, Law A, Della Camera G, Hannon G, Peacock B, Mehn D, Ponti J, Geiss O (2021) Measuring particle size distribution and mass concentration of nanoplastics and microplastics: addressing some analytical challenges in the sub-micron size range. J Colloid Interface Sci 588:401–417

Carreón RV, Cortázar-Martínez O, Rodríguez-Hernández AG, Serrano de la Rosa LE, Gervacio-Arciniega JJ, Krishnan SK (2024) Ionic liquid-assisted thermal evaporation of bimetallic Ag–Au nanoparticle films as a highly reproducible SERS substrate for sensitive nanoplastic detection in complex environments. Anal Chem 96(15):5790–5797

Chern CS (2006) Emulsion polymerization mechanisms and kinetics. Prog Polym Sci 31(5):443–486

Coelho SC, Estevinho BN, Rocha F (2021) Encapsulation in food industry with emerging electrohydrodynamic techniques: electrospinning and electrospraying—a review. Food Chem 339:127850

Collinson DW, Sheridan RJ, Palmeri MJ, Brinson LC (2021) Best practices and recommendations for accurate nanomechanical characterization of heterogeneous polymer systems with atomic force microscopy. Prog Polym Sci 119:101420

Danielsen SP, Beech HK, Wang S, El-Zaatari BM, Wang X, Sapir L, Ouchi T, Wang Z, Johnson PN, Hu Y, Lundberg DJ (2021) Molecular characterization of polymer networks. Chem Rev 121(8):5042–5092

Deshmukh R, Wagh P, Naik J (2016) Solvent evaporation and spray drying technique for micro-and nanospheres/particles preparation: a review. Drying Technol 34(15):1758–1772

Drzeżdżon J, Jacewicz D, Sielicka A, Chmurzyński L (2019) Characterization of polymers based on differential scanning calorimetry based techniques. TrAC, Trends Anal Chem 110:51–56

Easton CD, Kinnear C, McArthur SL, Gengenbach TR (2020) Practical guides for x-ray photoelectron spectroscopy: analysis of polymers. J Vacuum Sci Technol A 38(2)

Ethirajan A, Landfester K (2010) Functional hybrid materials with polymer nanoparticles as templates. Chem European J 16(31):9398–9412

Faust JM, Hamzehlou S, Leiza JR, Asua JM, Mhamdi A, Mitsos A (2021) Closed-loop in-silico control of a two-stage emulsion polymerization to obtain desired particle morphologies. Chem Eng J 414:128808

Garg M, Pamme N (2024) Strategies to remove templates from molecularly imprinted polymer (MIP) for biosensors. TrAC, Trends Anal Chem 170:117437

Geyer R (2020) A brief history of plastics. Mare plasticum-the plastic sea: combatting plastic pollution through science and art, 31–47.

Gomide RAC, de Oliveira ACS, Rodrigues DAC, de Oliveira CR, de Assis OBG, Dias MV, Borges SV (2020) Development and characterization of lignin microparticles for physical and antioxidant enhancement of biodegradable polymers. J Polym Environ 28:1326–1334

Gulmine JV, Janissek PR, Heise HM, Akcelrud L (2002) Polyethylene characterization by FTIR. Polym Testing 21(5):557–563

Habib U, Khan ZI, Mohamad ZB (2024) Compatibility and miscibility of recycled polyethylene terephthalate/polyamide 11 blends with and without Joncryl® compatibilizer: a comprehensive study of mechanical, thermal, and thermomechanical properties. Iran Polym J 33(9):1313–1326

Imoisili PE, Jen T-C (2021) Facile extraction and characterization of silica nanoparticles from corn stalk by sol-gel hydrothermal methods. IOP Conf Ser: Mater Sci Eng 1107(1):012030

Imoisili PE, Jen TC (2023a) Synthesis and characterization of bioplastic films from potato peel starch; effect of glycerol as plasticizer. Mater Today: Proc

Imoisili PE, Jen TC (2023b) Effect of physical modification on the tensile and thermal properties of plantain fibre polymer composite. In: ASME international mechanical engineering congress and exposition, vol 87615. American Society of Mechanical Engineers, p V004T04A004

Imoisili PE, Jen TC (2024a) Mechanism and kinetics of water absorption of plantain fibre reinforced bio-composites. In: Materials science forum, vol 1115. Trans Tech Publications Ltd, pp 55–62

Imoisili PE, Jen TC (2024b) Effect of mercerization on the crystallographic, macromolecular, and thermal properties of plantain fibers for fiber reinforced composite. In: Materials science forum, vol 1115. Trans Tech Publications Ltd, pp 63–70

Imoisili PE, Ukoba KO, Jen T-C (2020) Green technology extraction and characterisation of silica nanoparticles from palm kernel shell ash via sol–gel. J Mater Res Technol 9(1):307–313

Imoisili PE, Adeleke O, Makhatha ME, Jen TC (2023) Response surface methodology (RSM)-artificial neural networks (ANN) aided prediction of the impact strength of natural fibre/carbon nanotubes hybrid reinforced polymer nanocomposite. Eng Sci 23:852

Ismail MF, Islam MA, Khorshidi B, Tehrani-Bagha A, Sadrzadeh M (2022) Surface characterization of thin-film composite membranes using contact angle technique: review of quantification strategies and applications. Adv Coll Interface Sci 299:102524

Ivleva NP (2021) Chemical analysis of microplastics and nanoplastics: challenges, advanced methods, and perspectives. Chem Rev 121(19):11886–11936

Kaur A, Bajaj B, Kaushik A, Saini A, Sud D (2022) A review on template assisted synthesis of multi-functional metal oxide nanostructures: status and pro

Keirouz A, Wang Z, Reddy VS, Nagy ZK, Vass P, Buzgo M, Ramakrishna S, Radacsi N (2023) The history of electrospinning: past, present, and future developments. Adv Mater Technol 8(11):2201723

Kelly TL, Wolf MO (2010) Template approaches to conjugated polymer micro-and nanoparticles. Chem Soc Rev 39(5):1526–1535

Kim KW, Park B, Kim J, Jo C, Kim JK (2023) Recent progress in block copolymer s oft-template-assisted synthesis of versatile mesoporous materials for energy storage systems. J Mater Chem A 11(14):7358–7386

Lamaoui A, Mani V, Durmus C, Salama KN, Amine A (2024) Molecularly imprinted polymers: a closer look at the template removal and analyte binding. Biosens Bioelectron 243:115774

Laoutid F, Lafqir S, Toncheva A, Dubois P (2021) Valorization of recycled tire rubber for 3D printing of ABS-and TPO-based composites. Materials 14(19):5889

Li G, He Y, Han W, Yu Y, Zhu L, Si T, Sun Y (2021a) An improved solvent evaporation method to produce poly (lactic acid) microspheres via foam-transfer. Int J Biol Macromol 172:114–123

Li M, Yu H, Xie Y, Guo Y, Cheng Y, Qian H, Yao W (2021b) Fabrication of eugenol loaded gelatin nanofibers by electrospinning technique as active packaging material. LWT 139:110800

Li Y, Zhu J, Cheng H, Li G, Cho H, Jiang M, Zhang X et al (2021c) Developments of advanced electrospinning techniques: a critical review. Adv Mater Technol 6(11):2100410
Li X, Lin Y, Liu M, Meng L, Li C (2023) A review of research and application of polylactic acid composites. J Appl Polym Sci 140(7):e53477
Lin ST, Kimble L, Bhattacharyya D (2017) Polymer blends and composites for biomedical applications. Biomater Implants Scaffolds 195–235
Liu W, Ullah B, Kuo CC, Cai X (2019) Two-dimensional nanomaterials-based polymer composites: fabrication and energy storage applications. Adv Polym Technol 2019(1):4294306
Liu B, Liu Z, Wu H, Pan S, Sun Y, Xu Y (2021) Insight into simultaneous selective removal of nitrogen and phosphorus species by lanthanum-modified porous polymer: performance, mechanism and application. Chem Eng J 415:129026
Lovell PA, Schork FJ (2020) Fundamentals of emulsion polymerization. Biomacromol 21(11):4396–4441
Makhatha ME, Imoisili PE, Jen TC (2024) Crystallographic, macromolecular, thermal, and mechanical properties of Bambusa balcooa fibers: effect of two-step chemical modification. Polym Bull 1–17
Munir MU, Phan DN, Khan MQ (2022) Recycling of nanomaterials by solvent evaporation and extraction techniques. In: Nanomaterials recycling. Elsevier, pp 209–222
Ntobeng MK, Imoisili PE, Jen T-C (2020) Synthesis and photocatalytic performance of Ag-TiVOX nanocomposite, J King Saud Uni Sci 32:3103–3110. https://doi.org/10.1016/j.jksus.2020.08.021
Nurazzi N, Asyraf MRM, Rayung M, Norrrahim MNF, Shazleen SS, Rani MSA, Shafi AR, Aisyah HA, Radzi MHM, Sabaruddin FA, Ilyas RA (2021) Thermogravimetric analysis properties of cellulosic natural fiber polymer composites: a review on influence of chemical treatments. Polymers 13(16):2710
Nwanna EC, Imoisili PE, Jen T-C (2020) Fabrication and synthesis of SnOX thin films: a review. Int J Adv Manuf Technol 111(9–10):2809–2831
Nwanna EC, Imoisili PE, Bitire SO, Jen T-C (2021) Biosynthesis and fabrication of copper oxide thin films as a P-type semiconductor for solar cell applications. Coatings 11(12):1545
Okafor CE, Onovo AC, Imoisili PE, Kulkarni KM, Ihueze CC (2021) Optimal route to robust hybridization of banana-coir fibre particulate in polymer matrix for automotive applications. Materialia 101098
Pavlenko V, Żółtowska S, Haruna AB, Zahid M, Mansurov Z, Supiyeva Z, Jesionowski T et al (2022) A comprehensive review of template-assisted porous carbons: modern preparation methods and advanced applications. Mater Sci Eng R Rep 149:100682
Pramudianto A (2024) Managing plastic waste in the sea to support the international agreements for the achievement of the 14th sustainable development goals (sdgs-14). J Law Sustain Dev 12(7):e453–e453
Pulingam T, Foroozandeh P, Chuah JA, Sudesh K (2022) Exploring various techniques for the chemical and biological synthesis of polymeric nanoparticles. Nanomaterials 12(3):576
Rai PK, Lee J, Brown RJ, Kim KH (2021a) Micro-and nano-plastic pollution: behavior, microbial ecology, and remediation technologies. J Clean Prod 291:125240
Rai PK, Kumar V, Sonne C, Lee SS, Brown RJ, Kim KH (2021b) Progress, prospects, and challenges in standardization of sampling and analysis of micro-and nano-plastics in the environment. J Clean Prod 325:129321
Rajeshkumar L (2022) Biodegradable polymer blends and composites from renewable resources. In: Biodegradable polymers, blends and composites. Woodhead Publishing, pp 527–549
Rane AV, Kanny K, Abitha VK, Thomas S (2018) Methods for synthesis of nanoparticles and fabrication of nanocomposites. In: Synthesis of inorganic nanomaterials. Woodhead publishing, pp 121–139
Saad MA, Abdurahman NH, Yunus RM (2020) Eco-friendly surfactant to demulsification water in oil emulsion: synthesis, characterization and application. Chem Data Collect 30:100582
Saxena P, Shukla P, Gaur MS (2021) Thermal analysis of polymer blends and double layer by DSC. Polym Polym Compos 29(9_suppl):S11-S18

Shafqat SR, Bhawani SA, Bakhtiar S, Ibrahim MNM, Shafqat SS (2023) Template-assisted synthesis of molecularly imprinted polymers for the removal of methyl red from aqueous media. BMC Chem 17(1):46

Sisanth KS, Thomas MG, Abraham J, Thomas S (2017) General introduction to rubber compounding. In: Progress in rubber nanocomposites. Woodhead Publishing, pp 1–39

Su Y, Wang F, Wu S, Fan Y, Bai W, Wang S, Sun H, Zhu Z, Liang W, Li A (2021) Template-assisted preparation of conjugated microporous polymers membranes for selective separation. Sep Purif Technol 259:118203

Szczęch M, Szczepanowicz K (2020) Polymeric core-shell nanoparticles prepared by spontaneous emulsification solvent evaporation and functionalized by the layer-by-layer method. Nanomaterials 10(3):496

Tan D, Wang Y, Tong Y, Chen GQ (2021) Grand challenges for industrializing polyhydroxyalkanoates (PHAs). Trends Biotechnol 39(9):953–963

Taware AS, Rathod PB, Katariya AP, Tagad CK, Wagh PS, Sonar JP, Deshmukh SU, Kanagare AB (2024) Technological advancement in the development of nano fertilizers for sustainable agriculture. J Soil Sci Plant Nutrition 1–17

Thomas RM, Sambhudevan S, Hema S, Reghunadhan A (2023) PTT/rubber, thermoplastic and thermosetting polymer blends and IPNs. In: Poly trimethylene terephthalate: based blends, IPNs, composites and nanocomposites. Springer Nature Singapore, Singapore, pp 49–77

Vayshbeyn LI, Mastalygina EE, Olkhov AA, Podzorova MV (2023) Poly (lactic acid)-based blends: a comprehensive review. Appl Sci 13(8):5148

Vélez-Escamilla LY, Contreras-Torres FF (2022) Latest advances and developments to detection of micro-and nanoplastics using surface-enhanced Raman spectroscopy. Part Part Syst Charact 39(3):2100217

Veskova J, Sbordone F, Frisch H (2022) Trends in polymer degradation across all scales. Macromol Chem Phys 223(13):2100472

Walker TR, Fequet L (2023) Current trends of unsustainable plastic production and micro (nano) plastic pollution. TrAC, Trends Anal Chem 160:116984

Xu Z, Deng W, Wang X (2021) 3D hierarchical carbon-rich micro-/nanomaterials for energy storage and catalysis. Electrochem Energy Rev 4:269–335

Xu F, Miao D, Li W, Jin J, Liu Z, Shen C, Zhang J, Shuai J, Li X (2023) Specificity and competition of mRNAs dominate droplet pattern in protein phase separation. Phys Rev Res 5(2):023159

Yabu H, Higuchi T, Ijiro K, Shimomura M (2005) Spontaneous formation of polymer nanoparticles by good-solvent evaporation as a nonequilibrium process. Chaos: Interdisciplinary J Nonlinear Sci 15(4)

Yang DL, Faraz F, Wang JX, Radacsi N (2022) Combination of 3D printing and electrospinning techniques for biofabrication. Adv Mater Technol 7(7):2101309

Yu CH, Betrehem UM, Ali N, Ali F, Nawaz S, Sajid M, Yang Y, Chen T, Bilal M (2022) Design strategies, surface functionalization, and environmental remediation potentialities of polymer-functionalized nanocomposites. Chemosphere 306:135656

Zhang T, Xu G, Blum FD (2023) Eco-friendly room-temperature polymerization in emulsions and beyond. Polym Rev 63(4):852–865

Zhao G, Liu J, Liu M, Han X, Peng Y, Tian X, Liu J, Zhang S (2020) Synthesis of molecularly imprinted polymer via emulsion polymerization for application in solanesol separation. Appl Sci 10(8):2868

Zheng C, Xu X, He F, Li L, Wu B, Yu G, Liu Y (2010) Preparation of high-quality organic semiconductor nanoparticle films by solvent-evaporation-induced self-assembly. Langmuir 26(22):16730–16736

Zielińska A, Carreiró F, Oliveira AM, Neves A, Pires B, Venkatesh DN, Durazzo A, Lucarini M, Eder P, Silva AM, Santini A (2020) Polymeric nanoparticles: production, characterization, toxicology and ecotoxicology. Molecules 25(16):3731

Zimmerman JB, Anastas PT, Erythropel HC, Leitner W (2020) Designing for a green chemistry future. Science 367(6476):397–400

Chapter 7
Application of Nano ZnO-Blended Polymeric Membrane

C. Joel, R. Biju Bennie, and S. Daniel Abraham

Abstract The incorporation of zinc oxide (ZnO) nanoparticles into polymeric membranes has emerged as a promising strategy to address the limitations of conventional membrane systems in separation, filtration, and biomedical applications. This chapter presents an overview of nano ZnO-blended polymeric membranes, describing their synthesis methodologies, structural and morphological characteristics, physicochemical properties, and various applications. This chapter focuses on the role of ZnO nanoparticles in enhancing membrane performance, including improved hydrophilicity, mechanical reinforcement thermal and UV stability, as well as significant antimicrobial and photocatalytic functionalities. Fabrication techniques such as solution casting, phase inversion, electrospinning, in situ polymerization, and sol-gel synthesis are discussed. The applications including wastewater treatment, biomedical engineering (e.g., wound dressings, drug delivery systems), environmental remediation, food packaging, and protective coatings are also discussed. The chapter also addresses current challenges related to nanoparticle agglomeration, scalability, and long-term stability and suggests future directions for the development of next-generation multifunctional membrane materials.

Keywords ZnO · Polymeric membranes · Nanocomposites · Photocatalysis · Antimicrobial activity

7.1 Introduction

The Greek numerical prefix ‘nano’, which denotes a billionth, and the word technology are combined to form the term nanotechnology. As a result, nanotechnology is usually thought of as being smaller than 100 nm (Ratner and Ratner 2003). Renowned

---

C. Joel (✉) · R. Biju Bennie
Postgraduate Department of Chemistry, St. John’s College, Tirunelveli, Tamil Nadu, India
e-mail: joeldcc@yahoo.co.in

S. Daniel Abraham
Department of Chemistry, Madras Christian College (Autonomous), Chennai, Tamil Nadu, India

P. M. Visakh (ed.), *Engineering Applications of Polymer based Nano Blends*, Advanced Structured Materials 235, https://doi.org/10.1007/978-3-031-90979-5_7

physicist Richard Feynman predicted the theoretical potential of nanotechnology in his well-known and memorable speech in 1959. "There's plenty of room at the bottom," through which he described the potential for very small-scale manipulation and controlling of objects (Feynman 1959). In 1986, K. Eric Drexler popularized the word "Nanotechnology" in his book "Engines of the Creation: The Coming Era of Nanotechnology" (Drexler 2006). He suggested using molecules or particles that can self-assemble to create devices that are only a few nanometers wide.

With the advent of nanotechnology, material science and engineering have advanced quickly, especially in the design and production of hybrid materials. Among these developments, polymeric membranes combined with nanoparticles are a noteworthy advance. By combining nanoparticles into polymer matrices, these membranes are produced, creating hybrid systems that take advantage of both the inherent benefits of polymers and the special properties of nanomaterials (Zhu et al. 2018; Goh et al. 2015; Zhang et al. 2016). This combination has made it possible to create materials with specific qualities, tackling difficult problems in a range of scientific and industrial fields.

Because of their high surface area-to-volume ratios and nanoscale dimensions, nanomaterials have remarkable promise for improving the functionality of polymeric membranes. These materials come in a variety of nanostructures, each with unique chemical, structural, and electrical properties, including metal oxides, carbon-based nanomaterials, quantum dots, and others. These nanoparticles can be combined with polymers to create membranes that have better interfacial interactions, increased structural integrity, and more functional diversity. Techniques such as direct dispersion, surface modification of nanomaterials for improved compatibility, or in situ polymerization to guarantee even distribution and integration within the polymer matrix can all be used in the blending process (Kango et al. 2013).

Polymeric membranes have been the focus of research for a long time due to their chemical diversity, lightweight nature, and usefulness for filtration and separation activities (Liu et al. 2020). However, particularly in harsh operating environments, conventional polymeric membranes frequently encounter difficulties like fouling, heat deterioration and mechanical strength restrictions (Ulbricht 2006). One intriguing method to get over these restrictions is to incorporate nanoparticles into the polymer framework. Researchers can now construct materials with precisely adjustable molecular-level properties, which add a new dimension to membrane performance and design. The ultimate properties of the membrane are largely determined by the nanoscale interactions between the polymer chains and the distributed nanoparticles.

A multidisciplinary approach is used in the synthesis and development of nanomaterials-blended polymeric membranes, using ideas from materials engineering, nanotechnology and polymer science. Achieving a uniform dispersion of nanoparticles within the polymer matrix is an essential step in this process since agglomeration can seriously impair the performance of the membrane and decrease the efficiency of the nanomaterials (Ng et al. 2013; Hiremath et al. 2021). In order to overcome these obstacles and improve compatibility and interaction with polymer matrices, developments in chemical functionalization and surface modification of nanomaterials have proven crucial. Furthermore, the production process and the selection of polymers and nanomaterials enable membranes to be tailored to satisfy certain performance requirements.

The development of these cutting-edge materials depends heavily on characterization techniques. To investigate the morphology, dispersion and structural integration of nanomaterials inside the polymer matrix, techniques like heat analysis, spectroscopy and electron microscopy are used. These revelations aid in comprehending the fundamental principles that control the behavior of systems involving nanomaterials and polymers and direct the improvement of synthesis procedures. As a result, the final membranes are designed to have the ideal blend of structural, chemical and physical characteristics, which qualifies them for a variety of specific uses.

Nanotechnology has the ability to revolutionize conventional material systems, as demonstrated by nanomaterials-blended polymeric membranes (Abraham et al. 2023). Researchers are working on a new class of advanced materials that push the limits of usefulness and efficacy by fusing the remarkable properties of nanoparticles with the advantages of polymers. These membranes offer huge possibilities for innovation and discovery in this rapidly expanding field, reflecting the strong connection between materials science and engineering.

Membrane technology is essential to many industries, such as chemical processing, medical filtration and water purification (Baker 2023; Zhang et al. 2024). Because of their improved physical, chemical and antibacterial qualities, blended membranes that incorporate nanoparticles have recently drawn interest. The incorporation of nano zinc oxide (ZnO) into polymeric membranes is one such development. Nano ZnO-polymeric blended membranes combine the flexibility and adaptability of polymeric materials with the beneficial characteristics of ZnO nanoparticles, including enhanced mechanical strength, UV resistance and antibacterial activity (Zhang et al. 2024; Tian et al. 2023; Sahu et al. 2021). Applications in environmental protection, medical devices and water treatment have been made possible by this combination of technologies.

This chapter provides a detailed exploration of nano ZnO-polymeric blended membranes, focusing on their structure, properties, preparation methods, and practical applications. The analysis also examines how ZnO nanoparticles impact membrane functionality and the challenges associated with scaling and deploying these materials in real-world applications.

7.2 Nano ZnO-Polymeric Blended Membrane

Blended membranes made of metal oxide polymers are a novel and versatile family of materials with uses in energy storage, gas separation and water treatment. These membranes are made of metal oxide nanoparticles incorporated into a polymeric matrix, creating a synergistic blend that incorporates the benefits of both materials. Mechanical strength, flexibility and a continuous phase for transportation are all offered by the polymeric matrix (George and Thomas 2001). Polyvinylidene fluoride (PVDF), polysulfone (PSF), polyethersulfone (PES), and polydimethylsiloxane (PDMS) are examples of common polymer blends used in membrane technology (Khulbe et al. 2010). Properties including hydrophilicity, thermal stability and antifouling are improved by the metal oxides. Aluminum oxide (AlO_3), silicon dioxide (SiO_2), zinc oxide (ZnO), and titanium dioxide (TiO_2) are a few examples.

In purification, metal oxides improve water flux and decrease fouling by increasing water affinity. The rigid structures of the metal oxide reinforce the polymer matrix, increasing the longevity of membranes. Certain oxides, such as ZnO and TiO_2, have antibacterial properties that make them useful for filtration and medical applications (Naseem and Durrani 2021). Selectivity for particular molecules or ions can be improved by using nanoparticles to adjust the surface characteristics and pore size of membranes. Usually, the nanoparticles interact at the molecular level by aligning on the surface of the polymer matrix or dispersing within it. By altering surface chemistry and enhancing structural integrity, these interactions can enhance performance in demanding conditions. Metal oxides are reactive filtering agents because they can also catalyze processes that break down contaminants.

Microorganisms, organic pollutants and heavy metals are among the contaminants that can be removed with this method (Abdullah et al. 2022; Khulbe and Matsuura 2018; Salim et al. 2021). Additionally, it is used to separate gases from mixtures, such as CO_2, O_2 or H_2 (Turk et al. 2001). It benefits from improvements in ionic conductivity and is utilized in fuel cells and batteries (Agarwal et al. 2013). It is utilized in scaffolding and drug delivery (Gholamali and Yadollahi 2021), where controlled permeability and biocompatibility are crucial.

Despite their benefits, there are still difficulties with minimizing agglomeration and maintaining homogeneous metal oxide dispersion inside the polymer matrix. To get beyond these challenges, research is currently underway to improve synthesis methods including sol–gel procedures and in situ polymerization. The development of scalable, affordable production techniques and the investigation of innovative combinations suited to particular industrial requirements are key to the future of metal oxide polymeric blended membranes.

One class of metal oxide nanoparticles that is well known for its distinct chemical and physical characteristics is zinc oxide (ZnO) nanoparticles (NPs). ZnO nanoparticles are being researched extensively for use in electronics, optics, medicine and environmental remediation because of its large surface area, size-dependent properties and multifunctional nature (Sharma et al. 2022). These particles, which usually have sizes between 1 and 100 nm, have characteristics that set them apart from their bulk counterparts. These characteristics include increased UV absorption, antibacterial activity and photocatalytic activity. At ambient temperature, the broad bandgap of ZnO, a direct bandgap semiconductor, is around 3.37 eV. Its optical and electrical characteristics are also influenced by its high exciton binding energy (60 meV). Structurally, ZnO can crystallize in three forms: Hexagonal wurtzite (most stable and common form), Cubic zinc blende, and Rock salt (under high-pressure conditions) (Klingshirn 2007a, b). ZnO nanoparticles typically exhibit the hexagonal wurtzite structure. The surface of these nanoparticles often contains a high density of oxygen vacancies and defects, which significantly impact their reactivity and photoluminescence properties.

ZnO nanoparticles can be synthesized using various physical, chemical and biological methods, with the choice of method influencing their size, shape and properties. ZnO nanoparticles stand out for their excellent photoluminescence and UV-absorbing properties. Because of these properties, they can be used in sunscreens, UV-blocking cosmetics and optoelectronic devices like lasers and light-emitting diodes. Since ZnO can produce reactive oxygen species (ROS) when exposed to UV light, it acts a strong photocatalyst. Self-cleaning surfaces, air purification and water purification all make advantage of this feature (Fig. 7.1).

ZnO nanoparticles exhibit strong antimicrobial effects against a wide range of bacteria and fungi. The mechanisms include ROS generation, disruption of microbial membranes and leaching of Zn ions. This makes them useful in medical applications such as wound dressings, antibacterial coatings and packaging materials. The semi-conducting nature and high electron mobility make it a key material in gas sensors, biosensors and transparent conductive films for touch screens and solar cells. ZnO nanoparticles are being explored for drug delivery, cancer therapy and bioimaging

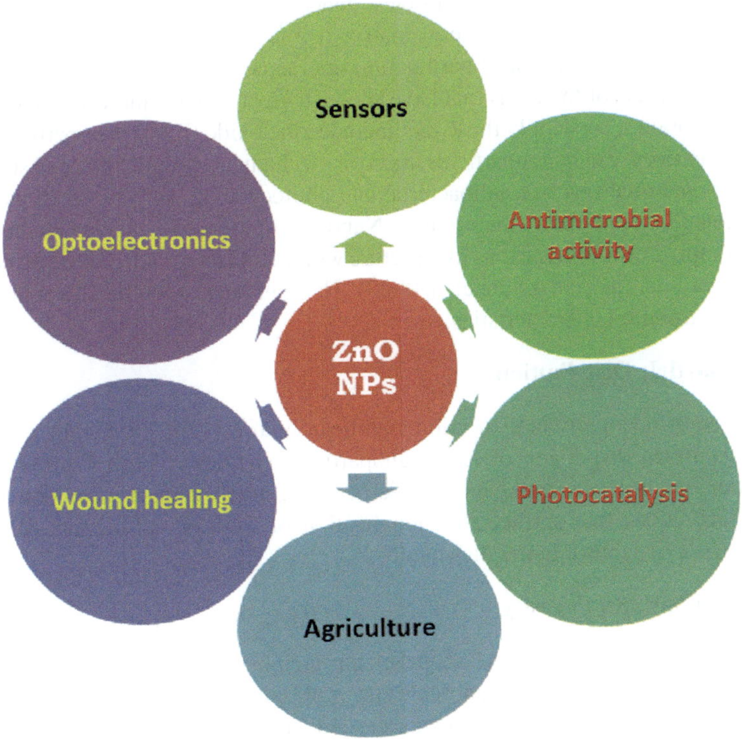

Fig. 7.1 Diverse applications of ZnO NPs

due to their biocompatibility, surface functionality and ability to interact with biological systems. ZnO nanoparticles play a significant role in the degradation of organic pollutants in water and air. Their ability to break down contaminants under UV light makes them an essential material in environmental cleanup efforts.

Despite their attractive characteristics, ZnO nanoparticles have limitations like they can be cytotoxic and induce oxidative stress in biological systems, which raise questions regarding their safety in environmental and medicinal applications. It is still difficult to produce homogenous ZnO nanoparticles with consistent characteristics on a large scale. Under some circumstances, ZnO nanoparticles are prone to aggregation and dissolution, which impairs their functionality.

7.2.1 Structure of Nano ZnO-Polymeric Blended Membranes

Nano ZnO-polymeric blended membranes are typically synthesized by incorporating ZnO nanoparticles within a polymer matrix. The choice of polymer matrix is crucial as it affects the mechanical stability, pore structure and filtration properties of the membrane. ZnO nanoparticles, generally sized between 1 and 100 nm, are embedded within the polymeric matrix or on the surface of the membrane, enhancing its properties. Water purification, gas separation and medicinal applications are just a few of the areas that have shown a great deal of interest in nanocomposite membranes, especially those that include zinc oxide (ZnO) nanoparticles into polymer matrices. ZnO nanoparticles improve mechanical strength, thermal stability and functional qualities like antibacterial and photocatalytic activity when they are incorporated into polymeric membranes. Nano ZnO-polymeric blended membranes have a composite arrangement as its structure where ZnO nanoparticles are dispersed within a polymer matrix (Asture et al. 2023). These membranes typically exhibit the following structural characteristics:

- **Nanoparticle Distribution**

The degree of nanoparticle dispersion significantly affects the membrane's performance. Uniform dispersion of ZnO nanoparticles ensures enhanced surface area interaction and reduces aggregation, which can negatively impact membrane properties. Techniques like surface functionalization of ZnO or optimizing synthesis parameters help achieve better dispersion.

- **Morphology and Porosity**

ZnO nanoparticles influence the shape of the membrane and frequently result in higher porosity. Because of its nanoscale size, ZnO can interact with polymer chains and change the pore shape of the membrane. As a result, separation processes have better water flux and selectivity.

- **Interfacial Interaction**

Strong interfacial bonding between ZnO nanoparticles and the polymer matrix ensures mechanical stability. Functional groups on the ZnO surface can form hydrogen bonds or covalent interactions with polymer chains, enhancing compatibility.

- **Surface Features**

ZnO nanoparticles at the membrane surface impart functional properties such as hydrophilicity, photocatalytic activity and antimicrobial effects. These features are critical for applications like fouling resistance in filtration or antimicrobial coatings.

7.2.2 Composition of Nano ZnO-Polymeric Blended Membranes

The composition of these membranes varies depending on the specific polymer and ZnO nanoparticles used, as well as the intended application. Commonly used polymers (Fig. 7.2) in ZnO-polymeric membranes includes, Polyvinylidene fluoride (PVDF) which is known for its chemical resistance, thermal stability and mechanical strength. Polyethersulfone (PES) provides excellent thermal and chemical stability, often used in ultrafiltration and nanofiltration membranes (Tsehaye et al. 2018). Polymethyl methacrylate (PMMA) offers good optical transparency and compatibility with ZnO. Chitosan is a biodegradable and biocompatible polymer which is ideal for biomedical applications. Polyvinyl alcohol (PVA) is frequently used due to its hydrophilicity and ability to form strong bonds with ZnO.

The polymer regulates the membrane's porosity and permeability and offers flexibility and structural support. Furthermore, these polymers are very resistant to environmental and chemical deterioration, which is crucial for long-term uses. Solution casting, phase inversion, electrospinning and direct mixing are some methods for incorporating ZnO nanoparticles into the polymer matrix. Each method has an impact on how the nanoparticles are distributed throughout the membrane, which changes characteristics like mechanical strength, porosity and hydrophilicity.

7.2.3 Properties of Nano ZnO-Polymeric Blended Membranes

The blending of ZnO nanoparticles with a polymeric matrix imparts unique properties to these membranes, making them suitable for diverse applications.

Because of their well-established capacity to prevent microbial development, ZnO nanoparticles are especially helpful in sterilization-intensive applications like water purification systems and medical devices. The mechanism of antimicrobial action

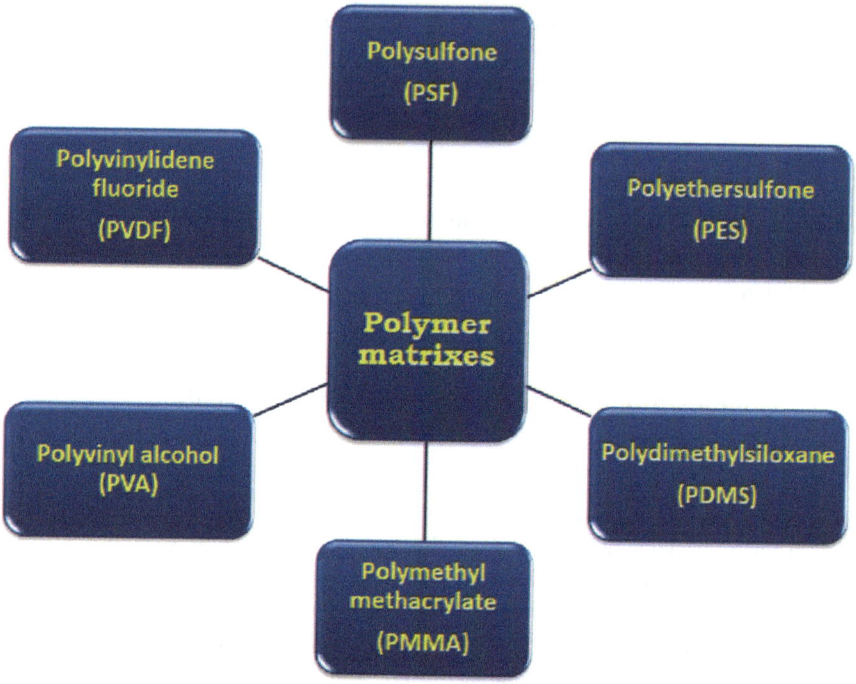

Fig. 7.2 Representation of various types of polymer matrixes

involves the generation of reactive oxygen species (ROS), which can damage bacterial cell walls and disrupt cellular function, thus providing a self-cleaning property to the membrane (Ogunsona et al. 2020). ZnO nanoparticles act as reinforcing agents within the polymer matrix, improving mechanical strength, tensile strength and elasticity. The membrane's lifetime and many possible uses are expanded due to its enhanced flexibility, which also makes it more resistant to stress and operational wear and tear. By absorbing UV radiation, ZnO nanoparticles can stop UV exposure from degrading the polymer matrix. This is especially helpful in outdoor or high-radiation settings because UV deterioration is a prevalent problem that shortens the lifespan of polymeric materials (Muzata et al. 2023). The hydrophilic properties of ZnO NPs can increase the membrane's water affinity, increasing its permeability and fouling resistance. This characteristic is particularly important in filtration purposes because fouling, which is brought on by particles building up on the membrane surface, lowers efficiency and raises maintenance expenses. Nano ZnO-polymeric membranes are useful in water treatment applications because of the photocatalytic activity under UV light, which can break down organic contaminants. When ZnO is exposed to light, it produces reactive oxygen species (ROS) that can degrade pollutants, decreasing the need for chemical cleaning agents and producing a secondary purification effect.

7.2.4 *Fabrication Techniques*

The fabrication of ZnO with polymer membranes is important due to its synergistic properties, offering enhanced performance in various applications. This integration combines the flexibility of polymers with the multifunctionality of nano ZnO, enabling advanced material solutions (Fig. 7.3).

- **Solution Casting and Phase Inversion**: In solution casting, ZnO nanoparticles are dispersed in a polymer solution, which is then cast into a film. The phase inversion method, on the other hand, involves casting the polymer solution with ZnO particles in a coagulation bath, leading to the formation of a porous membrane structure. This method is advantageous in controlling the membrane's pore size distribution, which is critical for filtration applications.
- **Electrospinning**: Electrospinning creates nanofibrous membranes by applying a high voltage to a polymer solution. ZnO nanoparticles can be added to the polymer solution before spinning, resulting in a membrane with high porosity and large surface area. Electrospun ZnO-polymeric membranes have promising applications in air filtration, tissue engineering and wound dressing.
- **In situ Polymerization**: This method involves mixing ZnO nanoparticles with monomers, which are then polymerized to form the membrane. In situ polymerization allows for uniform dispersion of ZnO particles and can result in a robust polymer-nanoparticle interface, enhancing the overall properties of the membrane.

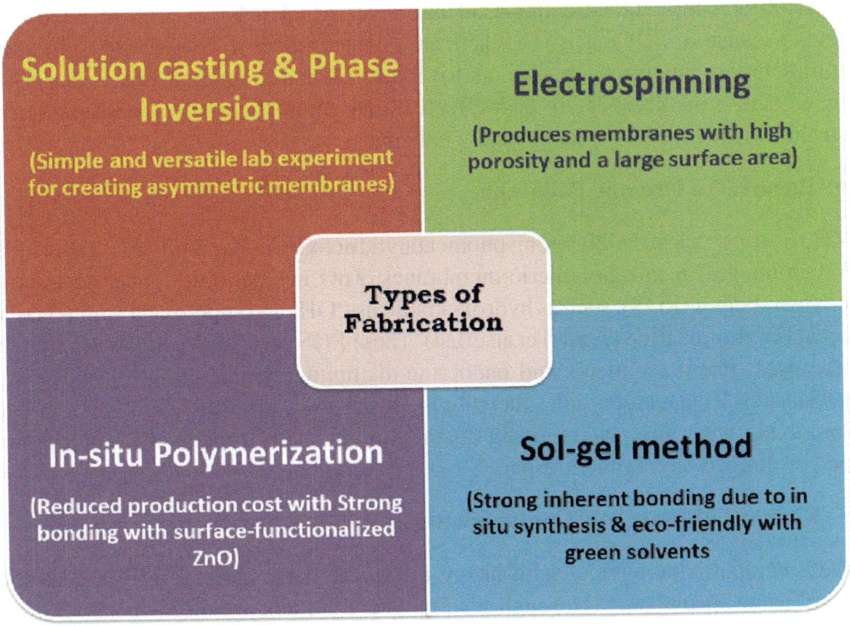

Fig. 7.3 Schematic representation of fabrication of polymeric membranes

- **Sol–Gel Method**: ZnO nanoparticles are synthesized within the polymer matrix via a sol–gel reaction. This method ensures a homogeneous distribution of ZnO throughout the polymer, which can be advantageous for membranes that require consistent antimicrobial and photocatalytic properties.

7.3 Application of Nano ZnO-Blended Polymeric Membrane

7.3.1 Water Treatment and Filtration

Nano-ZnO-blended polymeric membranes play a significant role in water treatment and filtration due to their multifunctional properties (Fig. 7.4). The incorporation of ZnO nanoparticles enhances the membrane's antimicrobial activity, preventing biofouling and prolonging operational life. Their photocatalytic properties aid in the decomposition of hazardous compounds and organic pollutants. These membranes have better hydrophilicity, which increases water flow and lowers energy consumption (Yin and Deng 2015; Ravi et al. 2020; Nqombolo et al. 2018). Through adsorption and catalytic reactions, nano-ZnO also helps remove dyes and heavy metals from wastewater. Their robust performance and high salt rejection rates make them effective in desalination procedures. These membranes' improved mechanical strength and thermal stability guarantee longevity under challenging operating circumstances. Nano-ZnO is perfect for creating drinkable water in isolated or disaster-affected locations because of its capacity to inactivate microorganisms. They are used in ultrafiltration and nanofiltration systems for industrial and municipal wastewater treatment. These membranes are cost-effective and environmentally friendly, offering sustainable solutions to global water challenges.

- **Removal of Organic Pollutants**

ZnO nanoparticles exhibit strong photocatalytic activity under ultraviolet (UV) light. When integrated into polymeric membranes, ZnO nanoparticles generate reactive oxygen species (ROS), such as hydroxyl radicals ($^{\cdot}OH$) and superoxide anions (O_2^{-}), upon UV illumination (Ansari et al. 2024). These ROS break down organic pollutants like dyes, pharmaceuticals and endocrine-disrupting chemicals into less harmful substances. ZnO nanoparticles increase the hydrophilicity of the membrane surface, improving water flux and reducing organic fouling. This leads to efficient filtration and prolonged membrane lifespan.

- **Removal of Inorganic Contaminants**

ZnO nanoparticles interact with heavy metal ions such as lead (Pb^{2+}), cadmium (Cd^{2+}), and arsenic (As^{3+}) through adsorption or ion exchange mechanisms (Dhiman and Kondal 2021). When ZnO surfaces are functionalized, their affinity for particular metal ions is increased, which helps remove harmful impurities. ZnO nanoparticles

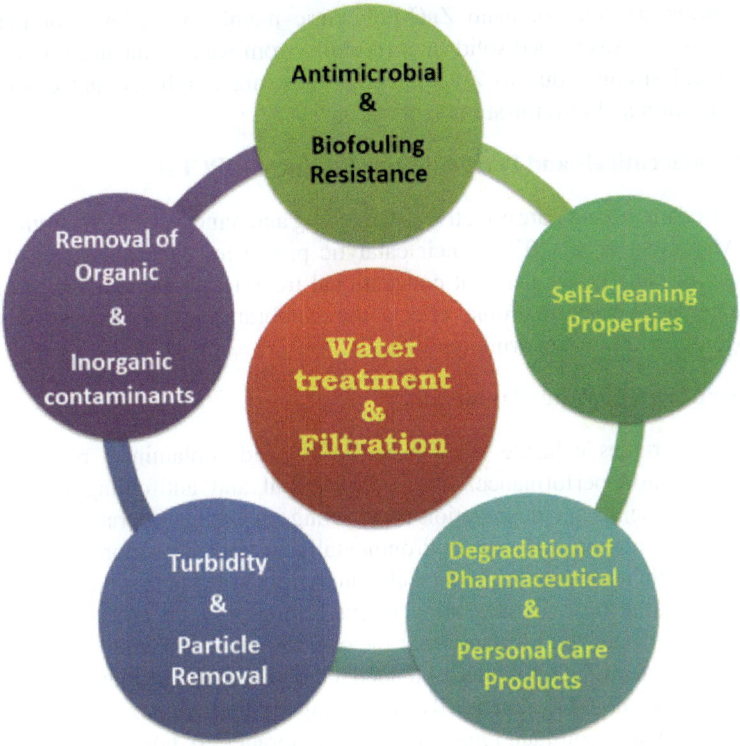

Fig. 7.4 Application of nano ZnO-Blended polymeric membrane in water treatment and filtration

enhance water permeability and salt rejection in reverse osmosis (RO) membranes and nanofiltration, which makes them useful in brackish and seawater desalination procedures.

- **Antimicrobial and Biofouling Resistance**

ZnO nanoparticles are inherently antimicrobial due to their ability to generate ROS, release Zn^{2+} ions, and disrupt microbial membranes. When embedded in polymeric membranes, they provide long-term protection against bacterial and fungal growth, which are major causes of biofouling in water treatment systems. It is mainly used in the prevention of biofilm formation in membrane bioreactors (MBRs) and ultrafiltration systems (Warsinger et al. 2018).

- **Self-cleaning Properties**

ZnO nanoparticles impart photocatalytic self-cleaning capabilities to membranes. UV light activates the ZnO, decomposing organic fouling agents on the membrane surface. This reduces cleaning frequency and operational costs.

- **Turbidity and Particle Removal**

The porous structure of nano ZnO-polymeric membranes allows for the effective removal of suspended solids and turbidity from water. Enhanced porosity and mechanical strength due to ZnO integration ensure durability and efficiency in microfiltration and ultrafiltration applications.

- **Pharmaceuticals and Personal Care Products (PPCPs)**

ZnO-based membranes are effective in degrading and removing PPCPs from wastewater (Majumder et al. 2020). Their catalytic properties enable the breakdown of complex organic molecules that conventional treatment methods may struggle to eliminate. Besides antimicrobial effects, the membranes act as a physical barrier to pathogens, ensuring safe drinking water.

- **Advantages in Water Treatment**

ZnO nanoparticles enhance water permeability and contaminant rejection, optimizing filtration performance. The antimicrobial and antifouling properties of ZnO reduce membrane degradation and fouling, extending operational life. ZnO-polymeric membranes offer an environmentally sustainable option for water treatment due to their potential to reduce chemical usage in cleaning and maintenance. While the initial cost of fabrication might be higher, reduced fouling, cleaning frequency and extended lifespan lower the overall operational costs.

In general, nano ZnO-polymeric blended membranes are a revolutionary approach to water filtration and treatment. By addressing challenges like fouling, microbial contamination and pollutant removal, these membranes are poised to play a critical role in ensuring access to clean and safe water worldwide.

7.3.2 Biomedical Applications

Nano ZnO-polymeric blended membranes have gained prominence in biomedical applications (Fig. 7.5) due to their unique combination of mechanical strength, biocompatibility and multifunctional properties such as antimicrobial activity and photocatalytic behavior. These membranes are especially helpful in drug delivery systems, biosensors, tissue engineering, and wound healing applications. Nano ZnO-polymeric membranes are employed in the biomedical industry as antimicrobial coatings for medical equipment, drug delivery, and wound dressings (Asture et al. 2023). They are beneficial in reducing hospital-acquired infections because of their biocompatibility and capacity to inhibit bacterial development. Because of their improved stability and adjustable porosity, ZnO-polymeric membranes can also help with controlled drug release.

These membranes function as efficient dressings during wound healing, preventing microbial infections and encouraging tissue regeneration. Because ZnO nanoparticles may produce reactive oxygen species (ROS) and release Zn^{2+} ions, they have antibacterial capabilities that prevent bacteria and fungi from growing on

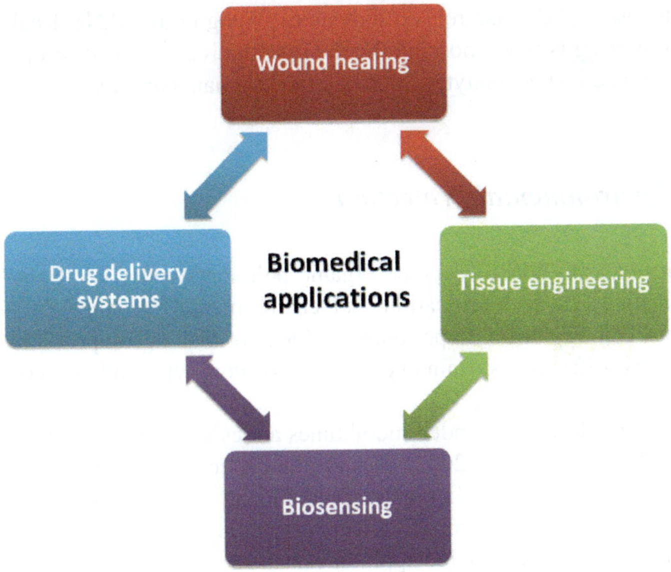

Fig. 7.5 Biomedical applications of nano ZnO-blended polymeric membrane

the surface of wounds. Furthermore, the membranes' hydrophilic properties improve moisture retention, fostering an environment that is conducive to cell migration and proliferation.

Nano ZnO-polymeric membrane mechanical and structural characteristics make them appropriate scaffolds for promoting cell growth and adhesion in tissue engineering (Rahman 2020). While ZnO nanoparticles offer antibacterial protection to reduce infections during tissue regeneration, their porosity shape enables effective nutrition and oxygen exchange. Growth factors or other biologically active compounds that promote tissue regeneration and cellular differentiation can also be provided by functionalizing these membranes.

As drug delivery devices, these membranes offer controlled and extended release of therapeutic compounds. Drugs and ZnO nanoparticles can interact to produce certain release characteristics, enabling effective delivery to the targeted site. Additionally, ZnO nanoparticles can improve the durability of medications that are encapsulated, preventing them from degrading. The photocatalytic capabilities of ZnO enable light-responsive drug release, which permits on-demand administration in reaction to external stimuli (Dristant et al. 2023; Sabbagh and Kim 2022; Weng et al. 2023).

ZnO nanoparticles incorporated in polymeric membranes have a large surface area and semiconducting characteristics, which are used in biosensing to create sensitive and specific detection tools. These membranes are used in sensors that track pH levels, glucose and disease-related indicators (Wang et al. 2024). Real-time diagnostic monitoring is made possible by the quantitative electrical or optical signals that are produced when analytes interact with ZnO nanoparticles.

7.3.3 Environmental Protection

The ability of ZnO NPs to degrade organic pollutants under UV light has potential applications in air purification and environmental cleanup. ZnO-polymeric membranes can trap airborne pollutants, which are then decomposed by photocatalytic activity, offering a sustainable solution for improving indoor and outdoor air quality.

Nano ZnO-polymeric blended membranes are essential for preserving the environment (Manikandan et al. 2022) because they effectively resolve challenges with waste management, air quality enhancement, and water purification (Fig. 7.6). These membranes are excellent at reducing environmental pollution because of their improved qualities, which include great mechanical strength, chemical stability and functional diversity. The removal of organic pollutants, heavy metals and microbiological contaminants from wastewater is one important use for these membranes in water treatment. Under UV light, the photocatalytic activity of ZnO nanoparticles incorporated in the polymer matrix allows hazardous organic molecules to be broken down into less hazardous compounds. This characteristic is very helpful for breaking down dyes, insecticides and medication residues that are challenging to get rid of using traditional techniques.

These polymeric membranes are efficient at adsorbing and immobilizing heavy metal ions such as lead, cadmium and arsenic in addition to removing organic pollutants. Strong interactions with these ions are made possible by the enormous surface area of ZnO nanoparticles, which allows for their capture and subsequent removal from contaminated water sources. Additionally, the membranes have antibacterial activity, which inhibits the formation of biofilms and lowers the possibility of microbial fouling, providing the filtration systems' long-term effectiveness and robustness.

Fig. 7.6 Application of nano ZnO-blended polymeric membrane in environmental protection

Because of this characteristic, they can also be used in membrane bioreactors, which help clean industrial and municipal effluent.

Nano ZnO-polymeric membranes are used in filtration systems that absorb particulate matter and volatile organic compounds in order to improve the quality of the air. When exposed to light, these membranes' ZnO photocatalytic qualities enable them to convert airborne contaminants like formaldehyde and benzene into harmless compounds (Scaffaro and Citarrella 2023; Grasso et al. 2022; Dutta et al. 2024). In urban and industrial settings, where air pollution poses serious health and environmental hazards, this application is especially crucial.

These kinds of membranes are used in waste management to help recover and recycle valuable materials from industrial effluents. They are employed, for example, in the separation of oil–water mixtures, successfully eliminating oil impurities from wastewater produced by the food processing and petrochemical sectors (Ismail et al. 2020; Mousa et al. 2022; Coelho et al. 2024). High separation efficiency and endurance under demanding conditions are guaranteed by their hydrophilic and oleophobic qualities, which are enhanced by the presence of ZnO nanoparticles.

7.3.4 Food Packaging

Nano ZnO-polymeric blended membranes are increasingly being utilized in food packaging due to their ability to enhance the safety, shelf life and quality of packaged products. These membranes combine the mechanical and barrier properties of polymers with the antimicrobial and UV-blocking capabilities of zinc oxide nanoparticles, offering a multifunctional approach to food preservation. Additionally, UV-blocking properties of ZnO protect packaged goods from degradation due to light exposure, making these membranes ideal for sensitive food items (Fig. 7.7).

The antimicrobial activity of ZnO nanoparticles plays a crucial role in inhibiting the growth of bacteria, fungi and other pathogens that can cause spoilage or contamination. By releasing Zn^{2+} ions and generating reactive oxygen species, these nanoparticles effectively disrupt microbial cell walls, ensuring that the food remains fresh for a longer period.

These membranes offer superior protective characteristics against moisture, oxygen and UV rays, which are significant causes of food spoiling, in addition to antimicrobial protection (Guerraf et al. 2024; Zhao et al. 2023; Shen et al. 2023). By decreasing the permeability of the polymer matrix, ZnO nanoparticle integration stops oxygen and water vapor from penetrating, which might cause oxidation or microbial growth. At the same time, its capacity to block UV rays keeps delicate food items like dairy and oils safe from photodegradation, maintaining their nutritional value and flavor (Irede et al. 2024).

The possibility for active packaging applications is a further advantage of employing nano ZnO-polymeric membranes in food packaging. By functionalizing these membranes, bioactive substances such as antioxidants or antimicrobials can be released into the packaging environment, extending shelf life and actively preventing

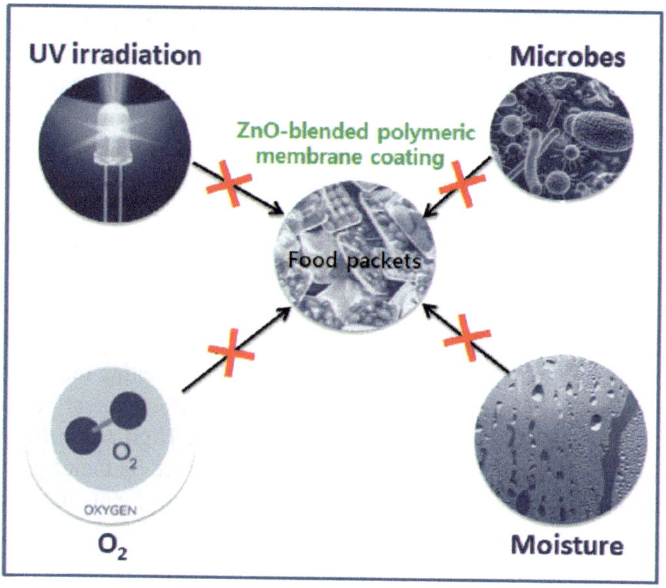

Fig. 7.7 Food packaging application of nano ZnO-blended polymeric membrane

spoiling. For example, the membrane's regulated release of antimicrobial agents can be tailored to the requirements of certain food items, ensuring effective and contamination prevention.

These membranes also help make food packing more environmentally friendly. Their capacity to prolong shelf life lowers food waste, and when combined with eco-friendly polymers, they may be biodegradable or recyclable, meeting the increasing demand for sustainable packaging materials.

7.3.5 Protective Coatings and Textiles

Nano ZnO-polymeric blended membranes have emerged as valuable materials for protective coatings and textiles, owing to their multifunctional properties such as antimicrobial activity, UV-blocking ability and mechanical durability (Fig. 7.8). In protective coatings, these membranes are applied to surfaces to enhance resistance to environmental factors, including microbial contamination, UV radiation and chemical exposure (Rivero et al. 2015; Singh et al. 2023). The incorporation of ZnO nanoparticles into the polymer matrix imparts significant UV-shielding properties, protecting underlying materials from photodegradation. This feature is particularly useful for coatings applied to outdoor structures, vehicles and electronic devices, where prolonged UV exposure can lead to material deterioration (Shah et al. 2022; Alam et al. 2021; Zhu et al. 2022).

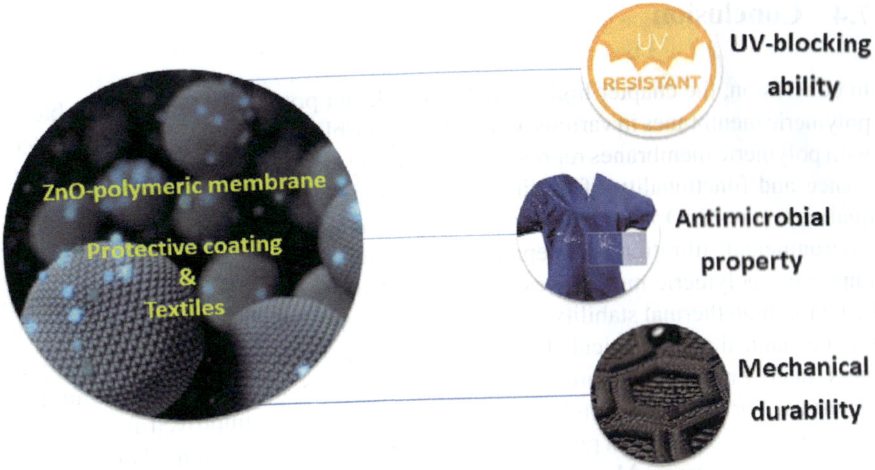

Fig. 7.8 Application of nano ZnO-blended polymeric membrane in protective coatings and Textiles

The antimicrobial properties of ZnO nanoparticles make these membranes highly effective in preventing the growth of bacteria, fungi and other pathogens on coated surfaces. This functionality is especially beneficial in environments where hygiene is critical, such as healthcare facilities, food processing plants and public transportation. By disrupting microbial cell walls and generating reactive oxygen species, ZnO-polymeric coatings help maintain cleanliness and reduce the risk of infections.

Nano ZnO-polymeric blended membranes are utilized in textiles to create fabrics with superior functional and protective qualities. Because these membranes offer antimicrobial protection, they can be used in military uniforms, sporting and health-care settings. Even after extended usage, the antibacterial function guarantees that textiles stay odor-free and hygienic. Furthermore, they protect the skin from harmful UV rays, which makes them perfect for outdoor apparel and accessories.

The polymer matrix's mechanical strength and durability guarantee that these membranes' resilience will hold up against washing and wear. Additionally, ZnO's self-cleaning qualities, which are made possible by the photocatalytic breakdown of organic matter, increase their usefulness in low-maintenance coatings and fabrics. Furthermore, these membranes help textiles resist stains and water, increasing their usefulness and durability.

Irede EL, Awoyemi RF, Owolabi B, Aworinde OR, Kajola RO, Hazeez A, Raji AA, Ganiyu LO, Onukwuli CO, Onivefu AP, Ifijen IH (2024) Cutting-edge developments in zinc oxide nanoparticles: synthesis and applications for enhanced antimicrobial and UV protection in healthcare solutions. RSC Adv 14(29):20992–21034

Ismail NH, Salleh WNW, Ismail AF, Hasbullah H, Yusof N, Aziz F, Jaafar J (2020) Hydrophilic polymer-based membrane for oily wastewater treatment: a review. Sep Purif Technol 233:116007

Kango S, Kalia S, Celli A, Njuguna J, Habibi Y, Kumar R (2013) Surface modification of inorganic nanoparticles for development of organic–inorganic nanocomposites—a review. Prog Polym Sci 38(8):1232–1261

Khulbe KC, Matsuura T (2018) Removal of heavy metals and pollutants by membrane adsorption techniques. Appl Water Sci 8:1–30

Khulbe KC, Feng C, Matsuura T (2010) The art of surface modification of synthetic polymeric membranes. J Appl Polym Sci 115(2):855–895

Klingshirn C (2007a) ZnO: from basics towards applications. Physica Status Solidi (b) 244(9):3027–3073

Klingshirn C (2007b) ZnO: material, physics and applications. ChemPhysChem 8(6):782–803

Liu H, Cao C, Huang J, Chen Z, Chen G, Lai Y (2020) Progress on particulate matter filtration technology: basic concepts, advanced materials, and performances. Nanoscale 12(2):437–453

Majumder S, Chatterjee S, Basnet P, Mukherjee J (2020) ZnO based nanomaterials for photocatalytic degradation of aqueous pharmaceutical waste solutions—a contemporary review. Environ Nanotechnol Monitor Manage 14:100386

Manikandan S, Subbaiya R, Saravanan M, Ponraj M, Selvam M, Pugazhendhi A (2022) A critical review of advanced nanotechnology and hybrid membrane based water recycling, reuse, and wastewater treatment processes. Chemosphere 289:132867

Mousa HM, Fahmy HS, Ali GA, Abdelhamid HN, Ateia M (2022) Membranes for oil/water separation: a review. Adv Mater Interfaces 9(27):2200557

Muzata TS, Gebrekrstos A, Orasugh JT, Ray SS (2023) An overview of recent advances in polymer composites with improved UV-shielding properties. J Appl Polym Sci 140(14):e53693

Naseem T, Durrani T (2021) The role of some important metal oxide nanoparticles for wastewater and antibacterial applications: a review. Environ Chem Ecotoxicol 3:59–75

Ng LY, Mohammad AW, Leo CP, Hilal N (2013) Polymeric membranes incorporated with metal/metal oxide nanoparticles: a comprehensive review. Desalination 308:15–33

Nqombolo A, Mpupa A, Moutloali RM, Nomngongo PN (2018) Wastewater treatment using membrane technology. Wastewater and Water Quality 29:30–40

Ogunsona EO, Muthuraj R, Ojogbo E, Valerio O, Mekonnen TH (2020) Engineered nanomaterials for antimicrobial applications: a review. Appl Mater Today 18:100473

Rahman MM (2020) Polyurethane/zinc oxide (PU/ZnO) composite—synthesis, protective property and application. Polymers 12(7):1535

Ratner MA, Ratner D (2003) Nanotechnology: A gentle introduction to the next big idea. Prentice Hall Professional

Ravi J, Othman MHD, Matsuura T, Bilad MRI, El-Badawy TH, Aziz F, Ismail AF, Rahman MA, Jaafar J (2020) Polymeric membranes for desalination using membrane distillation: a review. Desalination 490:114530

Rivero PJ, Urrutia A, Goicoechea J, Arregui FJ (2015) Nanomaterials for functional textiles and fibers. Nanoscale Res Lett 10:1–22

Sabbagh F, Kim BS (2022) Recent advances in polymeric transdermal drug delivery systems. J Control Release 341:132–146

Sahu DK, Ghosh G, Rath G (2021) Nanofibers in drug delivery. Nanopharmaceutical advanced delivery systems, pp 99–123

Salim SH, Al-Anbari RH, Haider AJ (2021) Polymeric membrane with nanomaterial's for water purification: a review. In: IOP conference series: earth and environmental science, vol 779, no 1. IOP Publishing, p 012103

Scaffaro R, Citarrella MC (2023) Nanofibrous polymeric membranes for air filtration application: a review of progress after the COVID-19 pandemic. Macromol Mater Eng 308(9):2300072

Shah V, Bhaliya J, Patel GM, Deshmukh K (2022) Advances in polymeric nanocomposites for automotive applications: a review. Polym Adv Technol 33(10):3023–3048

Sharma DK, Shukla S, Sharma KK, Kumar V (2022) A review on ZnO: fundamental properties and applications. Mater Today: Proc 49:3028–3035

Shen Y, Seidi F, Ahmad M, Liu Y, Saeb MR, Akbari A, Xiao H (2023) Recent advances in functional cellulose-based films with antimicrobial and antioxidant properties for food packaging. J Agric Food Chem 71(44):16469–16487

Singh P, Ali SW, Kale RD (2023) Antimicrobial nanomaterials as advanced coatings for self-sanitizing of textile clothing and personal protective equipment. ACS Omega 8(9):8159–8171

Tian J, Paterson TE, Zhang J, Li Y, Ouyang H, Asencio IO, Hatton PV, Zhao Y, Li Z (2023) Enhanced antibacterial ability of electrospun PCL scaffolds incorporating ZNO nanowires. Int J Mol Sci 24(19):14420

Tsehaye MT, Velizarov S, Van der Bruggen B (2018) Stability of polyethersulfone membranes to oxidative agents: a review. Polym Degrad Stab 157:15–33

Turk BS, Merkel T, Lopez-Ortiz A, Gupta RP, Portzer JW, Krishnan GN, Freeman BD, Fleming GK (2001) Novel technologies for gaseous contaminants control (No. AC26–99FT40675–01). National Energy Technology Laboratory (NETL), Pittsburgh, PA, Morgantown, WV, and Albany, OR (United States)

Ulbricht M (2006) Advanced functional polymer membranes. Polymer 47(7):2217–2262

Wang KN, Li ZZ, Cai ZM, Cao LM, Zhong NN, Liu B, Zhou K, Huo FY, Cai B, Bu LL (2024) The applications of flexible electronics in dental, oral, and craniofacial medicine. NPJ Flexible Electronics 8(1):33

Warsinger DM, Chakraborty S, Tow EW, Plumlee MH, Bellona C, Loutatidou S, Karimi L, Mikelonis AM, Achilli A, Ghassemi A, Padhye LP (2018) A review of polymeric membranes and processes for potable water reuse. Prog Polym Sci 81:209–237

Weng Z, Xu Y, Gao J, Wang X (2023) Research progress of stimuli-responsive ZnO-based nanomaterials in biomedical applications. Biomater Sci 11(1):76–95

Yin J, Deng B (2015) Polymer-matrix nanocomposite membranes for water treatment. J Membr Sci 479:256–275

Zhang Y, Wu B, Xu H, Liu H, Wang M, He Y, Pan B (2016) Nanomaterials-Enabled Water and Wastewater Treatment. Nanoimpacticles 3:22–39

Zhang Y, Zhang J, Li J, Shen J, Zhang X (2024) Thermosensitive ZnO-PMMA/PEG microgels: a smart sunscreen with adapting SPF to ambient temperature fluctuations. Mater Today Chem 37:102039

Zhao Y, Li B, Zhang W, Zhang L, Zhao H, Wang S, Huang C (2023) Recent advances in sustainable antimicrobial food packaging: insights into release mechanisms, design strategies, and applications in the food industry. J Agric Food Chem 71(31):11806–11833

Zhu J, Hou J, Zhang Y, Tian M, He T, Liu J, Chen V (2018) Polymeric antimicrobial membranes enabled by nanomaterials for water treatment. J Membr Sci 550:173–197

Zhu Q, Chua MH, Ong PJ, Lee JJC, Chin KLO, Wang S, Kai D, Ji R, Kong J, Dong Z, Xu J (2022) Recent advances in nanotechnology-based functional coatings for the built environment. Mater Today Adv 15:100270

Chapter 8
Rubber-Based Nanoblends

Süleyman Gökhan Çolak, Ahmet Güngör, Melis Özge Alaş Çolak, Utku Bulut Şimşek, and Asiye Güngör

Abstract Recently, rubber materials mixed with tiny, nanoscale fillers have really taken off in many fields. This chapter looks at several common rubbers—natural rubber, SBR, EPDM, NBR, and silicone rubber—and shows how adding things like graphene, carbon nanotubes, or metal-oxide particles can bump up properties such as strength, conductivity, and heat resistance. We talk through practical challenges too, for example, how hard it is to get those fillers to spread evenly. Then we move on to real-world uses: from car parts and vibration dampers to flexible circuits, energy storage devices, and EMI shielding panels. Sustainability comes up as well, because today's research needs to think about both performance and the planet. All in all, rubber nanocomposites are set to play an even bigger role, especially as we push for greener, smarter materials.

S. G. Çolak (✉)
Department of Biomedical Engineering, Faculty of Engineering and Natural Sciences, Iskenderun Technical University, Hatay, Turkey
e-mail: sgokhan.colak@iste.edu.tr

A. Güngör
Faculty of Engineering and Natural Sciences, Sabanci University, Istanbul, Turkey

Center of Excellence for Functional Surfaces and Interfaces for Nano-Diagnostics (EFSUN), Sabancı University, Orhanlı, Istanbul, Turkey

M. Ö. A. Çolak
Sabancı University Nanotechnology Research and Application Center (SUNUM), Sabanci University, Istanbul, Turkey

Department of Chemical Engineering, Mersin University, Mersin, Turkey

U. B. Şimşek
Department of Chemical Engineering, Bogazici University, Istanbul, Turkey

Department of Machine, Anadolu BIL Vocational School, İstanbul Aydın University, İstanbul, Turkey

A. Güngör
Izmit Vocational High School, Kocaeli University, Kartepe Kocaeli, Turkey

P. M. Visakh (ed.), *Engineering Applications of Polymer based Nano Blends*, Advanced Structured Materials 235, https://doi.org/10.1007/978-3-031-90979-5_8

Keywords Rubber nanocomposites · Natural rubber · SBR · EPDM · NBR · Silicone rubber · Carbon-based materials · Metal oxides · Flexible electronics · Energy storage · EMI shielding · Sustainability

8.1 Introduction to Rubber

8.1.1 History of Rubber

Rubbers are an important material and could be classified as an elastomer, which is produced through natural and artificial processes. Natural rubber has been used since ancient times. It was initially found by the natives of South America and extracted as latex from trees. The discovery of South America by European explorers in the fifteenth century brought the existence of rubber to the attention of the Western world. In the early nineteenth century, Charles Goodyear invented the vulcanization technique, and this considerably improved the features of rubber. Ever since then, during the Industrial Revolution, rubber became more and more widely used. Later, in the twentieth century, synthetic rubbers were synthesized from raw materials sourced from petroleum derivatives, which reduced demand for natural rubber (Chandrasekaran 2017).

8.1.2 Chemical Structure of Rubber

Natural rubber is an elastomer obtained by the polymerization of latex obtained from the rubber tree, Hevea brasiliensis, mainly as polyisoprene $(C_5H_8)_n$. Its polymer structure was formed by long-chain hydrocarbons, which are responsible for the material's key properties, such as elasticity and flexibility. The linking of the isoprene units, 2-methyl-1,3-butadiene, in a cis manner within natural rubber makes the material more flexible. The synthetic rubbers have similar polymer structures but can be prepared by the polymerization of different monomers, which makes it possible variations in their chemical, thermal, and mechanical properties (Vayyaprontavida Kaliyathan et al. 2020).

8.1.3 General Properties of Rubber

The most characteristic property of rubber is its elasticity: the possibility of eventual return to the initial shape of a material after the action of some deforming force. Because of this, it finds good applications in engineering at very high levels of flexibility and resistance to impact. Other salient properties of rubber include exceptional abrasion resistance, chemical resistance, and electrical insulating properties. Due to the fact that it is a good electrical insulator, natural rubber has become one of the most used materials in cables and electrical equipment. It resists water pretty well and also can remain flexible at low temperatures, which makes this material very appropriate for broad industrial and commercial applications (Marković and Visakh 2017).

8.1.4 Types of Rubber

Rubber can be broadly categorized into two main types: natural rubber and synthetic rubber (Sethulekshmi et al. 2022).

EPDM (Etilen Propilen Dien Monomer)

EPDM is a synthetic rubber obtained by polymerization of ethylene and propylene with a small amount of non-conjugated diene. It exhibits excellent resistance to heat, oxidation, ozone, and aging. Properties of EPDM are usually improved by vulcanization. On the other hand, adding nanofillers reduces the production cost and enhances physical and chemical properties of composites. Different nanofillers, including carbon nanotubes, graphene, and nanoclay, have been added to the EPDM matrix with a view to improving thermal and electrical conductivity. The applications of such nanocomposites have been extended to electromagnetic interference shielding, sensors, nuclear applications, and thermal insulation, among many other areas (Costa et al. 2024a).

Most notably, EPDM has been attracting wide interest as a thermal energy storage material. Nanofillers distribution enhances the mechanical and morphological properties of such composites, which in turn improves their thermal stability. EPDM/paraffin systems can be applied in applications that balance building energy consumption. The improvement in the mechanical performance of composites is based on the homogeneous distribution of the nanofiller in the EPDM matrix; this contributes to a decrease in crack formation and degradation processes within the material. Other benefits of cost-effective materials in the course of nanocomposite preparation will have enormous scalability benefits. Environmental sustainability, coupled with production efficiency by using nanofillers, makes nanocomposites of EPDM greener (Costa et al. 2024a).

The cost advantages based on durability that the EPDM nanocomposites provide, therefore, ensure their high growth in industrial and technological uses. Their flexibility and long-term performance in energy storage, thermal insulation, and across various engineering applications resulted in an increase in their usage.

Natural Rubber (NR)

Improvements in natural rubber are usually achieved by the addition of nano-sized reinforcing fillers, which enhance its durability, abrasion resistance, and UV stability, among other performance properties. Fillers alter the course of vulcanization and curing kinetics, and changes in cross-link density further improve the mechanical and viscoelastic properties of the product. Among the most used fillers in the rubber industry is carbon black, which has very good strength in vulcanized rubbers and enhances abrasion and UV deterioration resistance. However, manufacturing carbon black requires heavy oil or natural gas and results in the release of greenhouse gases, along with other harmful emissions. Furthermore, carbon black may spill into the environment through waste rubber products and thus pose ecological and health threats due to its carcinogenic and poisonous nature. Consequently, both business and academics have been evaluating carbon black substitutes (Sinclair et al. 2019).

In the past two years, a number of nanomaterials have been proposed as effective fillers in rubber. These include nano-clay, carbon nanotubes, graphene, and cellulose nanofibers/nanocrystals, CNF/CNC, which are derived from renewable resources. CNF/CNC has outstanding properties of stress transfer and reinforcement owing to its highly extended surface area, high aspect ratio, and ability to form strong interfacial bonds with various polymer matrices. However, CNF/CNC is inherently hydrophilic, which makes it a big challenge to incorporate it into the hydrophobic polymer matrices and to induce good fiber-polymer bonding. Thus, surface functionalization becomes an important task. A number of publications have demonstrated the potential of functionalized CNF/CNC as reinforcement fillers in rubber. For example, Trovatti et al. added BC nanofibers and polystyrene-decorated BC into natural rubber latex, improving the tensile modulus and strength of the resulting nanocomposites. This esterification with polystyrene enhanced the compatibility of BC with the hydrophobic NR, improving the interface bonding and dispersion (Sinclair et al. 2019).

Surface functionalization methods are also important. For example, Parambath Kanoth et al. grafted mercaptan (–SH) functional groups onto cellulose nanocrystals to make them active for use both as cross-linking agents and reinforcing fillers in natural rubber. Covalent bonds between CNC and the matrix of natural rubber were obtained through UV-initiated free radical reactions, hence enhancing cross-link density and improving tensile strength.

Nitrile Rubber (NBR)

NBR with polarity exhibits high oil resistance, but its tensile strength is low and, hence, has to be increased by incorporating reinforcing fillers into it. In general, properties of filler-reinforced rubber composites depend upon filler dispersion, filler-rubber interaction, and cross-link density. To improve the properties of rubbers, nanofillers are used, and among them, clay has a tiny size, an excellent aspect ratio, and low cost, hence, it is one of the most potential nanomaterials (He et al. 2023).

Organic quaternary ammonium salts, silane coupling agents, and antioxidants have also been used in rubber/clay nanocomposites to improve the filler-rubber interactions, which enhance the performance of the materials. Various methods have been studied with the hope of overcoming the dispersion issues of clay in various rubber matrices, including melt mixing, solution blending, and latex compounding. Curiously, despite the fact that cross-link density is one of the most influential factors on composite properties, studies related to the cross-link density of rubber/clay nanocomposites are rather scarce. The conventional vulcanization, efficient vulcanization, and semi-efficient vulcanization systems represent the most common methods in general use in the rubber business. So far, the CV method has been widely used for building cross-link networks in rubber/clay nanocomposites; however, it may limit the further development of mechanical qualities (He et al. 2023).

Recently, coordination cross-link networks with metal salts have been applied to rubber and have attracted considerable attention. Coordination cross-linking systems replaced conventional vulcanization processes with the addition of metal cations in nitrile-containing rubbers or polymer blends. Metal salts work not only as a coordination cross-linking agent but also as a reinforcing filler, leading to good mechanical qualities. In this work, the NBR/clay nanocomposites were prepared by mixing clay gel with solid rubber, then a metal salt, namely copper sulfate pentahydrate (CSP), was added to form NBR/clay/CSP nanocomposites. These nanocomposites showed coordination crosslinking between the nitrile groups (–C≡N–) with copper ions (Cu^{2+}), presenting mechanical properties dependent on curing time, which is hardly reported or discussed so far.

Neoprene (CR–Polychloroprene)

Polychloroprene was the first commercially produced synthetic rubber. In 1930, a research team, led by Wallace H. Carothers at DuPont and attempting to synthesize synthetic rubber, succeeded in developing it. It is marketed under the brand name Neoprene. Synthesis methods of radical emulsion polymerization of chloroprene are similar to those adopted for styrene-butadiene rubber. Initiators usually used include potassium or ammonium persulfate. Nowadays, polychloroprene is manufactured starting from chloroprene monomer, produced from butadiene by halogenation; originally, polychloroprene was prepared through the dimerization of acetylene (C_2H_2). After a brief history of polychloroprene, for example, one may recognize that DuPont introduced the material under the name Neoprene in 1931 and was producing it for decades. Indeed, in 2015, DuPont transferred its Neoprene business to Denka Performance Elastomers. Today, other producers, including ExxonMobil (Chlorobutyl®) and Arlanxeo (Baypren®), manufacture this elastomer (Mandlekar et al. 2022).

The most fundamental reason underlying polychloroprene's success is an excellent combination of environmental durability with flexibility. The material is very resistant to dynamic applications that is, continuous stretching and bending with the buildup of heat. Simultaneously, resistance to sunlight, UV, ozone, oil, and gasoline is excellent. High gas barrier properties make this material desirable in the widest circle of applications.

Polychloroprene is hardened by vulcanization and usually cross-linked with sulfur. The most common are the sulfur-based vulcanizates, but there are sulfur-free modified versions existing for cross-linking and regulation of molecular weight—W-type CR. They contain structural units of chloroprene in 1,4-*trans* that directly affect their crystalline structure and mechanical properties. These structural units provide certainty that the material will retain its crystalline structure under conditions of stress. Other structural units include 1,4-*cis*, 3,4, and 1,2, conferring low crystallinity and mechanical strength (Mandlekar et al. 2022).

Another key variable inherent in the manufacturing process of polychloroprene involves temperature. CR made at high temperatures exhibits a low crystallization rate due to their polymer chains being highly disordered. These CRs will be suitable for applications that need elastomeric properties. The 1,2 repeat units of polychloroprene are also very important since the chlorine atom easily exerts electrophilic substitution reactions by its attack on the carbon atoms present in the allylic position. This property is essential to the CR cross-linking process.

In light of this, the characteristic properties of polychloroprene especially flexibility, resistance to environmental conditions, and dynamic applications have given it a peculiar position among all types of rubber. Advances in production technologies from the past to the present have increased the usability of polychloroprene in different sectors and its wide distribution under various manufacturer names and commercial brands.

Inflatable Structures

Inflatable structures usually consist of a thin layer of one or several materials. They are able to be, therefore, inflated by various gases like air, helium, or hydrogen. One very important advantage of these structures is the good gas barrier properties present. Besides the low-cost inflatable objects, different types of advertising items, most inflatable structures are made of fabric-covered or laminated construction. The fabric layer adds durability to the structure, while the polymeric coating contributes to prevention of gas permeation, improving resistance to external factors. This optimal selection of materials is necessary, as these structures have to be lightweight and strong; only such combinations can make them crucial for performance (Mandlekar et al. 2022).

The important advantages of inflatable structures are that they are lightweight and store compactly when not in use. Military applications of the same structures have a weight reduction of about 50%. This is a big advantage logistically. Their low cost of production and low maintenance make these structures more appealing. They have found wide applications in the military and aerospace industries, opening up considerable avenues both for cost as well as functionality in various defense-related technologies. Examples include lightweight aircraft such as aerostats and zeppelins for air threat detection and other military purposes. Besides their logistic advantages, the capability of these balloons to carry a load for a long time is an added plus (Mandlekar et al. 2022).

After the Second World War, elastomers started to be widely applied in inflatable constructions due to their flexibility, rather low gas and vapor permeability, and other properties. General-purpose elastomers have also had a couple of drawbacks, such as low heat resistance and poor ozone and oxygen resistance. Later on, the development of more specialized elastomers went in the direction of improvement in properties like weatherability, resistance to organic solvents, and low gas permeability. Also, the properties of elastomers depend on the nature of their chemical structure. For instance, those with a saturated chemical structure, such as EPDM and IIR, are more durable in comparison with those containing unsaturated structures, which include SBR and natural rubber.

By now, the benefits of using different elastomers for different applications are apparent. Given the general permeation of elastomers through the industrial and defense sectors whenever inflatable structures are used, the strategic value of such materials can be appreciated. There is one elastomer, however, that does provide flexibility at low temperatures and which exhibits good mechanical durability and resistance to gas permeation: thermoplastic polyurethane, as used in lightweight aircraft-accommodating structures (aerostats) and related applications. Such material is also replaced by patented combinations like CR and CSM combinations. It is for such innovations that materials science does not stay for a single moment without further development and improvement.

Conclusion: Most of the research on elastomers and inflatable structures pertains to their application in defense and space technologies; however, those technologies have enormous potential for use in a wider range of industries and commercial fields. Future research will be aimed at coming up with lighter, stronger, and cheaper materials.

Synthetic Rubbers

- **Polybutadiene Rubber**: Synthetic rubber composed of polybutadiene, an elastomer produced by the chemical conjunction of many butadiene molecules into huge molecules or polymers. It has typical characteristics such as high abrasion resistance, low heat generation, and crack formation resistance.
- **Butyl Rubber**: Special kind of rubber that provides one of the highest classes of resistance to gas permeation. The great elasticity of the material, combined with this characteristic, facilitates producing airtight products, including inner tubes, sports balls, and sealants.
- **Nitrile Rubber**: Artificial rubber with better oil- and chemical-resistant characteristics than normal natural rubber. Alongside this, Nitrile rubber can bear much higher temperatures compared to normal ones.
- **Chlorosulfonated Polyethylene Rubber**: Synthetic rubber from polyethylene. Also known as Hypalon, the proprietary trade name of DuPont for CSM. It shows excellent resistance to ozone, oxidation, sunlight, and weathering.

- **Silicone Rubber**: One of the most used rubbers due to its plasticity, biocompatibility, and resistance against very high and very low temperatures, fire, ozone, and UV radiation.
- **Polyurethane Rubber**: a tough rubber that possesses a high degree of resistance to abrasion and extrusion. However, O-ring polyurethane is not good for use in applications where good compression and high-temperature resistance are needed.
- **Polysulfides**: Synthetic rubber compounds that possess an excellent resistance towards oils, fuels, a large number of chemicals, steam, and ozone.
- **Fluororubbers**: The synthetic rubber compounds of fluors exhibit very good temperature stability from − 45 to 204 °C.

Type of rubber	Advantages	Disadvantages	Applications
Polybutadiene rubber	High abrasion resistance, low heat accumulation, resistance to cracking	Poor resistance to oils and chemicals, limited flexibility at low temperatures	Tires, belts, hoses, seals
Butyl rubber	High gas impermeability, excellent elasticity, resistance to chemicals and oils	Limited flexibility at low temperatures, poor resistance to abrasion	Inner tubes, sports balls, sealants, adhesives
Nitrile rubber	Good resistance to oils and chemicals, high temperature resistance, good abrasion resistance	Limited flexibility at low temperatures, poor resistance to ozone and UV radiation	Seals, gaskets, O-rings, fuel hoses
Chlorosulfonated polyethylene rubber	Excellent resistance to ozone, oxidation, sunlight, and weathering, good resistance to chemicals and oils, high temperature resistance	Limited flexibility at low temperatures, poor resistance to abrasion	Roofing membranes, coatings, adhesives, sealants
Silicone rubber	High temperature resistance, good resistance to ozone and UV radiation, biocompatibility, plasticity	Poor resistance to abrasion, limited flexibility at low temperatures	Cookware, medical devices, adhesives, sealants
Polyurethane rubber	High abrasion resistance, good resistance to extrusion, high tensile strength	Limited flexibility at low temperatures, poor resistance to chemicals and oils	O-rings, seals, gaskets, adhesives
Polysulfides	Excellent resistance to oils, fuels, many chemicals, steam, and ozone	Poor resistance to abrasion, limited flexibility at low temperatures	Seals, gaskets, O-rings, fuel hoses

(continued)

(continued)

Type of rubber	Advantages	Disadvantages	Applications
Fluororubbers	Exceptional temperature stability from – 45 to 204 °C	High cost, limited flexibility at low temperatures	O-rings, seals, gaskets, fuel hoses

8.1.5 Applications and Importance of Rubber

It is an industrial raw material indispensable to a wide variety of fields, including the motor industry. The basic uses of this material are seen in tires, seals, hoses, and suspension systems. Electrical and electronics industries employ the material in electrical cables, insulation, and vibration-dampening pads. Rubber also finds significant application in the medical sector in latex gloves, catheters, and sundry other devices. Other very common uses of rubber include sports apparatuses, household utensils, and even in the construction business (Chandrasekaran 2017).

8.1.6 Rubber Blends and Improvement Methods

Rubber blends: These are obtained by mixing two or more types of rubber in a certain proportion. The blending of rubber is done to optimize one or some properties. For the formulation of the blends, usually, durability, elasticity, and performance improvement of the rubber is made. According to the type of used rubber, additives, and processing methods, the content of the blends will differ (Low et al. 2024; Urtekin et al. 2023).

Rubber blends use additives such as fillers, plasticizers, and stabilizers, among other chemical compounds. These are added to rubber to improve its processability and performance. As such, carbon black is used as the filler in improving the abrasion resistance of rubber, while various chemical stabilizers are added to extend the life of the rubber through improving its resistance to environmental elements (Urtekin et al. 2023).

8.1.7 Production Processes and Costs

Rubber production is also a multi-stage process in a sequence leading to the development of the properties and qualities of the final product. The basic steps of rubber manufacturing consist of harvesting, processing, compounding, and vulcanization. Natural rubber is harvested from the tapping of rubber trees, while synthetic rubber is manufactured from petrochemical feedstocks through polymerization processes. The cost of producing rubber is based on many factors such as the type of rubber, the method of production, and the location of production. Natural rubber tends to be higher in price owing to the intensive labor process of harvesting, while the cost of synthetic rubber will depend on the price of the petrochemical feedstocks (Wichaita et al. 2021).

8.1.8 Environmental and Health Impacts

Rubber is an important industrial crop that creates environmental and health impacts. Rubber tree cultivation contributes to deforestation and habitat loss, mainly in tropical climates where the trees grow. This has led to the loss of biodiversity with an associated disruption of natural ecosystems. The resultant wastewater, solid wastes, and air emissions are byproducts of the production process, with processing and vulcanization chemicals released into the environment as hazardous materials. Workers in the rubber industry may be exposed to harmful chemicals, such as solvents, plasticizers, and vulcanization accelerators, which could cause health problems, such as respiratory system diseases, skin diseases, and cancers. In addition, large amounts of waste result from both the manufacture and use of rubber products, including scrap rubber and used tires, and waste rubber products at the end of their life, which contribute to environmental pollution if not properly disposed of Marković and Visakh (2017).

8.1.9 Conclusion and Future Perspectives

The diversity and width of application give rubber and rubber blends an extremely high importance from industrial point of view. Advances in technology and new processing methods are continuing to further enhance the performance of rubber materials. Some of the points of interest for rubber material research in the near future are reduction in environmental impact, functional improvement of rubber materials, and new fields of application.

The main idea is the overview of the history of rubber, its chemical structure, general properties, and types applied, as well as applications and importance. Applications in these industries allow one to see what role rubber plays and in which directions it is further being developed.

8.2 Introduction to Rubber Blends

Rubber compounds are essential to modern-day materials research and in several industrial applications. Fundamentally, these materials comprise carefully matched mixtures of various rubber forms or rubber and other polymers to enhance specific performance attributes (Sisanth et al. 2017). As a single type of rubber frequently does not meet all application demands, blends are an essential solution to address these constraints by delivering advantages such as cost efficiency, superior mechanical qualities, increased processing attributes, and tailored performance (Mark et al. 2019).

A primary advantage of rubber blends is their capacity to combine the strengths of several rubber varieties. Natural rubber provides exceptional flexibility and minimal heat generation (Ibrahim and Dahlan 1998). Nevertheless, synthetic rubbers offer enhanced resistance to lubricants, ozone, and abrasion (Ibrahim and Dahlan 1998). By blending these two types in suitable proportions, a material can be synthesized that blends the advantages of each; a composite that is simultaneously flexible and durable. This synergistic action supports rubber compounds' adaptability and extensive applications (Ibrahim and Dahlan 1998).

The performance characteristics of rubber compounds depend on factors like the ratio of rubber types, the mixing procedure, and additives (Vayyaprontavida Kaliyathan et al. 2020). Mixing procedures are crucial for ensuring uniform ingredient distribution and consistency in the mixture. Common methods involve high heat and specialized mixers for this purpose. Other important additives include vulcanization accelerators, antioxidants, fillers, and processing aids that may adversely affect the expected characteristics of the product. Antioxidants protect the material from aging and degradation, while additives like carbon black improve the strength and wear resistance of the mixture (Vayyaprontavida Kaliyathan et al. 2020).

Rubber compounds have specialized usages in multiple sectors, including construction, automotive, footwear, and medical (Rodgers and Waddell 2005). Car tires are the most recognized product, balancing road grip, fuel efficiency, and wear resistance (Rodgers and Waddell 2005). Specially blended rubber compounds are also widely used in vehicle components like hoses, seals, and belts (Bandyopadhyay 2024). The applications of rubber compounds in the building and construction industries include sealants, vibration dampers, and roofing membranes. Some other uses of rubber compounds in the footwear industry are soles. Medical uses include surgical gloves, catheters, and drug delivery devices. The future of rubber compounds is bright with the advent of nanotechnology (Bandyopadhyay 2024). Nanoscale materials, such as graphene, carbon nanotubes, and nanoclays, have enhanced rubber compounds' mechanical strength and thermal conductivity. These nano-mixtures provide solutions to ultra-light, high-performance applications that conventional rubber compounds have long pursued. Applications of rubber compounds in materials science and engineering have been invaluable; the above advances further establish their potential. Current research into developing more durable, sustainable, and economical rubber will seek an optimal compromise between performance and environmental impact (Rodgers and Waddell 2005; Bandyopadhyay 2024).

8.2.1 Styrene Butadiene Rubber-Based Nanoblends

New research and development efforts in polymer material science have been focused on SBR-based nano-blends. SBR is an essential synthetic rubber widely used due to its outstanding abrasion resistance, good aging properties, and relatively low cost (Visakh 2017). The tensile strength and tear resistance could be unsatisfactory from the point of view of mechanical properties in some practical applications. Nanotechnology components with nano-size dispersion in the SBR matrix overcame these disadvantages (Šupová et al. 2011). These include carbon nanotubes (Soliman et al. 2012), graphene (Khan et al. 2022), nano clays (Rajasekar et al. 2009), silica (Wei et al. 2015), and various metal oxides (El-Sabbagh et al. 2017). When formulating SBR-based nano blends, the main objective is to sharply improve the mechanical and thermal properties by adding minimal nanoparticle quantities. The elevated surface area-to-volume ratio and distinctive characteristics of nanoparticles result in robust interactions with the SBR matrix and enhanced performance. Using carbon nanotubes in SBR can improve tensile strength, hardness, and wear resistance (He et al. 2017). Additionally, layered silicates, including nano clays and silica, can enhance gas-insulating characteristics and thermal stability (Vishvanathperumal and Anand 2020).

The nano-formulation of SBR-based mixtures requires careful optimization owing to the uniform dispersion of nanomaterials in the SBR matrix; otherwise, agglomeration or uneven dispersions reduce the potential benefits of nanoparticles in performance, abnormality, and overall qualities of the material. Consequently, solvent mixing, melt mixing, and in-situ polymerization achieve well-dispersed nanoblends. Furthermore, the surface modification of nanomaterials can enhance their compatibility and dispersion inside the SBR matrix (Chatterjee and Khobragade 2017).

Characterizing SBR-based nanoblends is essential for comprehending their structure–property relationships and enhancing their performance. Nanomaterials' distribution, morphology, and interaction in an SBR matrix are investigated using electron microscopy, X-ray diffraction, and spectroscopy methods. Besides, several tests were carried out to evaluate such nanoblends' mechanical behavior, thermal properties, and physical characteristics and determine their fulfillment of application requirements (Camargo et al. 2009). Research on SBR-based nano blends goes on at an impressive pace, with new nanomaterials, improvement in dispersion methodology, and understanding of structure–property relationships being in focus. Ecological impacts and sustainability are also paramount, hence the development of eco-friendly nanomaterials and processing techniques. This makes the field dynamic, having updated information on recent developments (Camargo et al. 2009; Stephen and Thomas 2010).

The SBR-based nano-blends have significantly advanced polymer material science. Through nanotechnology, these materials overcome the performance limits of conventional SBR and provide transformational opportunities in various industries. The current R&D works are targeted to translate the fullest potential with value addition toward performance and sustainability. This could ensure qualitative transformation in the automotive, aerospace, and biomedical sectors by developing more resistant, lighter, and ecologically friendly materials (Mishra and Hansora 2018).

8.2.2 *Ethylene Propylene Diene Rubber-Based Nanoblends*

Within the last years, nanocomposites based on EPDM have emerged as a promising frontier in the studies of high-tech material varieties. EPDM is a versatile synthetic rubber with outstanding resistance against weathering, ozone, and heat aging. Incorporating nanoparticles is regarded as one of the effective ways to further the characteristics of the EPDM matrix and expand its applications (Karpeles and Grossi 2000). It deals with studying EPDM-based nanocomposites because of various nanomaterials, processing techniques, improved properties, and the numerous applications of these innovative materials. EPDM is especially fit for outdoor applications due to its superior resistance against external conditions. Neither tensile strength nor tear resistance may be good enough and thus needs improvement by incorporating nanoparticles judiciously (Stephen and Thomas 2010; Visakh et al. 2013). Some special characteristics of graphene, silica, CNTs, nanoclays, and other metal oxide nanomaterials alter the performance of EPDM-based nanocomposites (Kumar and Kumar 2022). For EPDM-based nanoblends, the performance needs of the intended use determine the nanomaterial to use. For instance, carbon nanotubes (CNTs) are perfect for enhancing EPDM's electrical characteristics and reinforcing it due to their great mechanical strength and electrical conductivity (Spitalsky et al. 2010). Nanoblends based on EPDM can enhance their insulating characteristics and thermal conductivity by adding graphene, a two-dimensional material with remarkable strength and flexibility (Archibong et al. 2023). However, silica and metal oxides can enhance EPDM's thermal stability and flame retardancy, and nano clays provide cost-effective reinforcement with increased insulating characteristics (Ponnamma et al. 2013); for nano blends based on EPDM, intensive process optimization is required to achieve a consistent distribution of nanoparticles. Agglomeration or inadequate nanomaterial dispersion might result in reduced potential benefits and detrimental effects on material qualities. Obtaining nanoblends with good dispersion requires several methods, including melt mixing, solution mixing, and in situ polymerization. Through surface modification, nanomaterials can enhance their compatibility and dispersion with the EPDM matrix (Costa et al. 2024b; Kumar et al. 2024).

Characterization is critical to maximizing the performance of EPDM-based nano blends and understanding the structure–property correlations. Researchers use electron microscopy, X-ray diffraction, and spectroscopy methods to examine the nanomaterials' distribution, shape, and interactions within the EPDM matrix (Costa et al. 2024b). To assess the nanoblends' functionality and identify their potential uses, researchers conduct mechanical testing, thermal analyses, and assessments of insulating properties (Allahbakhsh et al. 2013). The qualities of EPDM-based nanoblends are vastly superior to those of pure EPDM. Adding nanoparticles to EPDM reinforces its modulus, tensile strength, tear, and abrasion resistance. Interestingly, another advantage of nano blends is their improved insulating properties against gases, liquids, and UV radiation, which has benefited various protective coating and sealing applications. Specific nanoparticles incorporated into EPDM further enhance thermal stability and flame retardation (Thomas and Maria 2016).

These versatile EPDM-based nanocomposites open up wide industries where they can be benefited. These improved properties have drawn wide applications for automobile gaskets, seals, hoses, and weather strips (Bhattacharya et al. 2020). Roofing membranes and waterproofing, among other applications that require resistance to vibrations, are the target of the EPDM-based nanocomposite in the construction industry (Bhattacharya et al. 2020). Nano-reinforced EPDM is also used in various other applications, such as insulating wires, cables, and manufacturing shoes and sports goods. Different aspects of research and development concerning EPDM-based nanoblends include the study of new nanomaterials, the development of new dispersion techniques, and structure–property relationships (Hagting 2015; Al-Shablle et al. 2022). These ongoing efforts ensure that the field remains dynamic and constantly evolving. Concerns regarding the long-term viability and ecological footprint of nano blends derived from EPDM are equally pressing. Sustainable nanomaterials and processing techniques are a current research focus, engaging materials science and polymer engineering professionals.

Finally, nano blends of EPDM are a huge step forward for polymer materials research. Using nanotechnology, these materials may surpass the performance constraints of pure EPDM, opening new possibilities for various uses. Continuous R&D efforts aim to maximize EPDM-based nano blends' performance and sustainability benefits across a range of sectors. These advancements pave the way for greener, stronger, and lighter materials that can reshape industries like biomedicine, aircraft, and automobiles.

8.2.3 Silicone Rubber-Based Nanoblends

Nanocomposites made of silicone rubber have recently gained much interest from researchers in the field of materials. Due to its biocompatibility, high and low-temperature strength, and remarkable flexibility, silicone rubber has extensive use in various contexts. However, there are some uses where silicone rubber's mechanical qualities—particularly its resistance to tears and abrasion—may not be enough.

One potential solution to these problems and further improving its performance is incorporating nanoparticles into the silicone rubber matrix (Mark 2006).

Since they include nanoparticles, silicone rubber-based nanocomposites exploit the benefits of such inclusion and create a synergistic effect. Nanomaterials, especially carbon nanotubes and graphene, nanoclays, silica, and various metal oxides, have thus been incorporated into silicone rubber to provide mechanical, thermal, and electrical reinforcements (Simon et al. 2008). For example, nanofillers based on carbon nanotubes result in a substantial improvement in tensile strength and hardness and an increase in the wear resistance of silicone rubbers. At the same time, nanoclays improve the insulation of properties against gases and enhance thermal stability (Kumar et al. 2021). This area of expansion in the properties of silicone rubber is one of the great opportunities that deserves considerable research and development attention.

Alternative Nanomaterials and Their Benefits

Nanocellulose, a renewable and biodegradable substance generated from plants, enhances mechanical capabilities and improves gas insulating characteristics when incorporated into the rubber matrix. Its reduced density and elevated surface area confer lightness and enhanced performance in rubber composites (Farid et al. 2019; Zhan et al. 2023).

Nanochitin/Nanochitosan: These biopolymers, derived from clam exoskeletons, exhibit biocompatibility, antimicrobial characteristics, and excellent film-forming capabilities. Their employment in rubber nanoblends is particularly significant for biomedical uses and food packaging (Sembiring et al. 2023; Yan et al. 2020).

Metallic nanoparticles like gold, silver, and copper add conductivity, antimicrobial properties, and catalytic activity to the rubber matrix, making them suitable for electronic gadgets, sensors, and biomedical devices (Chauhan et al. 2005; Shim et al. 2023). ***Quantum dots***, semiconductor nanocrystals, enhance fluorescence and sensing when incorporated into rubber nanoblends, finding applications in sensors, imaging, and light-emitting devices (Trung et al. 2012). ***Polyhedral Oligomeric Silsesquioxanes (POSS)***, hybrid nanomaterials, improve thermal stability, flame retardancy, and mechanical properties, opening up high-performance rubber applications (Liu et al. 2013a).

Molecularly engineered nanomaterials, such as dendrimers (Dvornic 2006) and polymeric micelles (Marsat et al. 2013), can be precisely designed to achieve specific properties in rubber nanoblends. Combining two or more nanomaterials, hybrid nano blends are being developed to offer superior characteristics and multi-functionality. The preparation of silicone rubber-based nanoblends requires precise optimization to ensure proper nanoparticle dispersion, as aggregation can reduce their benefits. Methods like solvent mixing, melt mixing, and in situ polymerization help achieve well-dispersed nanoblends. Surface modification of nanomaterials can also improve their compatibility and dispersion within the silicone rubber matrix (Mahmood et al. 2024).

Due to their exceptional mechanical qualities, thermal stability, and biocompatibility, silicone rubber-based nano blends possess extensive applications in sectors like automotive, aerospace, electronics, biomedical, and construction (Liu et al. 2021). These nanoblends are advancing in sophisticated domains such as flexible electronics and sensors, underscoring their versatility and influence (Liu et al. 2021). The research concentrates on identifying innovative nanomaterials, improving dispersion techniques, and elucidating structure–property correlations while prioritizing ecological consequences and sustainability. Silicone rubber-based nanoblends signify substantial advancement in polymer materials science (Liu et al. 2021). Utilizing nanotechnology, these materials exceed the constraints of silicone rubber and enable novel uses. Present research and development initiatives seek to improve functionality, durability, and sustainability across various industries.

8.2.4 Natural Rubber-Based Nanoblends

Natural rubber-based nanocomposites improve the performance of natural rubber, which is used in various industries due to its remarkable flexibility, increased tensile strength, and biodegradable nature. Using nanoparticles could help to lower restrictions, including limited abrasion resistance, poor heat resistance, and weathering sensitivity. These nanocomposites integrate the benefits of natural rubber with the specific properties of nanoparticles, resulting in a synergistic impact (Bokobza 2018). Apart from traditional nanomaterials, including carbon nanotubes, graphene, nanoclays, and silica, other nanomaterials can also be formed into the natural rubber matrix to offer different purposes to the material (Feldman 2017). Because of their hollow architecture, halloysite nanotubes are ideal media for adding functional molecules, including catalytic agents and medication delivery systems (Ismail et al. 2013). Natural rubber-based nanocomposites thus improve mechanical properties and provide other capabilities. Made with magnetite (Fe_3O_4) nanoparticles, magnetically responsive natural rubber nanocomposites can be produced fit for sensors, actuators, and electromagnetic shielding uses (Kong et al. 2010). Including zinc oxide (ZnO) nanowires gives natural rubber antibacterial and UV protection, making it fit for food packaging and medical equipment (Parvathi and Ramesan 2023). Recognized for their photocatalytic properties, titanium oxide (TiO_2) nanoparticles can be used in natural rubber nanocomposites for self-cleaning surfaces and air pollution control (Datta et al. 2016). Integrating lignin nanoparticles with natural rubber can create biodegradable nanocomposites, reducing material costs (Barana et al. 2016).

Particularly regarding its Young's modulus and tensile strength, nanocellulose improves the mechanical properties of natural rubber. Also, it increases its gas-insulating capacity, which qualifies it for uses like food packaging. An ideal alternative for biomedical uses and wound healing materials, nano chitin/nano chitosan generated from shellfish shells gives natural rubber biocompatibility, antibacterial properties, and great film-forming ability. Metallic nanoparticles such as gold, silver, and copper can provide conductivity, antibacterial properties, and catalytic action to the natural rubber matrix (Bras et al. 2010). Quantum dots (QDs) give natural rubber-based nanoblends fluorescence and sensing capabilities based on their unique optical and electrical properties for sensors, biosensors, and imaging uses. Possibilities improve natural rubber's mechanical properties, flame retardancy, and thermal stability (Edayadiyil et al. 2021).

8.2.5 *Nitrile Rubber Latex Blends*

Synthetic rubber made from copolymerizing acrylonitrile and butadiene monomers is nitrile rubber latex (NBR). This material's outstanding resistance to petroleum-based oils, fuels, and many chemicals makes it highly sought-after in industrial uses, attesting to its dependability. Among many uses, including paints, adhesives, and sealants, nitrile rubber latex is preferred as a water-based dispersion (Sadeghalvaad et al. 2019). The most essential quality of this material is its low-temperature elasticity retention ability. In recent years, nanotechnology-based approaches have been increasingly important in improving the characteristics of nitrile rubber latex (Sadeghalvaad et al. 2019).

Enhancing nitrile rubber latex with nanotechnology considerably increases the material's mechanical, thermal, and chemical characteristics. Incorporating nano-sized additives into the material matrix yields a more homogeneous structure and improved performance. Carbon nanotubes are among the most often occurring nanomaterials in nitrile rubber latex composites. These nanotubes boost the material's tensile strength and elastic modulus due to their considerable surface area and exceptional mechanical properties (Wang et al. 2022). By improving the electrical conductivity of a material, carbon nanotubes can be used to make it suitable for both conductive and antistatic uses (Wang et al. 2022). Graphene is a remarkable nanomaterial that garners interest in nitrile rubber latex composites due to its two-dimensional architecture and outstanding mechanical qualities. Particularly in high-performance uses, mixtures enhanced with graphene are preferred because of their better thermal conductivity and chemical resistance (Zhang and Cho 2019).

The main nanomaterial used in nitrile rubber latex composites is silica nanoparticles. These particles improve the abrasion resistance and tear strength of the material, enabling the manufacturing of durable goods with great longevity (Ghamarpoor and Jamshidi 2022). Moreover, silica nanoparticles improve the surface properties of the material, thereby improving adhesion and coating effectiveness. As was already indicated, these features increase the use of nitrile rubber latex mixtures, especially in the automotive and construction sectors (Sala et al. 2014).

Combining nitrile rubber latex with nanotechnology yields better performance traits than more traditional materials. These developments enable the employment of nitrile rubber latex in a broader spectrum of innovative applications. Nitrile rubber latex is quite crucial in the industrial sector, and its value will only increase with possible developments in nanotechnology and environmentally friendly research. Mixtures of nitrile rubber latex will remain rather crucial in the industrial sector (Chi Wang et al. 2018).

8.2.6 Polyurethane Rubber-Based Nanoblends

Polyurethane rubber, a versatile polymer, is used in various industries because of its remarkable properties, such as flexibility, abrasion, and chemical resistance. Its versatility in automotive, building, textiles, and healthcare makes it a preferred material. Particularly in the development of polyurethane rubber-based nanoblends, the potential of polyurethane rubber to provide personalized solutions through variable hardness levels and modification with diverse additives has spurred significant research in recent years. Designed using nanotechnology, nanoblends based on polyurethane rubber are meant to improve the material's mechanical, thermal, and chemical properties (Vaithylingam et al. 2017).

Carbon nanotubes improve the tensile strength and elastic modulus of materials as they are powerful and flexible. A study on nano blends that included carbon nanotubes in a polyurethane matrix showed a 50% increase in tensile strength relative to ordinary polyurethane rubbers (Zhang et al. 2023). Particularly in the automotive and aviation sectors, these nano mixes are helpful in situations where exceptional performance is needed.

Graphene oxide is a significant nanomaterial utilized in polyurethane rubber-based nanoblends. Due to its two-dimensional architecture and elevated surface area, graphene oxide enhances the material's thermal conductivity and chemical resistance (Liu et al. 2013b). Literature indicates that graphene oxide-doped polyurethane nanoblends retain dimensional stability at elevated temperatures and exhibit enhanced resilience to chemical corrosives. These characteristics promote the utilization of graphene oxide-doped nanoblends in electrical and energy storage applications (Liu et al. 2013b).

Nanoclay is a prevalent nanomaterial utilized in polyurethane rubber-based nanoblends. Nanoclay enhances the gas insulating characteristics and fire-retardant efficacy of the material. Research indicates that polyurethane nano blends using nano clay markedly diminish oxygen and water vapor permeability while enhancing fire resistance. These nano blends are particularly favored in the packaging and construction industries (Obazee et al. 2020).

Silica nanoparticles enhance mechanical strength and wear resistance in polyurethane rubber-based nanocomposites (Zhou et al. 2022). Nanoblends infused with silica yield enduring and resilient products by enhancing surface hardness and wear resistance. Literature indicates that polyurethane nano blends enhanced with silica nanoparticles exhibit higher performance, particularly in applications like floor coverings and industrial wheels. Thanks to the advantages of nanotechnology, polyurethane rubber-based nano blends thus show improved performance characteristics than conventional materials (Abdeen 2017). These developments provide polyurethane rubber-based nano blends with a broader range of innovative uses. Future developments in nanotechnology and sustainability-oriented research excite one since they are predicted to improve these materials' efficiency and ecological compatibility even more. In this regard, polyurethane rubber-based nano blends will likely remain important in the industrial sector (Al-Senani and Al-Qahtani 2024).

8.2.7 *Polybutadiene Rubber-Based Nanoblends*

The polybutadiene rubber industry is known for producing tires and other synthetic elastomer products with a low glass transition temperature and great flexibility. This substance is particularly common in automotive tires, shoes, and several other technological uses. Recently, nanotechnology-based solutions have drawn considerable interest in enhancing polybutadiene rubber's mechanical characteristics and durability (Parashiva et al. 2021).

Using many nanomaterials produces nanoblends based on polybutadiene rubber, a technique that significantly increases the material's performance. Usually enhanced with cutting-edge nanomaterials, including polymeric nanofibers, cellulose nanocrystals (Peng et al. 2019), and metal oxide nanoparticles, every nanomaterial offers unique benefits for the polybutadiene rubber matrix, improving the material's performance in many different uses (Soares 2017).

Metal oxide nanoparticles—including zinc oxide (ZnO) and titanium dioxide (TiO_2)—are widely employed in advanced nanomaterials in polybutadiene rubber-based nanoblends. These nanoparticles increase the UV resistance and thermal stability of the material. For example, ZnO nanoparticles increase the UV resilience of the polybutadiene rubber matrix (Wang and Chen 2006), extending the material's lifetime in outdoor applications. Using its photocatalytic activity, TiO_2 enhances the material's surface cleaning and antibacterial characteristics (Mohammed et al. 2023).

The literature shows that polybutadiene nanoblends treated with cellulose nanocrystals develop biodegradable properties, which makes them appropriate for ecologically benign uses. These nano blends are preferred in packaging and biological uses, especially (Peng et al. 2019).

As a result, polybutadiene rubber-based nanoblends display increased performance characteristics compared to conventional materials, thanks to the incorporation of advanced nanomaterials. These developments let polybutadiene rubber-based nanoblends find employment in more diverse creative fields. A cause of hope is the expectation of upcoming developments in sustainability-oriented research and nanotechnology, which will help to improve the ecological compatibility and efficacy of various materials even more. In this regard, polybutadiene rubber-based nano blends are expected to remain crucial in the industrial sector, guaranteeing the material's ongoing significance (Mishra and Hansora 2018; Hansora and Mishra 2017).

8.3 Role in Different Application Areas of Rubber-Based Nanoblends

8.3.1 Electromagnetic Shielding (EMI) Application

In recent decades, the rapid development of electronic gadgets and wireless communication technologies has led to their widespread usage in everyday life. This, in turn, has created a considerable rise in electromagnetic energy outputs across several frequency bands, leading to a critical problem of electromagnetic pollution (Al-Ghamdi et al. 2016). Therefore, avoiding exposure to electromagnetic radiation has become an essential sanitary practice in contemporary civilizations (Kruželák et al. 2021). So far, metals and alloys have been the most often utilized materials for electromagnetic interference (EMI) shielding. However, these materials suffer various problems, such as excessive weight, expense, lack of flexibility, sensitivity to environmental corrosion, and challenges in processing complicated forms (Kwon et al. 2014; Ameli et al. 2013). Rubber, on the other hand, is commonly utilized in electrical equipment and provides benefits such as low cost, flexibility, lightweight, and simplicity of manufacture. However, most rubber materials do not offer effective protection against electromagnetic radiation. Thus, the creation of elastomers with high electrical conductivity has become more vital, particularly in crucial industries like as electronics, aviation, medicine, and military. To answer this demand, ecologically safe, lightweight, and highly conductive nanomaterials (carbon materials like carbon nanotubes, graphene oxide, silver nanoparticles, iron oxide nanoparticles, etc.), and new technologies have been created and employed (Kwon et al. 2014).

Kwon et al. (2014) have synthesized flexible adhesive shields (FASs) made of silver flakes (Ag flakes), multi-walled carbon nanotubes decorated with nanoscale silver particles (nAg-MWNTs), and nitrile butadiene rubber (NBR) (Kwon et al. 2014). The addition of nAg-MWNTs to the Ag flake-NBR mixture significantly increased both conductivity and shielding effectiveness. The maximum obtained shielding effectiveness was approximately $\sim$ 75 dB at 1 GHz (Fig. 8.1).

Al-Ghamdi and colleagues (2016) developed conductive nanocomposites based on layered graphene nanosheets (GN) reinforced with an acrylonitrile butadiene rubber (NBR) matrix for EMI shielding efficacy at microwave frequencies spanning from 1 to 12 GHz. (Al-Ghamdi et al. 2016). They investigated how varying amounts of GN impact the morphology and mechanical properties of NBR/GN nanocomposites. Notably, an NBR/GN nanocomposite containing just 4% GN by weight demonstrated a shielding effectiveness (SE) of roughly 77 dB. This level of SE suggests that such a nanocomposite could serve as an effective shielding material against electromagnetic interference in areas such as radar systems, satellite broadcasting, vehicle detection technologies, defense tracking, weather satellites, home electronics, and beyond. In related work, Kruzelak et al. (2021) formulated rubber composites by incorporating manganese-zinc ferrite alone and in combination with carbon-based fillers like carbon black (CB) and carbon nanotube (CNT) into acrylonitrile-butadiene rubber

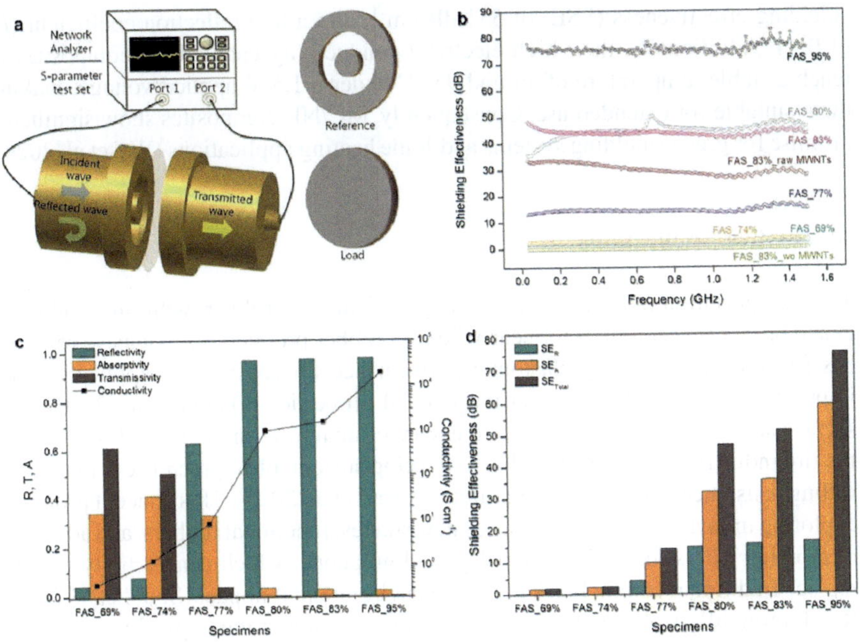

Fig. 8.1 The electromagnetic interference shielding effectiveness of FAS. **a** Schematic of the experimental set-up. A coaxial sample holder designed according to ASTM D4935-10, a network analyser, and an *S*-parameter test set were used. Schematics of the reference and load specimens (FAS on a PI film) are also shown. **b** Shielding effectiveness of FAS between 30 MHz and 1.5 GHz. **c** Reflectivity, absorptivity, and transmissivity of FAS are compared at 1 GHz (Agilent E5071C). The conductivity of specimen is also shown. **d** SE_R, SE_A, and SE_{Total} of FAS are compared. (A colour version of this figure can be viewed online) (Kwon et al. 2014)

(Kruželák et al. 2021). The composites showed an increase in electrical conductivity and permeability in the following sequence: ferrite composites, CB/ferrite composites, and CNT/ferrite composites. However, in the same order, their absorption shielding capacity decreased. Based on these findings, it was concluded that as electrical conductivity and permeability increase, the absorption shielding effectiveness within the 1 MHz-3 GHz frequency range declines. The introduction of carbon-based fillers enhanced the rubber matrix, improved the dispersion and distribution of ferrite, and thereby strengthened the mechanical properties of the composites. Since ferrites were found to absorb electromagnetic interference (EMI) well but had limited physical–mechanical reinforcement properties, combining ferrites with carbon-based fillers appears promising for future EMI shielding applications. In a related study, 2 nm Fe_3O_4 nanoparticles were securely attached to the exterior of intact CNTs through an in situ co-precipitation technique. These Fe_3O_4@CNTs were then incorporated into rubber to create green shielding materials. The magnetism and the conductive, segregated networks formed during latex blending led the rubber composites to achieve an impressive shielding effectiveness (SE) of 63.8 dB, specific shielding effectiveness (SSE) of 531 dB cm^{-1}, and a low reflection coefficient (R) of 0.24. Additionally, their high electrical conductivity enabled the composites to reach a stable temperature of around 95 °C under a 1.5 V applied voltage, making them suitable for extended use. Consequently, MC/NR composites show significant promise for green shielding systems and Joule heating applications (Wei et al. 2024).

Automotive Applications

The transportation industry is the primary consumer of rubber, with tires and associated products accounting for nearly 70% of rubber production. Extensive research has focused on enhancing elastomer performance for engineering applications by improving rubber's physical and mechanical properties (Oleiwi et al. 2010). The surface activity and particle size of fillers are critical in strengthening elastomers. In the tire industry, a key R&D goal is to develop sustainable, "green tires" that lower rolling resistance while enhancing safety (Ansari et al. 2024). This concept promotes exploring innovative materials, such as new silanes, functional rubbers, and advanced reinforcing fillers like silica, nanoclay, and graphene, which can partially or fully replace traditional carbon black to reduce environmental impact (Shoul et al. 2022). Reinforcing fillers play an essential role in tire manufacturing, typically comprising about 30% by weight, primarily in the form of carbon black (CB) and precipitated silica (SiO_2) (Limtrakul et al. 2022). Common elastomers in tire production include natural rubber (NR), polybutadiene rubber (PBR), and styrene-butadiene rubber (SBR) (Shoul et al. 2022).

Limtrakul and colleagues studied the effects of hybrid ratios of CB and SiO_2 on the properties of a passenger car tire (PCT) sidewall made from an NR and BR blend (Limtrakul et al. 2022). By preparing rubber compounds with varied hybrid filler ratios at a consistent 50 phr loading, they observed that higher SiO_2 content improved rubber-filler interaction, which in turn enhanced vulcanizate properties such as hardness, tensile strength, tear resistance, and fatigue resistance, along with curing efficiency and hysteresis behaviors (e.g., reduced heat buildup and increased dynamic set). The study also found that NR/BR vulcanizates reinforced with CB/SiO_2 hybrid fillers improved sidewall performance compared to CB-only vulcanizates. The findings also demonstrated that NR/BR vulcanizates reinforced with CB/SiO_2 hybrid filler showed improvements in sidewall performance compared to vulcanizates filled with only CB. Kang et al. have overcome the previously mentioned limitations of high-performance passenger car tire treads by adding rough-surfaced carbon nanotubes and soybean oil to a high-loaded silica-based elastomer composite. In another advance, Kang et al. addressed high-performance passenger tire tread limitations by integrating rough-surfaced carbon nanotubes (RS-CNT) and soybean oil into a high-loaded silica-based elastomer. Practical tire testing showed that the modified composite improved wet and dry braking, snow traction, and rolling resistance over unmodified silica-based materials. Additionally, wear resistance was boosted by over 20% while maintaining other properties at commercially viable levels (Kang et al. 2021). Industrially manufactured tires using the RS-CNT/soybean oil composite

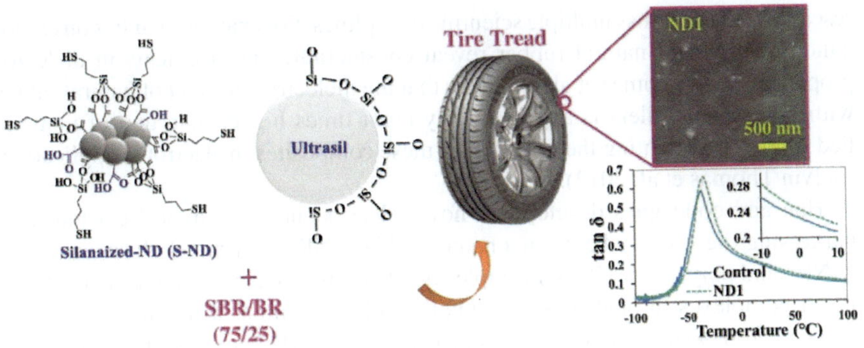

Fig. 8.2 Tire tread performance of silica-filled SBR/BR rubber composites incorporated with nanodiamond and nanodiamond/nano-SiO_2 hybrid nanoparticle (Salkhi Khasraghi et al. 2022)

underwent vehicle wear tests, with results showing that they provided an average of 122.5% more mileage than reference tires—128% more on front tires and 117% on rear tires.

Another study explored tire tread compound performance using nanodiamond (ND) and an ND/fumed nano-SiO_2 hybrid (Salkhi Khasraghi et al. 2022). A blend of SBR and cis-BR filled with high-dispersion silica served as the base compound (Fig. 8.2). Samples with ND showed increased curing and burn times compared to control compounds. With a 5 phr ND loading, significant improvements were observed in mechanical properties, thermal conductivity, and cross-link density, as shown by tube theory and swelling tests. Specifically, the 5 phr ND sample exhibited a 40% increase in Young's modulus and improved thermal stability, with temperature increases of around 40 °C and 10 °C at 10% and 50% weight loss, respectively. Furthermore, this ND loading resulted in an approximate 10% improvement in wet traction, a critical benefit for tire tread manufacturers.

Flexible Electronic Devices

Functional elastomers combine the inherent high elasticity of rubber materials with additional functional properties, opening up versatile applications in advanced technology fields (Han et al. 2019). A particularly interesting subset of these elastomers is electrically conductive elastomers, which combine flexibility with electrical conductivity, making them highly suitable for applications in soft electronic devices. These elastomers are utilized in a wide array of applications, including artificial skin, flexible sensors, stretchable circuits, electromagnetic shielding, flexible supercapacitors, and smart textiles, by incorporating conductive materials into their rubbery matrix (Jia et al. 2017; Vikulov et al. 2016; Ramuz et al. 2012). Although most elastomers are naturally insulating, integrating nanoscale conductive materials into their structures effectively promotes electron transport, thereby enabling conductivity (Han et al. 2019). Consequently, conductive rubber composites have garnered significant research interest across multiple scientific disciplines. For example, studies on carbon nanotube-modified natural rubber reveal considerable enhancements in dielectric properties, with Thomas et al. reporting that the dielectric constant of natural rubber with 10 phr nanofillers is approximately eight times higher than that of unmodified rubber, highlighting the potential of these composites in electronic applications (Selvin Thomas et al. 2012).

Han and colleagues developed a novel electroconductive hybrid elastomer by harnessing the synergy between a natural rubber (NR) matrix, cellulose nanofibers (CNF), and polyaniline (PANI) complexes. This was achieved by synthesizing CNF-PANI complexes through in-situ oxidative polymerization of aniline on the CNF surfaces, followed by homogeneous dispersion into NR latex. The resulting CNF-PANI/NR elastomers, with a 3D network structure, were then produced through latex coagulation. The addition of biodegradable CNFs provided not only mechanical reinforcement but also supported the formation of a conductive 3D network within the NR matrix. These bio-based elastomers displayed impressive mechanical properties, such as tensile strength reaching 9.5 MPa, elasticity over 500%, and a density of 1.15 g/cm^3, alongside excellent electrical conductivity. Furthermore, the elastomer integrated with 10 phr PANI allowed for flexible strain sensors capable of accurately detecting human motion. Elastomer-based electrodes with 25 phr PANI achieved a specific capacitance of 105 F/g at a current density of 0.3 A/g, with minimal capacitance degradation (under 20%) after 1000 cycles. This scalable, eco-friendly method using bio-based CNF and NR materials offers a promising approach to multifunctional elastomers with applications in flexible electronics, strain sensors, and electrodes (Fig. 8.3) (Han et al. 2019).

Parvathi et al. investigated chlorinated natural rubber (Cl-NR) and zinc ferrite ($ZnFe_2O_4$) nanocomposites, focusing on their optical, thermal, and electrical properties. They found that adding $ZnFe_2O_4$ nanoparticles decreased the band gap energy of the nanocomposites while improving thermal stability as the nanoparticle content increased. Additionally, the AC conductivity, dielectric constant, and dielectric losses rose with both temperature and frequency, showing non-ohmic behavior. DC conductivity also increased with $ZnFe_2O_4$ content and aligned with the McCullough model. These findings highlight the potential of these nanocomposites for use in flexible electronics, supercapacitors, and elastomeric sensors (Parvathi et al. 2022).

A study focused on the structural, morphological, thermal, and electrical properties of chlorinated natural rubber/polyaniline (Cl-NR/PIN) elastomer nanocomposites prepared with varying zinc oxide (ZnO) nanoparticle content. The inclusion of ZnO enhanced the thermal stability, with composites containing 5 wt% ZnO showing the highest electrical and dielectric performance. As the ZnO content increased, the dielectric constant improved considerably. Electrical relaxation was noted at lower frequencies, and conductivity rose with increasing temperature. Mechanical properties, including tensile strength and modulus, also improved with 5 wt% ZnO, indicating that Cl-NR/PIN composites have strong potential for flexible electronic applications (Parvathi and Ramesan 2023).

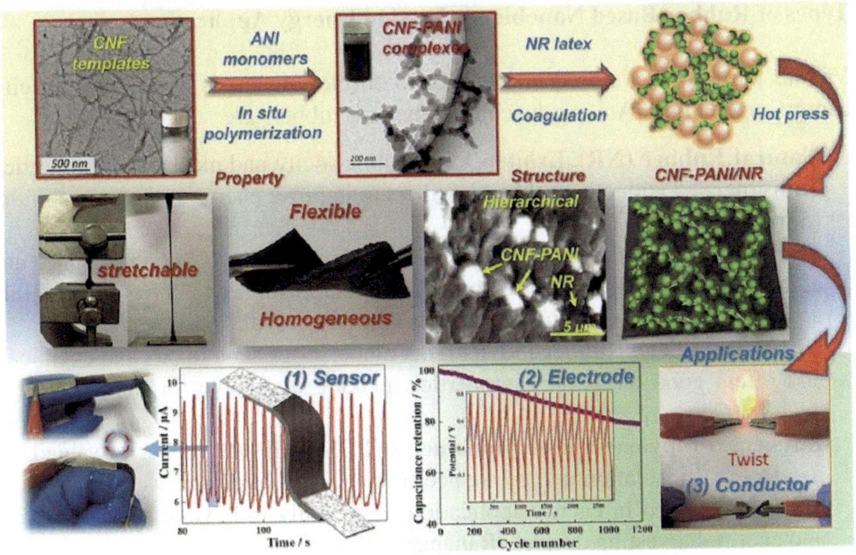

Fig. 8.3 Nanocellulose-templated assembly of polyaniline in natural rubber-based hybrid elastomers toward flexible electronic conductors (Han et al. 2019)

8.3.2 Rubber-Based Nanoblends for Energy Applications

Rubber blends are composite materials combining different types of elastomers or rubber polymers to achieve unique properties that surpass those of individual components (Daniel et al. 2024). Over the last several years, these materials have garnered significant attention in energy applications, particularly in areas like energy capturing, storage, and transformation (Ghosh et al. 2023). Rubber blends combined with functional additives can offer flexibility, durability, and enhanced electrical properties, making them ideal for energy devices like supercapacitors, batteries, and piezoelectric energy harvesters (Abdulwahid et al. 2023).

The intrinsic flexibility of rubber materials provides a mechanical advantage, the versatility of flexible supercapacitors makes them ideal for applications in wearable technology. Maintaining elasticity in rubber blends is essential for their performance in dynamic applications and robustness under mechanical deformation, while enhancing energy storage properties is a key area of exploration (Hao et al. 2024). Moreover, by modifying the composition of rubber blends, it's feasible to control dielectric, electrical, and thermal properties to match specific requirements, which are crucial for optimizing energy devices (Vipu Vinayak et al. 2024).

Types of Rubber-Based Nanoblends Used in Energy Applications

Various elastomers are used in blending to develop materials tailored for specific energy applications. Among the commonly used rubbers are:

Natural Rubber (NR): Exhibits excellent elasticity and mechanical properties. NR is blended with other rubbers or fillers to improve energy storage capabilities (Suriani et al. 2015).

Styrene-Butadiene Rubber (SBR): Known for its high abrasion resistance and aging stability, SBR is often blended with natural rubber to enhance durability and flexibility in energy storage devices (Pantrangi et al. 2024).

Nitrile Butadiene Rubber (NBR): This rubber is resistant to oils and chemicals and offers good mechanical properties, making it suitable for flexible energy devices (Cho et al. 2012).

Ethylene Propylene Diene Monomer (EPDM): EPDM rubber is valued for its electrical insulation and resistance to heat, oxidation, and weathering. It is increasingly used in energy applications, especially in flexible supercapacitors and energy-harvesting devices (Rahaman 2023).

By blending these rubbers with functional additives like graphene, carbon black, or other conductive nanomaterials, their performance in energy storage and conversion can be enhanced.

Additives and Conductive Fillers in Rubber-Based Nanoblends for Energy Applications

To enable rubber blends for use in energy storage and conversion devices, conductive fillers and additives are often incorporated. These fillers significantly influence the electrical and thermal conductivity of the rubber matrix, improving the performance of the composite in energy applications.

Carbon Black: A traditional conductive filler used in rubbers to improve electrical conductivity. Carbon black is often added to elastomers to produce rubber-based electrodes in supercapacitors and batteries (Anand et al. 2024).

Graphene: Graphene's exceptional surface area, high electrical conductivity, and notable mechanical strength render it an optimal filler material for rubber composites. The integration of graphene into rubber matrices significantly improves the composite's mechanical and electrical properties, thereby expanding its potential applications in advanced energy storage systems, such as supercapacitors, as well as in flexible electronic devices (Zhao et al. 2024a).

Carbon Nanotubes (CNTs): CNTs exhibit remarkable electrical conductivity and mechanical strength. Rubber blends incorporating CNTs show promise in stretchable energy storage devices, providing both flexibility and enhanced energy performance (Fan et al. 2022).

Metal Oxides: Rubber composites enhanced with metal oxides, such as titanium dioxide (TiO_2) or manganese dioxide (MnO_2), are widely utilized in energy storage applications due to their superior energy density and electrical conductivity. The inclusion of these metal oxides within rubber matrices enhances the material's electrochemical properties, making it highly suitable for advanced energy storage solutions (Gaikwad et al. 2024).

The interaction between rubber matrices and these conductive fillers is essential to the performance of rubber-based energy devices. Homogeneous dispersion of fillers within the rubber matrix ensures improved conductivity and mechanical integrity.

Applications of Rubber-Based Nanoblends in Energy Devices

Rubber blends are being explored for a variety of energy applications, including flexible supercapacitors, stretchable batteries, piezoelectric energy harvesters, and thermoelectric generators.

Flexible Supercapacitors

Supercapacitors are energy storage devices characterized by their ability to deliver rapid bursts of energy and charge/discharge cycles. The use of rubber blends as the substrate for supercapacitor electrodes enables flexibility and stretchability, making them ideal for wearable electronics (Zhao et al. 2024b).

The incorporation of conductive fillers, including graphene and carbon nanotubes (CNTs), into rubber matrices facilitates the fabrication of electrodes exhibiting enhanced conductivity, flexibility, and energy storage capacity. This modification significantly improves the electrochemical performance of rubber-based composites, making them viable for use in advanced energy storage applications (Yang et al. 2013). For instance, blending natural rubber (NR) with conductive fillers such as carbon black and graphene has resulted in supercapacitor electrodes with enhanced energy densities. The elasticity of rubber ensures that the electrodes maintain their integrity even under mechanical deformation, a critical feature for wearable supercapacitors (Niaz et al. 2024).

Eyssa et al. developed innovative sponge nanocomposites to improve energy storage performance, focusing on achieving both cost-effectiveness and high power output. Natural rubber (NR) and ethylene propylene diene monomer (EPDM) were combined and laminated onto cotton substrates, incorporating various nanofillers such as hexagonal boron nitride (h-BN), graphene oxide (GO), and vanadium oxide (V_2O_5). These nanocomposites were subsequently subjected to electron beam irradiation to optimize their properties for supercapacitor applications. The electrochemical properties of the composites were systematically evaluated through cyclic voltammetry, galvanostatic charge/discharge measurements, and electrochemical impedance spectroscopy. These techniques provided comprehensive insights into the composites' performance for energy storage applications. Among the samples tested, the NR/EPDM/h-BN nanocomposite demonstrated the highest tensile strength. Furthermore, the NR/EPDM/h-BN/GO composite exhibited enhanced porosity, flexibility, and durability, leading to improved electrochemical behavior attributed to increased charge-transfer efficiency. The adhesion between the composite and the cotton substrate was confirmed through Fourier-transform infrared (FTIR) spectroscopy, X-ray diffraction (XRD), and scanning electron microscopy (SEM) analyses. Additionally, the mechanical, thermal, and electrochemical properties of the NR/EPDM nanocomposites irradiated at 150 kGy were significantly enhanced compared to the non-irradiated samples. The NR/EPDM/h-BN/GO nanocomposite-based supercapacitor demonstrated a specific capacitance of 55 F/g at a current density of 1 A/g, maintaining approximately 97% of its initial capacitance after 1000 charge–discharge cycles at the same irradiation level (Eyssa et al. 2024).

Kamuralazam et al. propose an innovative methodology for the development of flexible, biodegradable, and cost-effective hydrogel electrolytes specifically designed for energy storage applications. The hydrogel electrolytes comprise a copolymer matrix formed from epoxidized natural rubber (ENR), acrylamide (AAm), and acrylic acid (AA), synthesized via free radical polymerization. To enhance the ionic conductivity, the hydrogels were immersed in 1 M sodium sulfate (Na_2SO_4) solutions for varying durations. This approach capitalizes on the synergistic effects of hydrogen bonding, electrostatic interactions within the hydrogel network, and interactions with the salt solution, resulting in an ionic conductivity of 19.4×10^{-2} S/cm, remarkable stretchability of up to 550%, and self-healing capabilities. When integrated into an electric double-layer capacitor (EDLC) using symmetrical porous carbon electrodes, the hydrogel achieved a specific capacitance of 55.65 F/g and demonstrated stable cycling performance over 3500 cycles with minimal degradation. Furthermore, the mechanical strength of the hydrogel was significantly enhanced following the self-healing process. This study highlights the critical importance of immersion time in optimizing the ionic conductivity and functional attributes of hydrogel electrolytes, thereby providing valuable insights for the advancement of energy storage technologies (Kamarulazam et al. 2023) (Figs. 8.4a, b and 8.5).

Kong et al. introduce a cost-effective method that utilizes natural rubber (NR) latex as a sacrificial template for the fabrication of three-dimensional porous MXene foam (MXF) designed for high-performance energy storage systems. MXenes are recognized for their exceptional electrical conductivity; however, they often experience performance degradation due to the aggregation and restacking of their two-dimensional nanosheets. To address this challenge, the authors employed uniform sub-macroparticles derived from NR latex to construct a porous architecture that effectively prevents nanosheet restacking while simultaneously facilitating efficient ion transport during charge–discharge cycles. This structural optimization significantly enhanced both the capacitive and rate performance of the electrodes. The MXF-70% foam demonstrated an impressive specific capacitance of 480 F/g at a scan rate of 2 mV/s, maintained high rate performance with 42.1% capacitance retention at 1 V/s, and exhibited exceptional cycling stability, retaining 97.1% of its initial capacitance after 10,000 cycles. The low cost and scalability of NR latex offer a promising pathway for the mass production of foam electrodes, positioning this approach as

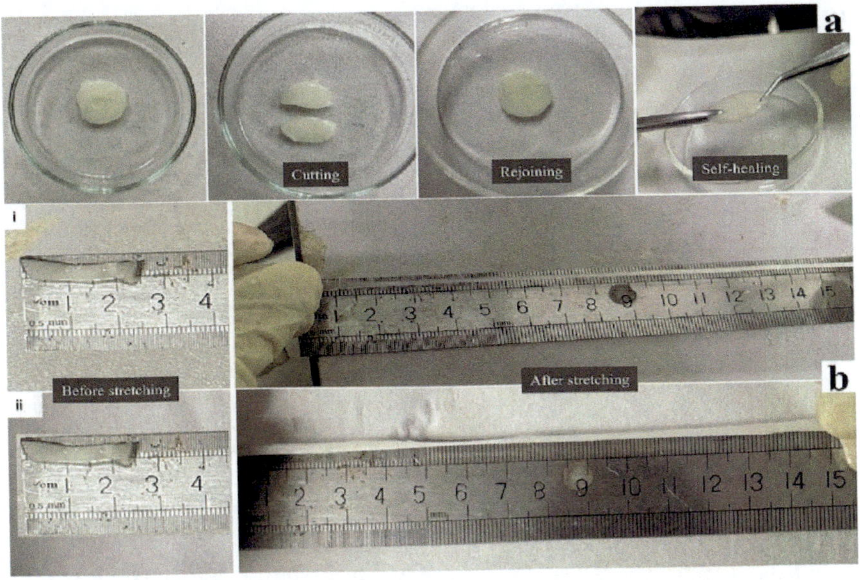

Fig. 8.4 Kamuralazam et al. **a** Images of a hydrogel sample. After cutting into two pieces and pressing together for 30 min at 60 °C, the hydrogels merged into a single piece **b** Images before and after stretching (i) S48 hydrogel sample (ii) S72 hydrogel sample

Fig. 8.5 Illustrates the fabricated device with light-emitting diode

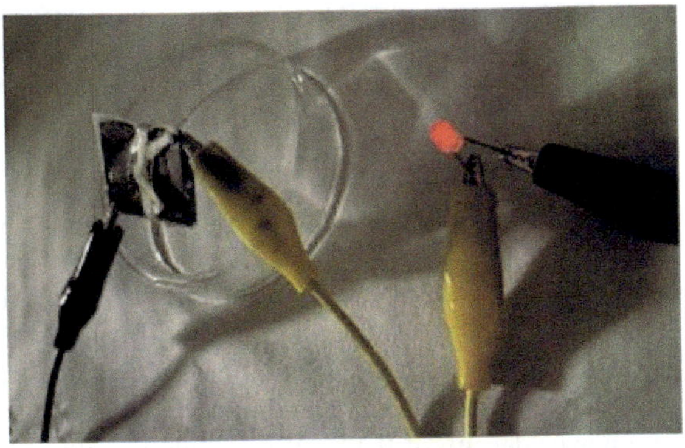

an effective solution for portable and integrated supercapacitor applications (Kong et al. 2022) (Fig. 8.6).

Han et al. introduce a novel strategy for the creation of highly compressible supercapacitors by integrating the current collector, active materials, and separator into a single cohesive device. In their work, natural rubber melamine foam (MF) is utilized as the structural framework, while $Ti_3C_2T_x$ nanosheets serve as the active materials. The fabrication process employs capillary action to infuse the MF structure with few-layer $Ti_3C_2T_x$ nanosheets, followed by freeze-drying to produce the MF/$Ti_3C_2T_x$ composite. This composite exhibits exceptional structural integrity,

Fig. 8.6 **a** Schematic illustration of the natural rubber and MXene preparation procedure. **b** X-ray diffraction patterns of $Ti_3Al_2C_2$ powders and pure MXene. **c** SEM image of $Ti_3C_2T_x$ MXene flakes deposited on a silicon wafer. **d** AFM image of $Ti_3C_2T_x$ MXene flakes deposited on a silicon substrate. The inset shows the height curves along the white line. **e** Digital image of MXF-70%. **f–h** SEM images of a cross-section of MXF-70% at different scanning multiples (Kong et al. 2022)

capable of enduring substantial compressive strains of up to 50% over 100 cycles. The electrochemical performance of the integrated supercapacitor was comprehensively assessed under varying compression strains. The resulting improvements in conductivity and the shortening of ion diffusion pathways promote efficient transport of ions and electrons, even at a compression strain of 60%. Consequently, the supercapacitor achieved a maximum volumetric specific energy of 0.37 mWh/cm^3 at a power density of 0.42 mW/cm^3, along with outstanding cycling stability over 2500 compression cycles (Guo et al. 2022).

Sun et al. introduce a novel compressible composite comprising MXene and nitrogen-doped carbon foam (MXene/NCF) designed for flexible supercapacitor applications. The composite is characterized by a unique three-dimensional (3D) hollow interconnected neuron-like structure, created through a one-step pyrolysis process. The MXene/NCF composite demonstrates a synergistic effect in which the nitrogen-doped carbon foam contributes additional pseudocapacitance via nitrogen atom doping while providing structural support to the MXene nanosheets. This architecture facilitates stable and efficient ion diffusion and electron transport, resulting in an increased number of contact sites alongside improved conductivity

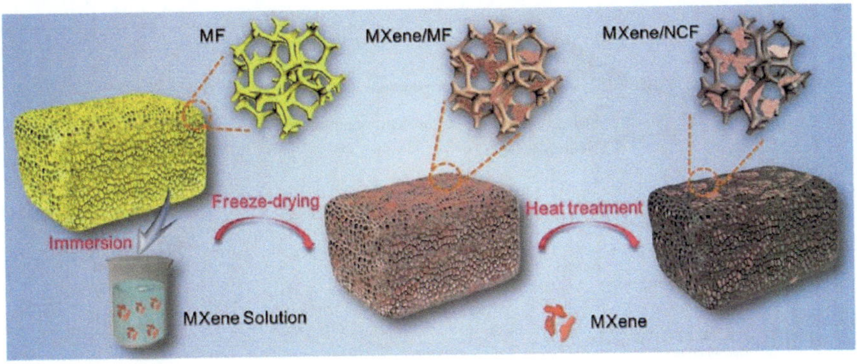

Fig. 8.7 MXene/N-doped carbon foam (MXene/NCF) (natural blends) (Sun et al. 2020)

and hydrophilicity. The freestanding MXene/NCF electrode exhibits outstanding electrochemical performance, achieving a gravimetric capacitance of 332 F/g and a volumetric capacitance of 3162 mF/cm^3, along with impressive rate performance, retaining 64% capacitance when transitioning from 0.5 to 100 A/g. Furthermore, it maintains an exceptional capacity retention of 99.2% after 10,000 cycles. Importantly, the all-solid-state supercapacitors based on MXene/NCF sustain high specific capacitance and remarkable rate performance, demonstrating remarkable stability under 60% strain without significant changes to their morphology or electrochemical properties. These superior characteristics suggest that MXene/NCF composites hold considerable promise for advancements in flexible supercapacitor technologies (Sun et al. 2020) (Fig. 8.7).

Stretchable Batteries

Rubber blends are also being explored in stretchable batteries for flexible electronics. These batteries require materials that can endure mechanical stretching while maintaining their electrochemical properties. Rubber blends combined with ionic conductors or lithium-based materials have been used to create stretchable and flexible electrodes for batteries. EPDM and SBR-based rubber blends have been shown to exhibit the necessary electrochemical stability, elasticity, and resistance to aging, making them promising candidates for next-generation flexible energy storage solutions (Kamarulazam et al. 2023).

Lu and colleagues have introduced an advanced laser-structured microarray electrode specifically designed for stretchable lithium-ion batteries (SLIBs), which hold great promise as a power source for wearable electronics. A key challenge in developing SLIBs is creating electrodes that maintain both stability and stretchability. In this study, an active material film is affixed to a planar stretchable current collector and is then ablated using an ultrafast laser, resulting in a microscale square array that significantly enhances the electrodes' stretchability. To assess the stretchability of

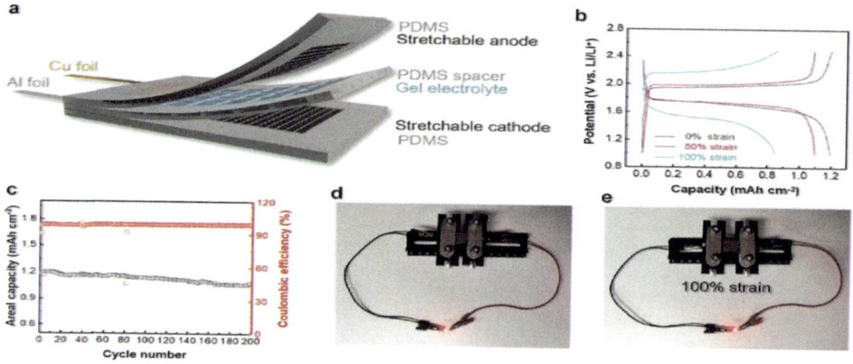

Fig. 8.8 Electrochemical performance of the stretchable batteries. **a** Schematic illustration of the structure of the stretchable battery. **b** Discharge/charge voltage profiles of the stretchable battery in the unstretched, 50% stretched, and 100% stretched states. **c** Cycling performance and coulombic efficiency over 200 cycles at 1C. Photographs of a light-emitting diode bulb operated by the stretchable battery in its **d** unstretched and **e** 100% stretched states (Lu et al. 2023)

these microarray electrodes under tensile stress, the researchers developed a one-dimensional elastic analytical model. This model illustrates that the microscale square array can remain securely bonded to the current collector, provided the shear stress remains below the adhesion force and the tensile stress does not exceed the tensile strength of the active material. The resulting electrodes feature a high active material loading of 10 mg cm^{-2} and demonstrate excellent electrochemical performance, even after 500 cycles at 100% strain. Furthermore, the SLIBs utilizing these electrodes exhibit a stable capacity of 1.2 mAh cm^{-2}, retaining over 70% of their initial capacity at 100% strain (Lu et al. 2023) (Fig. 8.8).

Emerging studies on next-generation flexible and wearable electronics have led to significant advancements in the development of durable, deformable energy devices that exhibit exceptional electrochemical performance. Liu and colleagues have created a high-performance, waterproof, customizable, and stretchable yarn zinc-ion battery (ZIB) that features double-helix yarn electrodes and a cross-linked polyacrylamide (PAM) electrolyte. The combination of the PAM electrolyte's high ionic conductivity and the unique helical design of the electrodes enables the yarn ZIB to achieve remarkable specific capacity and volumetric energy density values of 302.1 mAh g^{-1} and 53.8 mWh cm^{-3}, respectively, along with impressive cycling stability, retaining 98.5% of its capacity after 500 cycles. Notably, this quasi-solid-state yarn ZIB displays excellent knittability, substantial stretchability (up to 300% strain), and outstanding waterproof characteristics, maintaining 96.5% of its capacity after 12 h of immersion in water. Furthermore, the long yarn ZIB can be segmented into shorter pieces, each of which continues to operate effectively. Owing to its weavable and customizable properties, a 1.1-m-long yarn ZIB was divided into eight sections and woven into a textile that powered a flexible belt embedded with 100 LEDs and a 100 cm^2 flexible electroluminescent panel (Li et al. 2018) (Fig. 8.9).

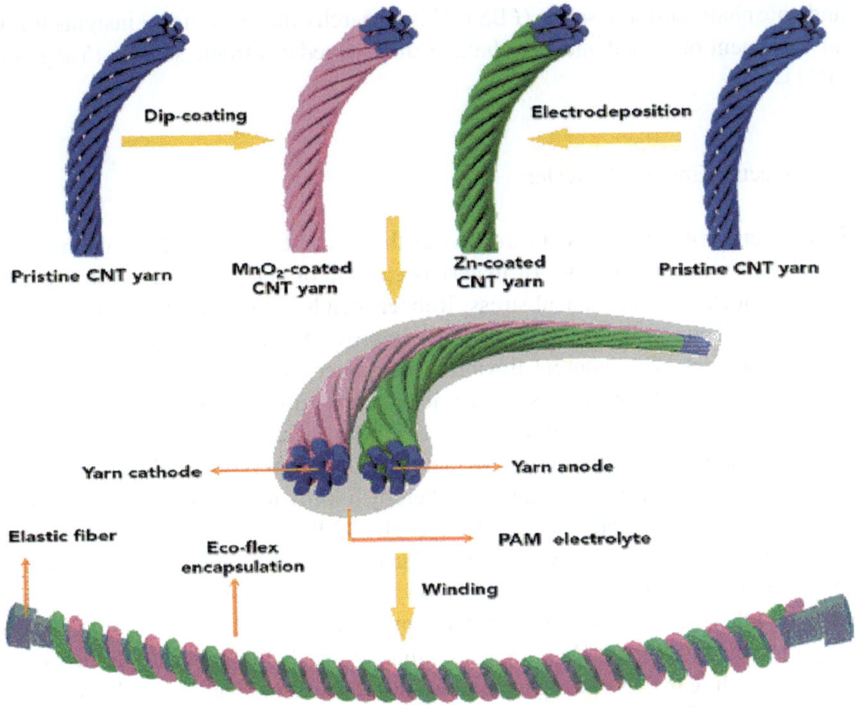

Fig. 8.9 Schematic diagram of fabrication and encapsulation of the yarn ZIB (Li et al. 2018)

Self-powered elastic conductors made from thermoelectric materials that can harness energy from the surrounding environment are crucial for electronic devices used in off-grid, maintenance-free conditions or scenarios where battery replacement is impractical. Among organic conductors, poly(3,4-ethylenedioxythiophene)-poly(styrenesulfonate) (PEDOT) is regarded as one of the leading candidates. However, its application in flexible devices is limited by brittleness and other irreversible characteristics when subjected to strain. To overcome this challenge, several polymer blends incorporating water-soluble polymers with PEDOT have been developed. Unfortunately, these blends frequently fail to strike an appropriate balance between electrical conductivity and mechanical performance, largely due to phase morphology issues that hinder the formation of a conductive network with PEDOT as the dispersed phase. Yang et al. detail a method where PEDOT is incorporated into natural rubber (NR), with NR serving as the dispersed phase. The resulting composite films, which contain 10 wt% PEDOT and are doped with dimethyl sulfoxide (DMSO), achieve conductivities of up to 87 S/cm while maintaining an impressive elongation at break of 490%. Notably, these films also demonstrate self-powered capabilities for sensing temperature and tensile strain. Furthermore, the study shows that the majority of irreversible strain and conductivity loss incurred under cyclic tensile strain can be effectively restored through post-treatment methods using water and phosphate-buffered saline (PBS). This research offers significant insights for the advancement of healed and stretchable self-powered electronic sensors (Yang et al. 2021).

Piezoelectric Energy Harvesters

Piezoelectric energy harvesters are devices that convert mechanical energy into electrical energy through the use of materials capable of generating an electric charge when subjected to mechanical stress. Rubber blends, especially those incorporated with piezoelectric fillers, can significantly improve the energy-harvesting efficiency of these devices. The inherent flexibility of rubber and its ability to endure repetitive mechanical deformations render it an ideal substrate for energy harvesters designed for dynamic operating conditions. The inclusion of fillers such as lead zirconate titanate (PZT) or barium titanate ($BaTiO_3$) can greatly enhance the piezoelectric response of rubber composites, thereby improving their overall performance in energy conversion applications (Zhou et al. 2020).

Sultana and colleagues have introduced an innovative method for developing flexible piezoelectric nanogenerators (FPNGs) by utilizing low-temperature hydrothermally synthesized wurtzite zinc sulfide nanorods (ZnS-NRs) in combination with polydimethylsiloxane (PDMS). Notably, this approach eliminates the necessity for electrical poling in these nanogenerators. The inclusion of conductive fillers, such as polyaniline (PANI) and multiwall carbon nanotubes (MWCNTs), significantly boosts the overall performance of the FPNGs. Specifically, the nanogenerator enhanced with PANI, designated as PZP-FPNG, achieves impressive electrical performance, delivering an open circuit voltage of approximately 35 V and a power density of around 2.43 $\mu W/cm^3$. The capacitor charging curves reveal a rapid transient response, indicated by a short time constant (τ), demonstrating that these nanogenerators can efficiently charge capacitors, reaching a voltage of 3 V in just 50 s. Moreover, the FPNGs are capable of powering multiple commercial light-emitting diodes (LEDs) and various portable electronic devices, such as wristwatches, calculators, and LCD screens. The experimental results underscore the potential application of PANI in the fabrication of piezoelectric-based nanogenerators utilizing low-temperature synthesized ZnS-NRs, emphasizing its importance for sustainable energy solutions. Additionally, the research investigates the feasibility of wireless piezoelectric signal detection, proposing the concept of a self-powered smart sensor (Sultana et al. 2015) (Fig. 8.10).

Liu and colleagues have developed an advanced skin-integrated rubbery electronic device designed to transform mechanical motions generated by the human body into electrical signals. This technology shows significant potential for use in biosensing and energy harvesting applications. The device utilizes a ternary piezoelectric rubber composite composed of graphene, lead zirconate titanate (PZT), and polydimethylsiloxane (PDMS), which is produced through a straightforward and cost-effective manufacturing process. Unlike a binary composite comprising only

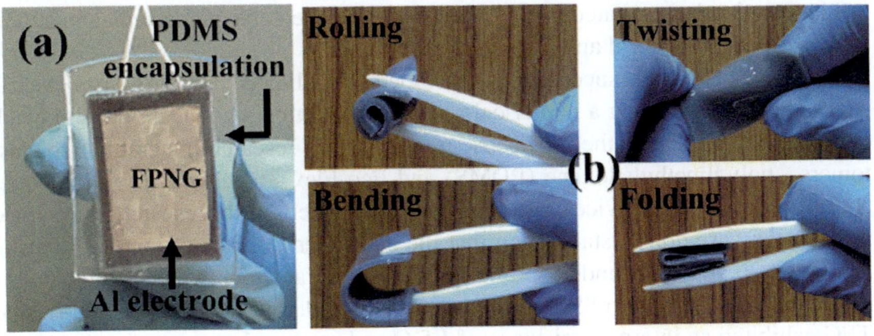

Fig. 8.10 **a** A digital photographic image of PDMS encapsulated PZP-FPNG with Al electrodes. **b** The flexibility of the PZP-FPNG is demonstrated by rolling, twisting, folding, and bending states (Sultana et al. 2015)

PZT and PDMS, the graphene-infused ternary composite exhibits a notable improvement in self-powered performance, achieving a peak power density of 972.43 μW/cm^3 during human movement. The combination of experimental and theoretical evaluations reveals that the graphene-enhanced PZT rubber maintains exceptional mechanical durability, effectively enduring bending, stretching, and twisting over thousands of cycles without significant degradation. The adaptation of device geometries, guided by optimized mechanical design principles, promotes better integration of these rubbery electronics with the human body. For example, annulus-shaped devices can be strategically placed on joints to ensure consistent power output, even under continuous and substantial deformation. This study highlights the promise of large-area, skin-integrated, self-powered electronics for energy harvesting and mechanical sensing applications related to human health (Liu et al. 2019).

Thermoelectric Generators

Rubber-based thermoelectric generators are designed to convert thermal energy into electrical energy by utilizing conductive rubber composites. By integrating materials with high thermoelectric efficiency, such as bismuth telluride (Bi_2Te_3), into rubber blends, researchers can develop flexible thermoelectric generators suitable for wearable devices. This combination of elastomeric flexibility and the thermoelectric properties of the fillers opens up new avenues in energy conversion technologies. These generators leverage the Seebeck effect, where a temperature gradient across the material generates an electric voltage. The incorporation of high-performance thermoelectric materials within a rubber matrix allows for the design of lightweight and flexible devices that can effectively harvest waste heat from the human body or the environment. Recent advancements have demonstrated significant improvements in thermoelectric performance through the optimization of filler content and composite structure. Additionally, studies show that such thermoelectric generators can be seamlessly integrated into clothing and other wearable items, enhancing their practicality in real-world applications.

Chen and colleagues successfully developed and validated flexible thermoelectric generators (TEGs) using a numerical approach, adopting a conventional sandwich configuration typical of thermoelectric modules. This research integrated two elastomers—polydimethylsiloxane (PDMS) and Eco-Flex—together with thin copper sheets to construct the devices. Utilizing the Taguchi method, the researchers assessed nine design variations, establishing that the elastomer reached a maximum tensile strength of 0.967 MPa under optimal processing parameters, which included stirring for 6 min, heating to 85 °C, and a total heating duration of 3 h. The fabricated TEG exhibited an internal resistance of 1.5 Ω and produced an open-circuit voltage of 0.011 V, resulting in an output power of 11 μW when subjected to a temperature differential of 75 °C. Notably, the incorporation of graphene into the PDMS matrix improved the elastomer's thermal conductivity by a factor of 9.6 at 370 K, consequently increasing the output power to 0.0515 W for the same temperature differential, which represents an enhancement of 4681 times. Furthermore, numerical simulations were conducted to assess the performance of the flexible TEG, demonstrating that the discrepancies between experimental results and numerical predictions were minimal, ranging from 4.6 to 5.2%. This finding reinforces the accuracy and reliability of the simulation outcomes. The innovative TEG technology holds significant promise for harvesting industrial low-temperature waste heat and biothermal energy, making it suitable for powering sensors in industrial settings as well as wearable devices for human and animal applications (Chen et al. 2024) (Fig. 8.11).

Zhang and co-workers introduce an innovative flexible thermoelectric device (TED) tailored for wearable electronic applications and the regulation of human body temperature, specifically addressing comfort concerns related to conventional rigid TEDs. Employing finite element analysis (FEA), the authors utilize porous polydimethylsiloxane (PDMS), which has low thermal conductivity, as a substitute for traditional organic elastomers, effectively minimizing heat transfer within the TED.

The findings indicate that the device achieves a maximum power density of 2.33 $\mu W/cm^2$ under natural convection conditions. When a temperature differential (ΔT) of 8.8 K is introduced, the power density escalates to 27.02 $\mu W/cm^2$. Additionally, at a current of 1.0 A, the flexible TED exhibits a cooling capability of up to 7 K under natural convection and a reduction of 2.8 K on the human wrist when exposed to 1-sun (1 kW/m^2) irradiation. This groundbreaking design not only boosts the efficacy of thermoelectric devices but also enhances their potential for integration into wearable technology (Zhang et al. 2021).

Fattahimoghaddam and associates conducted a study on the innovative integration of tellurium carbon-rich hybrid nanocables (NCs) into a polydimethylsiloxane (PDMS) matrix, crafting a composite optimized for solar-driven power generation. The unique interplay between tellurium nanowires (NWs) and the carbon-rich shell of the NCs significantly enhances photo-assisted thermoelectric energy conversion. When exposed to conditions simulating 1 sun illumination, the pure PDMS film

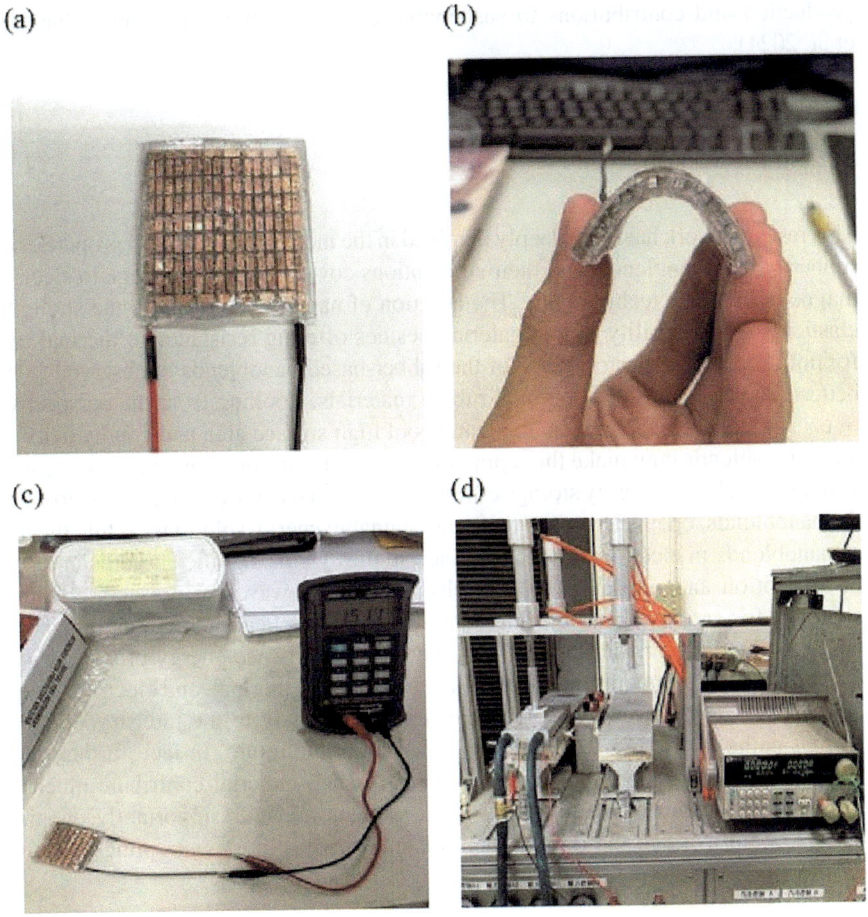

Fig. 8.11 **a** Top view and **b** bending of the completed flexible TEG and Flexible TEG module testing system of **c** internal resistance and **d** open circuit voltage (Chen et al. 2024)

generates an output voltage of 54.8 mV and a current of 17.2 mA. The introduction of carbonaceous nanospheres (NSs) into the PDMS matrix notably elevates these parameters, achieving 71.7 mV and 22.4 mA, which corresponds to an approximate power output of 1600 μW. Additionally, the incorporation of NCs results in an even more pronounced enhancement in power generation, with the NCs/PDMS composite attaining peak voltage and current values of 94.3 mV and 33.4 mA, respectively, leading to a power output of around 3150 μW. This significant increase is attributed to improved photothermal performance, arising from the synergistic effects of two distinct photothermal mechanisms: non-radiative recombination facilitated by the tellurium NWs and lattice relaxation within the carbonaceous shell. The resulting flexible NCs/PDMS composites capitalize on the synergistic properties of tellurium, carbon, and PDMS, indicating their potential for applications in solar-driven energy production and contributions to sustainable energy solutions (Fattahimoghaddam et al. 2024).

8.4 Conclusion

This research work has been deeply involved in the mechanical, thermal properties of rubber-based nanoblends, and their applications covering energy storage to electromagnetic interface technologies. The addition of nanofillers enhances the strength, elasticity, and durability of the material, besides offering resistance to thermal and friction forces. The performance of the rubber-based nanoblends is observed to be better as compared to conventional rubber materials. Looking from the perspective of energy storage applications, the features of high surface area and conductivity in these nanoblends may make them potential candidates in supercapacitor and battery systems. The higher energy storage efficiency, attributed to the electrical properties of the nanoblends, opens new ways toward sustainable energy solutions. While the use of nanoblends in electromagnetic interface studies yields significant gains in terms of absorption and insulation against electromagnetic waves, it presents effective solutions for various electronic and communication applications. The optimization of such materials can substantially be important to enhance energy efficiency and electromagnetic compatibility. In brief, the mechanical, thermal, and electrical properties of rubber-based nanoblends are of immense benefit to the industry, opening various paths for studies that will be considered in the future. In fact, further optimization and expansion of application areas of such material contribute much to raising the efficiency level of energy storage systems, and most importantly, opening quite new horizons in the development of innovative solutions in applications for electromagnetic interfaces.

References

Abdeen ZI (2017) Polyurethane rubber-based nanoblends: preparation, characterization and applications, pp 89–103

Abdulwahid RT, Aziz SB, Kadir MFZ (2023) Replacing synthetic polymer electrolytes in energy storage with flexible biodegradable alternatives: sustainable green biopolymer blend electrolyte for supercapacitor device. Mater Today Sustain 23:100472

Al-Ghamdi AA, Al-Ghamdi AA, Al-Turki Y, Yakuphanoglu F, El-Tantawy F (2016) Electromagnetic shielding properties of graphene/acrylonitrile butadiene rubber nanocomposites for portable and flexible electronic devices. Compos B Eng 88:212–219

Allahbakhsh A, Mazinani S, Kalaee MR, Sharif F (2013) Cure kinetics and chemorheology of EPDM/graphene oxide nanocomposites. Thermochim Acta 563:22–32

Al-Senani GM, Al-Qahtani SD (2024) Silica nanoparticle-reinforced polyurethane plastic immobilized with strontium aluminate nanofiller: ultraviolet protection, superhydrophobicity and photoluminescence. J Photochem Photobiol A Chem 456:115808

Al-Shablle M, Al-Waily M, Njim EK (2022) Analytical evaluation of the influence of adding rubber layers on free vibration of sandwich structure with presence of nano-reinforced composite skins. Arch Mater Sci Eng

Ameli A, Jung PU, Park CB (2013) Electrical properties and electromagnetic interference shielding effectiveness of polypropylene/carbon fiber composite foams. Carbon N Y 60:379–391

Anand A, Raosaheb BV, Reddy MVH, Aan MPS, Ekwipoo K, Kumar RSP, Vijayasankar AV, Binish CJ, Johns J (2024) Unveiling the synergistic effect of zinc oxide and carbon black nanoparticles on the dielectric properties of vulcanized natural rubber. Next Nanotechnol 6:100092

Ansari A, Mohanty TR, Sarkar S, Ramakrishnan S, Amarnath SKP, Singha NK (2024) Epoxy modified styrene butadiene rubber (SBR) in green tire application. Eur Polym J 213:113069

Archibong FN, Orakwe LC, Ogah OA, Mbam SO, Ajah SA, Okechukwu ME, Igberi CO, Okafor KJ, Chima MO, Ikelle II (2023) Emerging progress in montmorillonite rubber/polymer nanocomposites: a review. J Mater Sci 58:2396–2429

Bandyopadhyay S (2024) Fundamentals of rubber compounding. In: Rubber products: technology and cost optimisation, pp 1–52

Barana D, Ali SD, Salanti A, Orlandi M, Castellani L, Hanel T, Zoia L (2016) Influence of lignin features on thermal stability and mechanical properties of natural rubber compounds. ACS Sustain Chem Eng 4:5258–5267

Bhattacharya AB, Raju AT, Chatterjee T, Naskar K (2020) Development and characterizations of ultra-high molecular weight EPDM/PP based TPV nanocomposites for automotive applications. Polym Compos 41:4950–4962

Bokobza L (2018) Natural rubber nanocomposites: a review. Nanomaterials 9:12

Bras J, Hassan ML, Bruzesse C, Hassan EA, El-Wakil NA, Dufresne A (2010) Mechanical, barrier, and biodegradability properties of bagasse cellulose whiskers reinforced natural rubber nanocomposites. Ind Crops Prod 32:627–633

Camargo PHC, Satyanarayana KG, Wypych F (2009) Nanocomposites: synthesis, structure, properties and new application opportunities. Mater Res 12:1–39

Chandrasekaran C (2017) Anticorrosive rubber lining a practical guide for plastics engineers. William Andrew, imprint of Elsevier

Chatterjee A, Khobragade PS (2017) Styrene–butadiene rubber-based nanoblends (SBR-nB): preparation, characterization and applications, pp 209–248

Chauhan BPS, Sardar R, Latif U, Chauhan M, Lamoreaux WJ (2005) Nanoengineering of metallic solutions through silicone constructs †. Acta Chim Slov 52:361–370

Chen WH, Bai WC, Luo D, Bandala AA, Wang XD, Tuan Hoang A (2024) Optimization of material properties and performance of flexible thermoelectric generators with/without graphene. Energy Convers Manage: X. https://doi.org/10.1016/j.ecmx.2024.100741

Chi Wang A, Wu C, Pisignano D et al (2018) Sustainable rubbers and rubber additives. J Appl Polym Sci 135:45701

Cho HW, Hepowit LR, Nam HS, Kim SH, Lee YM, Kim JH, Kim KM, Ko JM (2012) Synthesis and supercapacitive properties of electrodeposited polyaniline composite electrode on acrylonitrile-butadiene rubber as a flexible current collector. Synth Met 162:410–413

Costa NL, Hiranobe CT, Cardim HP et al (2024a) A review of EPDM (ethylene propylene diene monomer) rubber-based nanocomposites: properties and progress. Polymers (Basel). https://doi.org/10.3390/polym16121720

Costa NL, Hiranobe CT, Cardim HP et al (2024b) A review of EPDM (ethylene propylene diene monomer) rubber-based nanocomposites: properties and progress. Polymers 16:1720

Daniel P, Coelho-Diogo C, Gaucher V, Stoclet G, Robin C, Lorthioir C (2024) Solid-state NMR of vulcanized natural rubber/butadiene rubber blends: Local organization and cross-linking heterogeneities. Magn Reson Lett 200155

Datta J, Kosiorek P, Włoch M (2016) Effect of high loading of titanium dioxide particles on the morphology, mechanical and thermo-mechanical properties of the natural rubber-based composites. Iran Polym J (English Edition) 25:1021–1035

Dvornic PR (2006) PAMAMOS: the first commercial silicon-containing dendrimers and their applications. J Polym Sci A Polym Chem 44:2755–2773

Edayadiyil JJ, Abraham J, Rajeevan S, George SC (2021) Synthesis and characterization of natural rubber/graphene quantum dot nanocomposites. J Polym Res 28:1–8

El-Sabbagh SH, Ahmed NM, Turky GM, Selim MM (2017) Rubber nanocomposites with new core-shell metal oxides as nanofillers. In: Progress in rubber nanocomposites, pp 249–283

Eyssa HM, Ali NM, Yousif NM (2024) Structure-property performance of irradiated sponge rubber blend nanocomposites containing h-BN, GO, V_2O_5, and adhesion on cotton fabric for positive electrode in flexible supercapacitors. Energ Technol. https://doi.org/10.1002/ente.202400100

Fan M, Li S, Wu L, Li L, Qu M, Nie J, Zhang R, Tang P, Bin Y (2022) Natural rubber toughened carbon nanotube buckypaper and its multifunctionality in electromagnetic interference shielding, thermal conductivity, Joule heating and triboelectric nanogenerators. Chem Eng J 433:133499

Farid M, Purniawan A, Susanti D, Rasyida A, Yulianto H, Ramadani M (2019) Effect of nanocellulose on acoustical and thermal insulation properties of silicone rubber composite. Mater Sci Forum 964:156–160

Fattahimoghaddam H, Kim IH, Jeong YW, An TK, Jeong YJ (2024) Exploring the potential of Te@Carbon–Rich hybrid nanocables/PDMS composites for Solar-Driven thermoelectric power generation. Compos Commun. https://doi.org/10.1016/j.coco.2024.101980

Feldman D (2017) Natural rubber nanocomposites. J Macromol Sci, Part A 54:629–634

Gaikwad P, Tiwari N, Kamat R, Mane SM, Kulkarni SB (2024) A comprehensive review on the progress of transition metal oxides materials as a supercapacitor electrode. Mater Sci Eng, B 307:117544

Ghamarpoor R, Jamshidi M (2022) Synthesis of vinyl-based silica nanoparticles by sol–gel method and their influences on network microstructure and dynamic mechanical properties of nitrile rubber nanocomposites. Sci Rep 12:1–15

Ghosh S, Majhi J, Sharma S, Priya K, Bandyopadhyay A (2023) A review on the development of electron and ion conductive polymer hydrogels and their composites for flexible and smart supercapacitors. J Energy Storage 74:109423

Guo R, Han X, Yuan P, He X, Li Q, Sun J, Dang L, Liu Z, Zhang Y, Lei Z (2022) Filling $Ti_3C_2T_x$ nanosheets into melamine foam towards a highly compressible all-in-one supercapacitor. Nano Res 15:3254–3263

Hagting E (2015) An investigation on the effect of clusters in thermo-reversible cross-linked rubbers

Han J, Lu K, Yue Y, Mei C, Huang C, Wu Q, Xu X (2019) Nanocellulose-templated assembly of polyaniline in natural rubber-based hybrid elastomers toward flexible electronic conductors. Ind Crops Prod 128:94–107

Hansora D, Mishra S (2017) Polybutadiene rubber-based nanoblends (PBR-nB): preparation, characterization and applications, pp 155–207

Hao JM, Yang YF, Shen YT, Zhou R, Zhang W, Chen H, Cheng WL (2024) Positive temperature coefficient material based on silicone rubber/paraffin/graphite/carbon nanotubes for wearable thermal management devices. Chem Eng J 493:152427

He Y, Gao J, Gong X, Xu J (2017) The role of carbon nanotubes in promoting the properties of carbon black-filled natural rubber/butadiene rubber composites. Results Phys 7:4352–4358

He S, Li J, Zhang X, Chen Z, Zhou H, Pang C, Weng Y, Lin J (2023) Nitrile butadiene rubber/clay nanocomposites cured and reinforced by copper sulfate pentahydrate. J Market Res 22:2338–2342

Ibrahim A, Dahlan M (1998) Thermoplastic natural rubber blends. Prog Polym Sci 23:665–706

Ismail H, Salleh SZ, Ahmad Z (2013) Properties of halloysite nanotubes-filled natural rubber prepared using different mixing methods. Mater des 50:790–797

Jia LC, Li YK, Yan DX (2017) Flexible and efficient electromagnetic interference shielding materials from ground tire rubber. Carbon N Y 121:267–273

Kamarulazam F, Bashir S, Pershaanaa M, Goh ZL, Surender G, Elumalai PNN, Farhana NK, Ramesh S, Ramesh K (2023) Stretchable, self-healable and highly conductive natural-rubber hydrogel electrolytes for supercapacitors: advanced wearable technology. J Energy Storage 71:108182

Kang CH, Bin JW, Kim HJ, Jung HT (2021) Highly enhanced tire performance achieved by using combined carbon nanotubes and soybean oil. J Appl Polym Sci 138:1–13
Karpeles R, Grossi AV (2000) EPDM rubber technology. In: Handbook of elastomers, pp 863–894
Khan A, Kian LK, Jawaid M, Khan AAP, Alotaibi MM, Asiri AM, Marwani HM (2022) Preparation of styrene-butadiene rubber (SBR) composite incorporated with collagen-functionalized graphene oxide for green tire application. Gels 8:161
Kong I, Hj Ahmad S, Hj Abdullah M, Hui D, Nazlim Yusoff A, Puryanti D (2010) Magnetic and microwave absorbing properties of magnetite–thermoplastic natural rubber nanocomposites. J Magn Magn Mater 322:3401–3409
Kong N, Lv K, Chen W, Guan J, Zhao P, Tao J, Zhang J (2022) Natural polymer template for low-cost producing high-performance $Ti_3C_2T_x$ MXene electrodes for flexible supercapacitors. ACS Appl Mater Interfaces 14:56877–56885
Kruželák J, Kvasniˇ AKH, Plavec R, Dosoudil R, Goˇralík M, Vilˇcáková J, Hudec I (2021) Absorption shielding effectiveness of rubber composites based on ferrite and carbon fillers. Polymers (Basel) 13:2937
Kumar P, Kumar P (2022) Natural nano-fillers materials for the Bio-composites: a review. J Indian Chem Soc 99:100715
Kumar V, Alam MN, Manikkavel A, Song M, Lee DJ, Park SS (2021) Silicone rubber composites reinforced by carbon nanofillers and their hybrids for various applications: a review. Polymers 13:2322
Kumar SP, Prabhakaran G, Vishvanathperumal S (2024) Influence of modified nanosilica on the performance of NR/EPDM blends: cure characteristics, mechanical properties and swelling resistance. J Inorg Organomet Polym Mater 34:3420–3442
Kwon S, Ma R, Kim U, Choi HR, Baik S (2014) Flexible electromagnetic interference shields made of silver flakes, carbon nanotubes and nitrile butadiene rubber. Carbon N Y 68:118–124
Li H, Liu Z, Liang G et al (2018) Waterproof and tailorable elastic rechargeable yarn zinc ion batteries by a cross-linked polyacrylamide electrolyte. ACS Nano 12:3140–3148
Limtrakul A, Sae-Oui P, Nillawong M, Sirisinha C (2022) Influence of carbon black/silica hybrid ratio on properties of passenger car tire sidewall. Periodica Polytech, Chem Eng 66:147–156
Liu Y, Shi Y, Zhang D, Li J, Huang G (2013a) Preparation and thermal degradation behavior of room temperature vulcanized silicone rubber-g-polyhedral oligomeric silsesquioxanes. Polymer (Guildf) 54:6140–6149
Liu HD, Liu ZY, Yang MB, He Q (2013b) Surperhydrophobic polyurethane foam modified by graphene oxide. J Appl Polym Sci 130:3530–3536
Liu Y, Zhao L, Wang L et al (2019) Skin-integrated graphene-embedded lead Zirconate Titanate rubber for energy harvesting and mechanical sensing. Adv Mater Technol. https://doi.org/10.1002/admt.201900744
Liu J, Yao Y, Chen S, Li X, Zhang Z (2021) A new nanoparticle-reinforced silicone rubber composite integrating high strength and strong adhesion. Compos Part A Appl Sci Manuf 151:106645
Low DYS, Mintarno S, Karia NR, Manickam S, Tan KW, Khalid M, Goh BH, Tang SY (2024) Nano-reinforced self-healing rubbers: a comprehensive review. J Ind Eng Chem. https://doi.org/10.1016/j.jiec.2024.05.002
Lu Y, Ru S, Li H, Wang G, Xu S (2023) Laser-structured microarray electrodes for durable stretchable lithium-ion battery. J Colloid Interface Sci 631:1–7
Mahmood K, Sani A, Sadiq H, Tang J, Li Q (2024) Modification of silicone rubber by nanocomposites for enhancing physicochemical properties: a review. Mater Sci Eng, B 310:117664
Mandlekar N, Joshi M, Butola BS (2022) A review on specialty elastomers based potential inflatable structures and applications. Adv Ind Eng Polym Res 5:33–45
Mark JE (2006) Some novel polymeric nanocomposites. Acc Chem Res 39:881–888
Mark JE, Erman B, Roland M (2019) Science and technology of RUBBER, 3rd edn
Marković G, Visakh PM (eds) (2017) Springer series on polymer and composite materials rubber nano blends preparation, characterization and applications

Marsat JN, Stahlhut F, Laschewsky A, Berlepsch HV, Böttcher C (2013) Multicompartment micelles from silicone-based triphilic block copolymers. Colloid Polym Sci 291:2561–2567
Mishra S, Hansora D (2018) Advancements in polymer nanotechnology. In: Novel applications in polymers and waste management, pp 93–121
Mohammed AA, Shnain ZY, Sukkar KA (2023) Improving the thermal and mechanical properties of reinforced polybutadiene rubber composite using nano-sized TiO_2. Egypt J Chem 66:253–258
Niaz F, Shah SS, Hayat K, Aziz MA, Liu G, Iqbal Y, Oyama M (2024) Utilizing rubber plant leaf petioles derived activated carbon for high-performance supercapacitor electrodes. Ind Crops Prod. https://doi.org/10.1016/j.indcrop.2024.119161
Obazee EO, Okieimen FE, Aigbodion AI, Bakare IO (2020) Effect of nanoclay reinforcement on the property of rubber seed oil polyurethane nanocomposites. In: Minerals, metals and materials series, pp 867–879
Oleiwi JK, Hamza MS, Abed MS (2010) Improving the properties of the tire tread by adding SiO_2 and Al_2O_3 to SBR rubber. Int J Appl Eng Res 5:1637–1652
Pantrangi M, Ashalley E, Hadi MK et al (2024) Flexible micro-supercapacitors: materials and architectures for smart integrated wearable and implantable devices. Energy Storage Mater 73:103791
Parashiva PC, Mohanty S, Gupta VK (2021) Modification of polybutadiene rubber: a review. Rubber Chem Technol 94:410–431
Parvathi K, Ramesan MT (2023) Tailoring the structural, electrical and thermal properties of zinc oxide reinforced chlorinated natural rubber/poly (indole) blend nanocomposites for flexible electrochemical devices. J Polym Res 30:1–15
Parvathi K, Bahuleyan BK, Ramesan MT (2022) Enhanced optical, thermal and electrical properties of chlorinated natural rubber/zinc ferrite nanocomposites for flexible electrochemical devices. J Macromol Sci Part A Pure Appl Chem 59:466–479
Peng C, Yang Q, Zhao W, Ren J, Yu Q, Hu Y, Zhang X (2019) Relationship between interface chemistry and reinforcement in polybutadiene/cellulose nanocrystal nanocomposites. Compos Sci Technol 177:103–110
Ponnamma D, Maria HJ, Chandra AK, Thomas S (2013) Rubber nanocomposites: latest trends and concepts, pp 69–107
Rahaman M (2023) Superior mechanical, electrical, dielectric, and EMI shielding properties of ethylene propylene diene monomer (EPDM) based carbon black composites. RSC Adv 13:25443–25458
Rajasekar R, Heinrich G, Das A, Das CK (2009) Development of SBR-Nanoclay composites with epoxidized natural rubber as compatibilizer. J Nanotechnol 2009:405153
Ramuz M, Tee BCK, Tok JBH, Bao Z (2012) Transparent, optical, pressure-sensitive artificial skin for large-area stretchable electronics. Adv Mater 24:3223–3227
Rodgers B, Waddell W (2005) The science of rubber compounding. In: Science and technology of rubber, pp 401–454
Sadeghalvaad M, Dabiri E, Zahmatkesh S, Afsharimoghadam P (2019) Improved curing conditions and mechanical/chemical properties of nitrile butadiene rubber composites reinforced with carbon based nanofillers. J Nanostruct 9:453–467
Sala RL, Arantes TM, Longo E, Leite ER, Paranhos CM, Camargo ER (2014) Evaluation of modified silica nanoparticles in carboxylated nitrile rubber nanocomposites. Colloids Surf A Physicochem Eng Asp 462:45–51
Salkhi Khasraghi S, Momenilandi M, Shojaei A (2022) Tire tread performance of silica-filled SBR/BR rubber composites incorporated with nanodiamond and nanodiamond/nano-SiO_2 hybrid nanoparticle. Diam Relat Mater 126:109068
Selvin Thomas P, Abdullateef AA, Al-Harthi MA, Atieh MA, De SK, Rahaman M, Chaki TK, Khastgir D, Bandyopadhyay S (2012) Electrical properties of natural rubber nanocomposites: effect of 1-octadecanol functionalization of carbon nanotubes. J Mater Sci 47:3344–3349

Sembiring S, Sembiring SM, Humaidi S, Riana M (2023) Study and characterization of tribofilm based on activated carbon-nanochitosan modified with silicone oil for application of self-healing coatings on metal surfaces. J Prisma Sains 11:972–981
Sethulekshmi AS, Saritha A, Joseph K (2022) A comprehensive review on the recent advancements in natural rubber nanocomposites. Int J Biol Macromol 194:819–842
Shim H, Hwang S, Byun E, Choi M, Kim B, Song S (2023) Inkjet patterning of plasmonic metal nanoparticles in silicone elastomers. Adv Mater Technol 8:2201614
Shoul B, Marfavi Y, Sadeghi B, Kowsari E, Sadeghi P, Ramakrishna S (2022) Investigating the potential of sustainable use of green silica in the green tire industry: a review. Environ Sci Pollut Res 29:51298–51317
Simon MW, Stafford KT, Ou DL (2008) Nanoclay reinforcement of liquid silicone rubber. J Inorg Organomet Polym Mater 18:364–373
Sinclair A, Zhou X, Tangpong S, Bajwa DS, Quadir M, Jiang L (2019) High-performance styrene-butadiene rubber nanocomposites reinforced by surface-modified cellulose nanofibers. ACS Omega 4:13189–13199
Sisanth KS, Thomas MG, Abraham J, Thomas S (2017) General introduction to rubber compounding. In: Progress in rubber nanocomposites, pp 1–39
Soares BG (2017) Rubber nanocomposites with metal oxides as nanofillers. In: Progress in rubber nanocomposites, pp 285–318
Soliman EM, Kandil UF, Taha MMR (2012) The significance of carbon nanotubes on styrene butadiene rubber (SBR) and SBR modified mortar. Mater Struct/Mater Constr 45:803–816
Spitalsky Z, Tasis D, Papagelis K, Galiotis C (2010) Carbon nanotube–polymer composites: chemistry, processing, mechanical and electrical properties. Prog Polym Sci 35:357–401
Stephen R, Thomas S (2010) Nanocomposites: state of the art, new challenges and opportunities. In: Rubber nanocomposites: preparation, properties, and applications, pp 1–19
Sultana A, Alam MM, Garain S, Sinha TK, Middya TR, Mandal D (2015) An effective electrical throughput from PANI supplement ZnS nanorods and PDMS-based flexible piezoelectric nanogenerator for power up portable electronic devices: an alternative of MWCNT filler. ACS Appl Mater Interfaces 7:19091–19097
Sun L, Song G, Sun Y, Fu Q, Pan C (2020) MXene/N-doped carbon foam with three-dimensional hollow neuron-like architecture for freestanding, highly compressible all solid-state supercapacitors. ACS Appl Mater Interfaces 12:44777–44788
Šupová M, Martynková GS, Barabaszová K (2011) Effect of nanofillers dispersion in polymer matrices: a review. Sci Adv Mater 3:1–25
Suriani AB, Nurhafizah MD, Mohamed A, Zainol I, Masrom AK (2015) A facile one-step method for graphene oxide/natural rubber latex nanocomposite production for supercapacitor applications. Mater Lett 161:665–668
Thomas S, Maria HJ (2016) Progress in rubber nanocomposites. Woodhead Publishig, an imprint of Elsevier
Trung NN, Luu QP, Son BT, Sinh LH, Bae JY (2012) Preparation and characterization of silicone resin nanocomposite containing CdSe/ZnS quantum dots. Polym Compos 33:1785–1791
Urtekin G, Ullah MS, Yildirim R, Ozkoc G, Kodal M (2023) A comprehensive review of the recent developments in thermoplastics and rubber blends-based composites and nanocomposites. Polym Compos 44:8303–8329
Vaithylingam R, Ansari MNM, Shanks RA (2017) Recent advances in polyurethane-based nanocomposites: a review. Polym Plast Technol Eng 56:1528–1541
Vayyaprontavida Kaliyathan A, Varghese KM, Nair AS, Thomas S (2020) Rubber–rubber blends: a critical review. Prog Rubber Plast Recycl Technol 36:196–242
Vikulov S, Di Stasio F, Ceseracciu L, Saldanha PL, Scarpellini A, Dang Z, Krahne R, Manna L, Lesnyak V (2016) Fully solution-processed conductive films based on colloidal copper selenide nanosheets for flexible electronics. Adv Funct Mater 26:3670–3677

Vipu Vinayak VJ, Deshmukh K, Murthy VRK, Pasha SKK (2024) Conducting polymer based nanocomposites for supercapacitor applications: a review of recent advances, challenges and future prospects. J Energy Storage 100:113551

Visakh PM (2017) Rubber nanoblends: state of the art, new challenges and opportunities, pp 1–13

Visakh PM, Thomas S, Chandra AK, Mathew AP (2013) Advanced structured materials advances in elastomers II composites and nanocomposites

Vishvanathperumal S, Anand G (2020) Effect of nanoclay/nanosilica on the mechanical properties, abrasion and swelling resistance of EPDM/SBR composites. SILICON 12:1925–1941

Wang J, Chen Y (2006) Application of nano-zinc oxide master batch in polybutadiene styrene rubber system. J Appl Polym Sci 101:922–930

Wang R, Sun Y, Hou Y, Tian Y, Zhang Y, Liu F, Han J (2022) Effect of ionic liquid modified carbon nanotubes on the properties of nitrile butadiene rubber and nitrile butadiene latex nanocomposites. J Appl Polym Sci 139:e53020

Wei H, Guo L, Zheng J, Huang G, Li G (2015) Effect of nanosilica-based immobile antioxidant on thermal oxidative degradation of SBR. RSC Adv 5:62788–62796

Wei Z, Cai Y, Zhan Y, Meng Y, Pan N, Jiang X, Xia H (2024) Ultra-low loading of ultra-small Fe_3O_4 nanoparticles on nonmodified CNTs to improve green EMI shielding capability of rubber composites. Small 20:1–10

Wichaita W, Promlok D, Sudjaipraparat N, Sripraphot S, Suteewong T, Tangboriboonrat P (2021) A concise review on design and control of structured natural rubber latex particles as engineering nanocomposites. Eur Polym J. https://doi.org/10.1016/j.eurpolymj.2021.110740

Yan Y, Ge F, Qin Y, Ruan M, Guo Z, He C, Wang Z (2020) Ultralight and robust aerogels based on nanochitin towards water-resistant thermal insulators. Carbohydr Polym 248:116755

Yang Z, Deng J, Chen X, Ren J, Peng H (2013) A highly stretchable, fiber-shaped supercapacitor. Angew Chem Int Ed 52:13453–13457

Yang Y, Zhao G, Cheng X, Deng H, Fu Q (2021) Stretchable and healable conductive elastomer based on PEDOT:PSS/natural rubber for self-powered temperature and strain sensing. ACS Appl Mater Interfaces 13:14599–14611

Zhan X, Chen J, Yang Z, Wu G, Kong Z (2023) Superhydrophobic film from silicone-modified nanocellulose and waterborne polyurethane through simple sanding process. Int J Biol Macromol 232:123431

Zhang Y, Cho UR (2019) Enhanced interfacial interactions of isocyanate-grafted graphene oxide/nitrile-butadiene rubber nanocomposites: mechanical and thermo-physical properties. Polym Compos 40:E1103–E1110

Zhang L, Han X, Du M, Shi Y, Zhang K (2021) Compliant three-dimensional thermoelectric generator filled with porous PDMS for power generation and solid-state cooling. Compos Commun. https://doi.org/10.1016/j.coco.2021.100793

Zhang J, Wang Q, Xue X, Li M, Sun X, Zhao J, Zhang W, Lu C (2023) Waste flame-retardant polyurethane foam/ground tire rubber/carbon nanotubes composites with hierarchical segregated structures for high efficiency electromagnetic interference shielding. Compos Part A Appl Sci Manuf 169:107530

Zhao X, Liu C, He J, Zhang W, Duan X, Liang C (2024a) Graphene rubber toward high content and energy saving enabled by spray drying. Diam Relat Mater 141:110594

Zhao R, Fang Y, Zhao Z, Song S (2024b) Ultra-stretchable, adhesive, fatigue resistance, and anti-freezing conductive hydrogel based on gelatin/guar gum and liquid metal for dual-sensory flexible sensor and all-in-one supercapacitors. Int J Biol Macromol 271:132585

Zhou H, Zhang Y, Qiu Y, Wu H, Qin W, Liao Y, Yu Q, Cheng H (2020) Stretchable piezoelectric energy harvesters and self-powered sensors for wearable and implantable devices. Biosens Bioelectron. https://doi.org/10.1016/j.bios.2020.112569

Zhou W, Ren S, Zhang F, Gao X, Song K, Fang H, Ding Y (2022) Reinforcement of boron–nitrogen coordinated polyurethane elastomers with silica nanoparticles. Polymer (Guildf) 256:125200

Chapter 9
Polylactic Acid and Their Blends for Biomedical Application

Ezequiel Perez and Raquel E. Martini

Abstract Polylactic Acid (PLA) is a biodegradable and compostable thermoplastic, derived from renewable resources, looked as an alternative to petroleum-based plastics due to its good processability and performance, lower environmental impact, and sustainable sourcing. The PLA applications have been expanded to a wide range of fields such as packaging, consumer goods, textiles, and agriculture, among others. Biodegradable medical devices are an important field for this biopolymer, mainly due to its combination of biocompatibility, hydrolytic degradation, tailorable performance, and well-established processing technologies. PLA can be processed as scaffolds, sutures, dressings, orthopedic implants, drug delivery devices, tissue engineering, among others. The limitations of PLA (thermal instability, brittleness, low melt strength and elongation at break, hydrophobicity, slow degradation rate, and poor cell affinity) could be improved by different techniques including copolymerization, crosslinking, or blending. It has been largely incorporated into PLA natural polymers (chitosan, collagen, elastin, and hyaluronic acid) and other biopolymer such as PHB, TPU, PGA, and PEG. The selection of blending procedure depends on the physicochemical properties of polymers, miscibility, physical structure, and desired final performance. The in-service properties of PLA and their blends could not be easily predicted due to the interdependence of characteristics (matrix properties, incorporated polymer, processing, and product requirements), and also varied trends have been largely reported. For this reason, the experimental performance of PLA-based blends should be exhaustively investigated to warrant the in-service requirements of biodegradable medical products.

Keywords Polylactic Acid (PLA) · Biodegradable polymer · Polymer blend · Experimental characterization · Biomedical applications

---

E. Perez (✉)
IPQA, Instituto de Investigación y Desarrollo en Ingeniería de Procesos y Química Aplicada, Universidad Nacional de Córdoba, CONICET, Av. Vélez Sarsfield 1611, X5016GCA Córdoba, Argentina
e-mail: emperez05@hotmail.com

R. E. Martini
Ciudad Universitaria, Córdoba, Argentina

P. M. Visakh (ed.), *Engineering Applications of Polymer based Nano Blends*, Advanced Structured Materials 235, https://doi.org/10.1007/978-3-031-90979-5_9

Nomenclature

AFM	Atomic force microscopy
AGM	Acetyl glycerolmonolaurate
ATC	Acetyl triethyl citrate
BF	Bamboo flour
DBS	Dibutyl sebacate
DMA	Dynamic mechanical analysis
DSC	Differential scanning calorimetry
EIS	Electrochemical impedance spectroscopy
EVA	Ethyl vinyl acetate
FAEs	Fatty acid esters
FTIR	Fourier transform infrared spectroscopy
HPLC	High performance liquid chromatography
MDI	Methyldiisocyante
NMR	Nuclear magnetic resonance
PBAT	Poly(butylene adipate-co-terephthalate)
PBOH	Poly(1,3-butanediol)
PBS	Polybutylene succinate
PBSL	Poly(butylene succinate co-L-lactate)
PBT	Poly(butylene terephthalate)
PCL	Poly(caprolactone)
PEG	Poly(ethylene glycol)
PEO	Poly(ethylene oxide)
PET	Polyethylene terephthalate
PGA	Poly(glycolic acid)
PHAs	Polyhydroxyalkanoates
PHB	Poly(hydroxybutyrate)
PLA	Poly(lactic acid)
POM	Polarized optical microscopy
PP	Polypropylene
PPD	Poly(p-dioxanone)
PS	Polystyrene
PTAT	Poly(tetramethyleneadipate-co-terephthalate)
ROP	Ring-opening polymerization
SEM	Scanning electron microscopy
TEM	Transmission electron microscopy
TGA	Thermo gravimetric analysis
TPS	Thermoplastic starch
TPU	Thermoplastic polyurethane
XRD	X-ray diffraction

9.1 Polylactic Acid

Polylactic Acid (PLA) is a biodegradable and compostable thermoplastic derived from renewable resources. It has gained significant attention as an alternative to petroleum-based plastics due to its good processability and properties, lower environmental impact, and sustainable sourcing (Hamad et al. 2018). PLA was discovered in the early 1800s by Pelouze who produced low molecular weight PLA by lactic acid condensation through a distillation process of water (Sin et al. 2013a). In 1932, almost a century later, Carothers (DuPont) produced a low molecular weight product by heating lactic acid under vacuum (Sin et al. 2013a). Due to the high cost of manufacture, mainly coming from the purification step, low molecular weight and low availability, the initial uses were limited to medical and pharmaceutical applications. (Sin et al. 2013a). The breakthrough ring-opening polymerization (ROP) of lactides developed by Cargill around the 1990s changed this situation (Xu and Song 2015). This method allows achieving higher molecular weight and narrower molecular weight distribution. Furthermore, different strategies of modification and customization of PLA can be carried out to get a wide range of properties including flexibility, strength, heat resistance, and barrier properties, resulting in an opportunity to develop innovative and high-value products. Thus, the PLA applications were expanded to different fields such as packaging, consumer goods, packaging, textiles, and agriculture, among others (Balla et al. 2021), producing a remarkably increase of PLA use. In 2023, the global PLA market was estimated in USD 1.5 billion and is expected to grow at a compound annual growth rate (CAGR) of 17.1% from 2023 to 2028 (https://www.marketsandmarkets.com/Market-Reports/polylactic-acid-pla-market-29418964.html). Between the different applications, biomedical devices are still an important field for this biopolymer, mainly due to their combination of biocompatibility, hydrolytic degradation in situ, tailorable properties, and well-established processing technologies (Casalini et al. 2019). In this sense, PLA is highly versatile and can be processed into various shapes, such as scaffolds, sutures, dressings, screws and pins for bone fixation, devices for controlled drug release, joint cartilage repair, muscle tissue engineering, tracheal replacements, and more. As production scales up, operational costs decrease, making PLA suitable for a wide range of products (Khouri et al. 2024).

For the mentioned properties, PLA has proven to be an effective bio-absorbable polymer for fixation devices like resorbable plates and screws. These devices, widely used in dissolvable suture meshes and orthopedic surgery, offer advantages over titanium and other metal implants by preventing bone erosion and eliminating the need for a second surgery to remove the implant, thus reducing medical costs. Additionally, bio-absorbable polymers support tissue recovery as they degrade, can stimulate tissue regeneration, and deliver drugs such as painkillers, anti-inflammatories, and antibiotics. Once resorbed, these plates also do not interfere with computed tomography scans, simplifying post-surgical imaging (Singhvi et al. 2019).

Chemical Structure and Synthesis of PLA

Poly(lactic acid) (PLA) belongs to the family of aliphatic polyesters commonly made from α-hydroxy acids. In this case, lactic acid (LA), or 2-hydroxypropanoic acid or α-hydroxypropanoic acid, is the only monomer present in their structure. This monomer has two optically active stereoisomers: dextro-(D-) and levo-(L-) (Avinc and Khoddami 2009). Based on this, three types of stereo forms of poly have been produced: poly (L-lactide), poly (D-lactide), and poly (DL-lactide), also known as meso-dilactide (Taib et al. 2023). The optically purity of isomers is important in determining the ultimate process/property relationships achievable in the resultant PLA. LA can be produced either by chemical synthesis by the hydrolysis of lactonitrile or by microbial fermentation routes from renewable resources like maize, sugarcane, potato, and corn. The major disadvantage of chemical synthesis is that this process leads to a racemic mixture, while the fermentative alternative produces optically pure L-or D-LA (Singhvi et al. 2019). The reaction/fermentation product is then purified by different techniques, such as nanofiltration and electrodialysis, ion exchange resin, hybrid short path evaporation, and reactive distillation (Taib et al. 2023). The PLA polymerization is then carried out following mainly two different routes: direct polycondensation of LA and ring-opening polymerization (ROP) of a cyclic dimer of lactic acid (lactide) (Fig. 9.1).

Direct polycondensation is carried out under high vacuum and high temperature. It yields a low to intermediate molecular weight polymer (Mw 10,000–20,000), brittle, glassy polymer with limited applications, because of the viscous polymer melt, the presence of impurities, water, the low concentration of reactive end-groups, and the "back-biting" equilibrium reaction that forms the six-member lactide ring (Avinc and Khoddami 2009; Taib et al. 2023; Garlotta 2001). The ring-opening polymerization (ROP) is the most common route to achieve high molecular weight polymers. The process consists of three steps: polycondensation, depolymerization of oligopolymers to produce LA cyclic dimer (lactide), and ROP of lactide in the presence of a catalyst. Generally, organometallic catalysts are used, including transition metals such as tin, aluminum, lead, zinc, bismuth, ion, and yttrium (Singhvi et al. 2019). The

Fig. 9.1 Polymerization routes to PLA

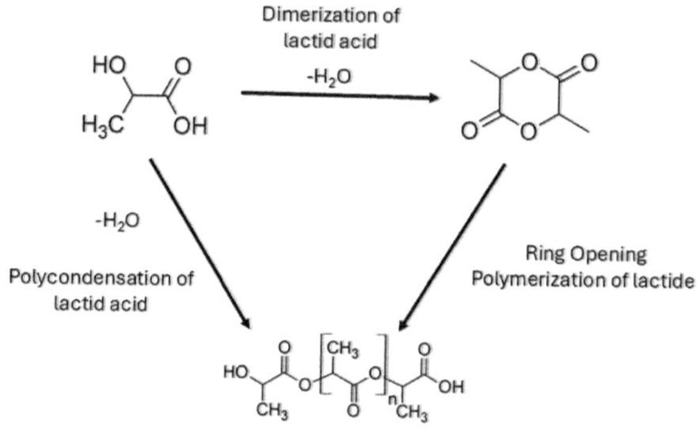

final properties of PLA produced via the lactide route can be modified by controlling the optical sequence of the polymer backbone. Different characteristics of the polymer could be obtained through the controlled production of the optical isomers of the lactic acid via combining the two forms of lactic acid through the three lactide intermediates (L-lactide, D-lactide, and meso-lactide) (Avinc and Khoddami 2009). Selecting L-chirality over D will determine the polymer's crystallinity, mechanical properties, and biodegradability (Silva et al. 2018). Thus, ROP is the most used methodology for PLA polymerization due to the short residence times, mild process conditions, absence of by-products, and high molecular weight product.

Tin (II) 2-ethylhexanoate [Sn(Oct)2] is a commercial catalyst, widely used industrially for the ring-opening polymerization of lactides, and provides high conversion and reaction rates (Stjerndahl et al. 2008). However, the residual amount of initiator present in the materials after the polymerization and its possible toxicity represents a limitation, especially for biomedical applications. The amount of tin in commercially used medical polymers is typically between 20 and 50 ppm, and it is essential that this amount is not exceeded (Stjerndahl et al. 2008). For this reason, much effort has been put in to developing more biocompatible metal-free organic catalysts for the use of PLA materials in biomedical, including the use of cationic or organic catalyst (Gadomska-Gajadhur and Ruśkowski 2020; Lohmeijer et al. 2006), as well as strategies to minimize or remove the remanent catalyst from the final product (Stjerndahl et al. 2008). The world's main producers of PLA are NatureWorks LLC (USA), Corbion® (The Netherland), Futerro (Galactic/Total) (HAD), Nantong JiudingBiologic (China), Shenzhen Bright China (China), Shanhai Tong-Jie-Liang (China), Zhejiang Hisun (China), Purac/Corbion (EU/Thailand), NatureWorks LLC (USA/Thailand), Pyramid Germany Teijin (Japan), among others.

Physicochemical Properties

High molecular weight poly(lactic acid) is a color-less, glossy, stiff thermoplastic polymer. The physical, mechanical, and thermal characteristics of polylactic acid are primarily influenced by its molecular weight distribution and stereochemistry composition. The variations of stereochemistry composition, controlled by the polymerization process of different optical monomers, allow tailoring the physical properties over a broad range (Khouri et al. 2024). The stereochemical composition and their distribution along the polymer backbone significantly influences the crystallization behavior of polylactic acid. Due to their structural regularity, pure PLLA or PDLA are semi-crystalline, while poly (DL-lactide) (PDLLA) is an amorphous polymer. The melting point of PLLA is at 170–200 °C with a T_g between 55 and 65 °C (Farah et al. 2016). Furthermore, PDLLA has T_g about 50–60 °C and has no melting point. A blend of poly(L-lactic acid) and poly(D-lactic acid) can lead to a polymeric stereocomplex with higher melting point than either L- or D-polymers alone (can be as high as 220 °C) (Avinc and Khoddami 2009) and with improved mechanical properties. The resulting crystallinity influences various physicochemical properties including mechanical properties, degradation, and/or solubility (Khouri et al. 2024; Taib et al. 2023). In this sense, the mechanical properties of lactic acid-based polymers can be extensively adjusted, allowing them to range from soft, flexible plastics to rigid, high-strength materials. Semi-crystalline PLA typically exhibits a tensile modulus of around 3 GPa, a tensile strength between 50 and 70 MPa, a flexural modulus of 5 GPa, a flexural strength of 100 MPa, and an elongation at break of approximately 4% (Farah et al. 2016). The amorphous PLA is soluble in most organic solvents such as tetrahydrofuran (THF), chlorinated solvents, benzene, acetonitrile, and dioxane, while crystalline PLA is soluble in chlorinated solvents and benzene at elevated temperatures (DeStefano et al. 2020). The main properties of PLA are shown in Tables 9.1 and 9.2.

Biodegradation

Biodegradability is a key feature of PLA and is an important factor for both the environment and biomedical field, particularly in non-removable or resorbable implants. The degradation of this biopolymer involves different mechanisms, such as hydrolytic, oxidative, thermal, microbial, enzymatic, chemical, and photodegradative processes, each contributing to its breakdown. Among them, hydrolysis plays a fundamental role in PLA degradation, producing the formation of oligomers and monomers of lactic acid by the cleavage of ester bonds (Khouri et al. 2024). LA is a naturally occurring organic acid because it is a metabolic by-product, then in the biomedical applications, it can be excreted from the body by the normal cellular activity through kidney filtration and urine or as carbon dioxide (Silva et al. 2018).

Table 9.1 Main physicochemical properties of PLA

Types of PLA	PLA	PLLA	PDLA	PDLLA
Crystalline structure	Semi-crystalline	Semi-crystalline	Crystalline	Amorphous
Glass transition temperature, T_g (°C)	60–65	50–65	50–65	50–60
Melting temperature, T_m (°C)	150–162	170–200	120–150	–
Density (g/cm^3)	1.21–1.25	1.24–1.30	1.248	1.25–1.27

Table 9.2 Commonly reported mechanical parameter values

	PLLA	PDLLA
Tensile test		
Elastic modulus (GPa)	3.5–4.2	3.6–4.1
Tensile strength (MPa)	47–66	40–44
Strain at break (%)	1.3–7	4.8–7.5
Flexural test		
Flexural modulus (GPa)	3.6–4.2	3.5–3.6
Flexural strength (MPa)	51–119	84–88
Strain at break (%)	1.6–4.7	4.1–4.8
Impact test		
Notched impact strength (J/m^2)	19–66	17–18

The intrinsic properties of polymer, such as the molecular weight, crystallinity, L-lactide content, presence of residual monomers, and oligomers, are the main factors influencing the biodegradation rate of this kind of polymer (Gorrasi and Pantani 2018; Shasteen and Bin Choy 2011). Due to the physical properties mentioned above, PLLA has better chemical stability, better withstands enzyme degradation, and normally a longer resorption time, while PDLA, on the other hand, breaks down inside the body relatively quickly (Karst and Yang 2006). It was found that the hydrolytic degradation rate of PLLA/PDLA blends is lower that of pure PLLA and PDLA monitored by number-average molecular weight (M_n) and sample weight loss (Tsuji and Tsuruno 2010). Also, the shape, porosity and size of specimen, and external factor, including pH and temperature, also affects the degradation reaction (Gorrasi and Pantani 2018). In this sense, the understanding of the dependence of kinetics on pH is of crucial importance in biomedical applications because pH values in the human body can range from strongly acidic (in the stomach, the pH is about 1–2) to almost neutral (in the blood, pH is 7.4), changing not only the rate of degradation, but also the mechanism (Gorrasi and Pantani 2018).

PLA also exhibits characteristically autocatalytic degradation. The carboxylic groups produced during cleavage of ester bond act as a catalyst for hydrolysis during the degradation of the PLA (Shasteen and Bin Choy 2011). Since the generated carboxyl groups are confined within the bulk of the PLA and cannot escape into the surrounding environment, their autocatalytic effect often leads to accelerated degradation within the interior of a PLA component. This causes PLA to typically degrade from the inside out, resulting in hollow and porous structures before it is fully degraded (Shasteen and Bin Choy 2011). Given that PLA degrades completely within the lifespan of the host, with a loss of tensile strength occurring over weeks and full degradation taking place over years (https://www.marketsandmarkets.com/Market-Reports/polylactic-acid-pla-market-29418964.html), managing this degradation rate has become a crucial factor in assessing its suitability for various biomedical applications. Thus, different strategies were developed to control the degradation according to the final application, including copolymerization, blending, incorporation of additives, or irradiation, among others (Shasteen and Bin Choy 2011). On the other hand, biocompatibility is related to the interaction of any medical part or component material with the tissue or living system, taking into account the toxicity and the immunological rejection. Thus, a polymeric implant is regarded as biocompatible if it is able to support a cell–biomaterial interaction and does not show an intolerable inflammatory reaction, it degrades in the implant site in the body to safe components that eventually fully eliminate from the body, without leaving traces, and it is free of microbial agents and pyrogens or traces of toxic chemicals introduced during the manufacturing, storage or transport (Ramot et al. 2016). Furthermore, of their biodegradability mentioned above with non-toxic degradation products, studies have shown that PLA exhibits favorable interactions with tissues, including minimal inflammation and cytotoxicity (Ramot et al. 2016). Nevertheless, possible toxicity can be produced because of accumulation of acidic degradation products, such as lactic acids. This risk increases as the degradation rate and size of implants increases. However, the acidity of lactic acid is in the range of pH4-5, which is generally rapidly neutralized by local body fluids (Ramot et al. 2016). On the other hand, instances of inflammatory issues associated with biodegradable PLA implants polymers are relatively uncommon, with a rate of less than 10% (Ramot et al. 2016).

9.2 Polylactic Acid and Their Blends

Despite PLA's advantages for medical applications, it has several limitations, including its semi-crystalline nature, brittleness, low melt strength, elongation at break below 10%, poor toughness, hydrophobicity, and slow degradation rate. Additional drawbacks such as slow crystallization, matrix cracking, thermal instability, low operating temperature, acidic by-products, and poor cell affinity further restrict its application in the biomedical field (Srivastava et al. 2024). Although its physical and biological properties can be influenced by factors such as stereochemistry, molecular weight, crystallinity, and blending ratios, and as mentioned above, different techniques can be employed for PLA bulk modification including copolymerization, crosslinking, and blending, in order to tailor its properties for specific applications (Srivastava et al. 2024; Saini et al. 2016). Polymer blending is a practical technique for physically modifying materials by combining the properties of different polymers, avoiding developing chemical strategies to produce a new polymer or copolymer, and tailoring a material with the desired properties via a thermodynamically driven mixing of two or more polymers. Furthermore to the tailoring of final properties, blending can also extend the engineering resins' performance by incorporation of less expensive polymers and improve the processability (Blend 2011). Many polymers from biological and petrochemical sources have been explored for PLA blends for use in a variety of applications like biomedical, packaging, textiles, and consumer products (Saini et al. 2016). Particularly, natural polymers like chitosan, collagen, elastin, and hyaluronic acid are preferable for biomedical applications such as tissue engineering, since they simulate the native materials found in biological systems (Saini et al. 2016; Torres-Hernández et al. 2018; Xie et al. 2022), while other polymer are used to enhanced the mechanical behavior such as PHB o TPU (Saini et al. 2016; Gao and Drozdov 2024; Mi et al. 2013) or enhance the biodegradability such as PGA and PEG (Saini et al. 2016; Takayama et al. 2014; Hu et al. 2003).

9.2.1 Stereoisomers Blend

The simpler PLA blend is those produced by blending the enantiomers PDLA and PLLA polymers. The percentage of poly(L-lactide) (PLLA) and poly(D-lactide) (PDLA) in polymer blends affect the crystal structure, melting point, and glass transition of PLA, and then the mechanical properties and the biodegradation rate (Karst and Yang 2006). Recent studies have shown that stereocomplex (SC) crystals, formed by poly(L-lactide) (PLLA) and poly(D-lactide) (PDLA), offer significant potential for modifying PLA materials. While high SC crystal content is typically used to create PLA-based engineering plastics, even a small amount can serve as nucleating agents and reinforcers, greatly enhancing PLA's crystallization rate and melt strength. Modification is usually achieved by adding PDLA to the PLLA matrix, with the performance of PDLA/PLLA blends depending on the characteristics of SC crystals. In asymmetrical PDLA/PLLA materials, the SC crystal formation and material properties are influenced by the PDLA's feed ratio and molecular weight (Yan et al. 2023).

9.2.2 *Polyglycolic Acid (PGA)*

Polyglycolide, or poly(glycolic acid) (PGA), is a biodegradable, thermoplastic polymer and the simplest linear, aliphatic polyester. Its chemical structure is shown in Fig. 9.2. It was first synthesized from petrochemical resources in 1940s and has been extensively studied in biomedical applications (Jem and Tan 2020). It can be prepared starting from glycolic acid by means of polycondensation or ring-opening polymerization. PGA presents good mechanical properties and biocompatibility. The glass transition temperature of PGA is between 35 and 40 °C, and the melting temperature is in the range of 220–230 °C. Because of high stereoregularity, PGA can be quickly crystallized (up to 52%) (Jem and Tan 2020). PGA has high mechanical properties compared with other biodegradable polymers, with a tensile strength about 60–99.7 MPa, a Young's modulus about 7 GPa and elongation at break 1.5–20% (Budak et al. 2020). PGA has a faster degradation rate and higher mechanical properties compared to PLA, mainly due to the higher hydrophobicity by the absence of -CH3 side groups in their polymer chain (Taib et al. 2023; Budak et al. 2020). Thus, the PLA blending with this polymer is a good alternative to control the rate of biodegradation and the final mechanical properties by varying the ratio of PLA/PGA in the blend (Srivastava et al. 2024; Takayama et al. 2014). The wide range of physicomechanical properties can be engineered for various specific applications, such as orthopedics tissue growth implants and fracture fixation devices (Pawar et al. 2014; Fossion 1993), a wound healing dressing (Cantón et al. 2010; Fang et al. 2019), and scaffolds (Choi et al. 1998; Singhal et al. 2007).

9.2.3 *Polycaprolactone (PCL)*

PCL, a biodegradable polyester (Fig. 9.2) like PGA and PLA, is an aliphatic hydrophobic, semi-crystalline polymer with a melting temperature between 59 and 64 °C (just above body temperature) and a glass transition temperature of – 60 °C. At physiological temperatures, PCL becomes rubbery, giving it high toughness and excellent mechanical properties, such as strength and elasticity, which vary based on its molecular weight (Dwivedi et al. 2020). The tensile strength for PCL is around

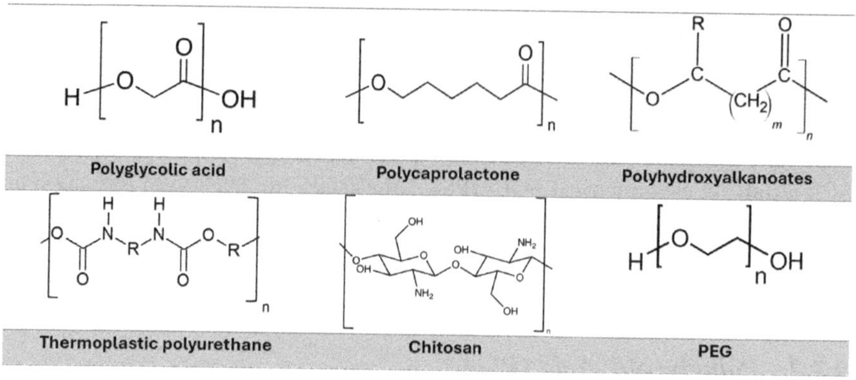

Fig. 9.2 Chemical structure of common polymers used for PLA blending

4–785 Mpa, Young modulus between 210 and 440 Mpa and elongation at break range is 20–1000% (Labet and Thielemans 2009). In the same way that PLA and PGA, the two main pathways to produce PCL are the polycondensation of a hydroxycarboxylic acid: 6-hydroxyhexanoic acid, and the ring-opening polymerization (ROP) of a lactone: ε-caprolactone (ε-CL), using different catalytic systems that were used (enzymatic, organic, and metal catalyst systems) (Labet and Thielemans 2009). This polymer is non-toxic and tissue-compatible. It has a slow degradation time (2–3 years), breaking down through microbial activity or hydrolysis of its aliphatic ester bonds under physiological conditions. Due to its five hydrophobic –CH2 groups in each repeating unit, PCL degrades more slowly than other polyesters, such as PGA and PLLA (Dwivedi et al. 2020). Thus, the blend with PLA is a good alternative to modulate the hydrophilicity and degradability behavior, especially for long-term applications, modifying also the mechanical and thermal properties of final product (Patrício et al. 2014, 2013; Chen et al. 2014; Patrício and Bártolo 2013; Yao et al. 2017).

9.2.4 Polyhydroxyalkanoatos (PHA)

PHA (polyhydroxyalkanoate) is a family of naturally occurring biopolyesters (Fig. 9.2) synthesized by various microorganisms, discovered by Lemogine in 1926. A PHA molecule typically consists of 600–35,000 (R)-hydroxy fatty acid monomer units containing a side chain, usually a saturated alkyl group, though it can also be unsaturated, branched, or substituted. About 150 different PHA monomers have been identified, and new types of PHA can be developed through the chemical or physical modification of naturally occurring PHA, or through the creation of genetically modified organisms (Tan et al. 2014). Thus, a diverse PHA properties can be tailored for various applications ranging from biodegradable packaging materials to medical products (Tan et al. 2014). Furthermore to its biodegradability, biocompatibility, chemical diversity, and production from renewable carbon sources, they can be considered as active compound and currently investigated as potential anti-HIV drugs, anticancer drugs, antibiotics, etc., increasing their applicability in the biomedical field (Tan et al. 2014). PHAs are usually produced by recombinant Escherichia coli for commercial use (Patrício and Bártolo 2013) but can be produced by a number of other microorganisms (i.e., Aeromonas, Azotobacter, Cupriavidus, Clostridium, Methylobacterium, Ralstonia, Pseudomonas, Syntrophomonas, etc.) (Meereboer et al. 2020). The synthesis of PHAs is important in biodegradation as the metabolic pathways are related to bio-assimilation. In general, poly (hydroxybutyrate) (PHB) is one of the most commonly observed forms of PHAs, as it is composed of packed monomers of R-3-hydroxybutyrate (R-3-HB) (Adnan et al. 2022). This polymer exhibits properties that are similar to several synthetic thermoplastics, including polypropylene. However, the high production costs of PHB in comparison with plastics derived from petrochemicals are one of the major problems for the extensive production and commercialization of this product. PHB is known for its excellent mechanical properties, with a tensile strength ranging from 20 to 40 MPa and a Young's modulus between 3.0 and 3.5 GPa (Zainuddin et al. 2023). It has a melting temperature in the range of 160–180 °C (Tang et al. 2014), and T_g around 5 °C (D'Amico et al. 2016). The hydrophobic behavior makes it more stable for reabsorption by the human organism, in comparison with PLA, being usually applied in long-term biomedical applications (Dominguez-candela et al. 2024). Thus, the blending of PHB with PLA is generally used with the aim of modified mechanical properties as well as biodegradation behavior (Gao and Drozdov 2024; Zhang and Thomas 2011; Abdelwahab et al. 2012).

9.2.5 *Thermoplastic Polyurethane (TPU)*

Thermoplastic polyurethane (TPU) as a class of PU has been widely employed in medical applications due to its biocompatibility and flexibility, and has been used in medical devices such as catheters, pacemaker leads, skin, and vascular grafts (Mi et al. 2013). Polyurethanes (PUs) date back to the late 1930s, when Otto Bayer and his team discovered their synthetic route using 1,6-diisocyanatohexane and 1,4-butanediol (BDO). Today, thermoplastic polyurethanes (TPUs) are one of the most significant applications of PUs. TPUs are synthesized from polyols and isocyanates (Fig. 9.2), resulting in alternating soft polyol segments and hard isocyanate segments. The soft segments form a flexible, non-crystalline matrix due to their low glass transition temperature (T_g), while the hard segments, with a high T_g or melting temperature (T_m), provide stiffness by acting as physical crosslinkers within the matrix (Al Nakib et al. 2022). TPUs are traditionally synthesized through various methods using different media, such as solvent-free, solution-based, or water-based approaches, depending on how the reactants are introduced (one-shot or pre-polymer process). In the pre-polymer process, polyols and isocyanates first react to form an isocyanate-terminated pre-polymer, followed by chain extension with a chain extender. Many

TPU elastomers are produced using NCO pre-polymer intermediates, enabling the full reaction of low-reactivity polyether diols without catalysts and allowing for the deliberate formation of segmented structures (Jung et al. 2022). Tertiary amines and organometallic compounds, particularly those containing Sn, Pb, and Fe, are usually used as catalyst in the reaction between isocyanates and compounds with unstable hydrogen atoms, such as alcohols, amines, water, carboxylic acids, and malonates (Król 2007). TPU attracts significant interest because they possess wide useful properties, including modulable flexibility, mechanical strength, elasticity, good abrasion resistance, and transparency. This material can present properties from those typical soft elastomers to hard plastics according to the type of polyol, which was during the synthesis of TPU. The melting temperature can vary from 70 to 220 °C, it T_g from − 48 to − 5 °C, while the ranges for mechanical properties are: Tensile strength 0.785–195 MPa, Young modulus 0.00500–1.90 GPa, and elongation at break 10.0–1100% (https://www.matweb.com/search/datasheet.aspx?matguid=1932586b674346e2a9d5cb4c7462dd33). Thus, combining PLA and TPU in various ratios allows for tailoring properties to meet the specific requirements of diverse tissue engineering scaffold applications, such as cartilage and bone healing and regeneration (Mi et al. 2013; Song et al. 2014; Hong et al. 2016; Dogan et al. 2017). TPUs are typically non-degradable, but they can become vulnerable to hydrolytic and oxidative degradation when exposed to vacuum conditions. This susceptibility poses challenges for long-term biomedical implants, leading to efforts in designing biodegradable polyurethanes for such applications (Allami et al. 2021).

9.2.6 Chitosan

Chitin and its deacetylated derivative, chitosan, are linear polysaccharides composed of varying amounts of (β1 → 4) linked residues of *N*-acetyl-2-amino-2-deoxy-D-glucose (glucosamine, GlcN) and 2-amino-2-deoxy-D-glucose (*N*-acetyl-glucosamine, GlcNAc). (Fig. 9.2). Chitin is one of the most abundant biopolymers, found in the exoskeleton of crustaceans, insect cuticles, algae, and fungal cell walls, while chitosan, less common in nature, is found in certain fungi (Mucoraceae). Traditionally, commercial chitosan has been produced by chemically deacetylating chitin from crustacean sources (Aranaz et al. 2021). In addition to chemical methods, biological approaches, such as enzymatic and fermentation methods, are also used to produce chitosan from crustacean by-products (Kou et al. 2021). However, chemical methods offer several advantages over biological approaches for chitosan production, as they involve simpler, faster processes and result in products with stronger bioactivity due to lower molecular weight and higher degrees of deacetylation (Kou et al. 2021). Chitosan is the only polycation in nature and its charge density depends on the degree of acetylation and pH of the media. The solubility of the polymer depends on the acetylation degree and molecular weight. Chitosan oligomers are soluble over a wide pH range, from acidic to basic ones (i.e., physiological pH 7.4). On the contrary, chitosan samples with higher Mw are only soluble in acidic aqueous media even at high deacetylation degrees. This lack of solubility at neutral and basic pH has hindered the use of chitosan in some applications under neutral physiological conditions (i.e., pH 7.4) (Aranaz et al. 2021). Chitin, chitosan, oligosaccharides, and derivatives exert many biological activities including antitumoral, antimicrobial, antioxidant, and anti-inflammatory activities, which could be used as therapeutic polymers (Aranaz et al. 2021). Besides these properties, chitosan is widely applied in biomedical due to its excellent oxygen barrier properties, antimicrobial activity, biodegradability, and non-toxicity properties (Srivastava et al. 2024). However, chitosan scaffolds, films, and beads demonstrate limited mechanical stability (Jung et al. 2018). These fragile scaffolds, designed for tissue regeneration, are prone to breaking, which could lead to complications for patients. Also, their lack of strength also hinders the formation of a cohesive matrix necessary for effective cell delivery. As drug delivery systems, the weak CS scaffolds may cause rapid drug release instead of the intended sustained release. The brittle nature of chitosan films and membranes makes them also unsuitable for handling in wound dressing applications (Jung et al. 2018). For this reason, the use of additives, or the blending of chitosan with natural polymers or other synthetics, like PLA, is necessary to enhance the mechanical performance (Hui et al. 2023; Olaiya et al. 2020).

Compared with PLA polymers, in vitro degradation of chitosan is much faster. Thus, the PLA/chitosan biocomposite can be mainly designed for non-loaded bone regeneration and sustain constant drug release (Srivastava et al. 2024). The interaction between PLA and Chitosan is due to hydrogen bonds that improve wettability and reduce the acid inflammation formed during the de-esterification of the chain. Additionally, it was demonstrated that chitosan encourages cell adhesion, proliferation, and metabolic activity (Torres-Hernández et al. 2018). Also, chitosan has been reported to neutralize the acidic effect created by the degradation of PLA and improve the ability of PLA to retain mechanical strength during degradation (Olaiya et al. 2020).

Chitosan presents a glass transition around 140–150 °C (Dong et al. 2004). This high value is inconvenient for its processing or blending by melting. In the same way, in solvent casting, there is no suitable co-solvent that can dissolve both chitosan and PLA polymers (Srivastava et al. 2024).

9.2.7 Polyethylene Glycol (PEG)

Polyethylene glycol (PEG), also called polyethylene oxide (PEO) or poly(oxyethylene) (POE), is a synthetic polyether (Fig. 9.2) known for being hydrophilic and biocompatible. Generally, polymers with a molecular weight below 20,000 g/mol are classified as PEGs, while those with a molecular weight above 20,000 g/mol are referred to as PEOs. PEG can be synthesized through the anionic polymerization of ethylene oxide using hydroxyl initiators, such as water, ethylene glycol, or other diols. It can also be produced via ring-opening polymerization of epoxyethane. Linear PEG has only two functional groups, which limits its potential for further derivatization with other residues or targeting ligands. To address this, commercial PEGs are available with varying degrees of polymerization and activated functional groups (D'souza and Shegokar 2016). PEG with molecular weight in the range of 100–700 are liquids at room temperature, those between 1000 and 2000 are soft solids, and PEGs with MW > 2000 are hard crystalline solids with melting points of around 63 °C (D'souza and Shegokar 2016). The glass transition temperature (T_g) of PEG varies from − 60 to − 75 °C, depending upon the molecular weight (Sheth et al. 2008). For instance, glass transition temperature for PEG 4000, 6000, and 20,000 is − 34.6 °C, − 22.7 °C, and − 22.37 °C with melting point of 67.7 °C, 58.7 °C, and 51.6 °C, respectively (D'souza and Shegokar 2016). The molecular weight of PEG also influences the mechanical properties of these polymers. For instance, the tensile strength decreases from around 2.50 to 1.24 MPa, when the molecular weight increases from 4000 to 35,000, while the young modulus decreases from around 2.42 to 0.98 GPa.

The high polarity of PEG significantly enhances its hydrophilicity, making it highly water-soluble—an exceptional characteristic compared to structurally similar polymers. PEG also exhibits excellent solubility in a wide range of organic solvents such as ethanol, acetonitrile, toluene, acetone, dichloromethane, hexane, and chloroform (D'souza and Shegokar 2016). The blending of PEG with PLA improves the PLA processing, tailoring the both surface and structural property, also modulating the degradation rate (Kumar et al. 2023; Serra et al. 2014; Kossentini Kallel et al. 2015).

Other Polymers

Additionally to the mentioned polymers, other macromolecules were used for PLA blending, such as Hyaluronic acid (HA) (Wu and Liao 2005), Poly(butylene adipate-coterephthalate) PBAT (Kumar et al. 2010; Al-Itry et al. 2014), Poly(vinyl alcohol) (PVA) (Chuaponpat et al. 2020; Restrepo et al. 2018), Polybutylene succinate (PBS) (Kun et al. 2012; Álvarez et al. 2020), among others.

9.3 Preparation of Polylactic Acid-Based Blends

Different methods can be used to prepare polymer blends. The selection of the most suitable method to prepare PLA blends depends on the physicochemical properties of polymers, their miscibility and the physical structure and mechanical properties of the end product. Most polymer pairs are immiscible, and therefore, their blends are not formed spontaneously. Moreover, the phase structure of polymer blends is not equilibrium and depends on the process of their preparation.

Melt Blending

Melt mixing is the most widespread method of polymer blend preparation in practice. The blend components are mixed in the molten state within a heated mixer, such as internal blender, single-screw extruder, and twin-screw extruder. The blending process using a twin-screw extruder is significantly more efficient, offering reduced blending times and increased productivity. This makes it an ideal choice for industrial applications, delivering superior performance and adaptability (Hamad et al. 2018). Therefore, melt blending is simple, cost-effective processing and available at an industrial level and allows for obtaining materials with the desired performance by changing the blend composition. In order to achieve a good polymer blending, avoiding degradation, the different processing parameters, such as temperature profile, screw speed, and mixing time, should be optimized during the extrusion procedure (Hamad et al. 2018). This technique is widely used to process blends of PLA with other thermoplastic polymers like PCL (Navarro-Baena et al. 2016; Fernández-Tena et al. 2024), PHB (Li et al. 2023; Vanovcanova et al. 2016), TPU (Mi et al. 2013; Cao et al. 2024), among others.

Solution Blending

In this technique, the polymers are dissolved in a common solvent and intensively stirred. The blend is separated by precipitation or evaporation of the solvent to form the end product, followed by evaporation of the solvent. The phase structure formed in the process is a function of blend composition, interaction parameters of the blend components, type of the solvent, and history of its separation (Mazumder et al. 2018). Polymer blends prepared by dissolution in solvent can be cast into films and porous materials, spun (wet, dry, or electrospun) and subjected to particulating techniques (e.g., emulsion–diffusion–evaporation, phase separation, spray-drying) to form particles of different size profiles. Due to the solvated state of polymers and the presence of large amounts of solvent, solution blending is not an applicable method for compression molding (Saini et al. 2016). The process offers advantages such as rapid mixing with low energy consumption and the ability to avoid undesirable chemical reactions. However, it is limited by the need to find a common solvent for the blend components and the challenge of removing large quantities of organic solvents, which may be expensive, toxic, or affect the bulk properties of the blend. As a result, solution blending is primarily used for polymer blend preparation on a laboratory scale. In industry, it is typically employed for producing thin membranes, surface layers, and paints (Mazumder et al. 2018). Furthermore, this alternative is useful to blend PLA with some natural polymers in order to avoid degradation and inter-chain chemical reactions that can occur during the melt blending of PLA, particularly for biomedical applications (Hamad et al. 2018). Different solvents are commonly used during the preparation of PLA blends via the solution blending method, including chloroform, dichloromethane (Vogel et al. 2009), dimethylformamide (Zhao et al. 2020), tetrahydrofuran (Peesan et al. 2006), and 1,4-dioxan (Meereboer et al. 2020). The type of solvent can play a crucial role in determining the final morphology of the PLA blend prepared via this method, as it was demonstrated for PLA/chitosan blends (Peesan et al. 2006). The as-spun fibers from H-chitosan/PLA blend solutions in chloroform, with H-chitosan content up to 50% w/w, were continuous and free of beads. In contrast, fibers spun from blend solutions in dichloromethane exhibited beaded structures, or in some cases, consisted entirely of beads at higher H-chitosan content (Peesan et al. 2006).

Mill Mixing and Fine Powder Mixing Technique

Milling and grinding serve as effective alternatives for blending components in various industrial applications. This method involves first grinding raw materials into a fine powder, followed by thorough mixing to ensure a uniform distribution at the micro-level. Achieving this uniformity is crucial for the quality of the final product (Mazumder et al. 2018). Once the powder mixture is prepared, it undergoes further processing, such as extrusion, to create the desired polymer blend. A variety of milling devices and grinders can be utilized for this process, with the Bunbury mixer being one of the most widely used pieces of equipment for the mechanical mixing of polymer ingredients. Its design is particularly well-suited for synthesizing high-quality polymer blends, ensuring consistent performance and optimal results (Mazumder et al. 2018). As example, this technique was applied in the preparation of PHA/PLA blends (Degradable 2014; Relinque et al. 2019; Astner et al. 2023).

Interpenetrating Polymer Network Technology

Interpenetrating polymer networks (IPNs) are a unique polymer blend formed by the interpenetration and entanglement of two polymers through the network (Yang et al. 2022), where at least one of which is synthesized and /or crosslinked in the immediate presence of the other(s). Thus, there are covalent bonds between the polymers within each separate network but not between the polymers in different networks. Interpenetrating polymer networks (IPNs) can be classified into three main types: (i) traditional IPNs, where two polymer networks are intertwined and interact with each other; (ii) homo-IPNs, which consist of two polymers with identical structures, each forming its own distinct network; and (iii) semi-IPNs, where one component is a crosslinked polymer network, and the other consists of linear polymer chains distributed within the network (Yang et al. 2022). Pseudo-interpenetrating polymer networks (PIPNs) are created through the partial penetration of one polymer into a network system. As a result, the polymer and network remain more distinct, preserving their individual properties to a greater extent. Additionally, the separation of the components is easier compared to semi-IPNs (Mazumder et al. 2018). An IPN can be distinguished from polymer blends, blocks, or grafts in two ways because an IPN swells, but does not dissolve in solvents, and creep and flow are suppressed (Blend 2011).

These systems are designed for the delivery of drugs at a predetermined rate and thus helps in controlled drug delivery. For instance, PLA blends with sodium alginate (Prabhakar et al. 2013) or PEG (Brown et al. 2005). Also, they can be used for scaffolds, as in the case of PCL/PLA semi-interpenetrating networks (Woodard et al. 2017). The selection of preparation technique will determine the final properties of the blend or end-use product. For instance, PCL/PLA scaffolds produced from solvent-cast blends demonstrate smaller pore size, improved mechanical properties, and enhanced cell adhesion compared to those produced by melt blending, even under identical processing conditions and geometry (Patrício et al. 2014). Similarly, chitosan/PLA films prepared by melt blending followed by hot-press presented higher secant modulus and the ultimate stress, lower water adsorption and different crystalline behavior in comparison with casted films (Cheerarot and Saikrasun 2023).

Compatibilization

Blending is a cost-effective and efficient method for creating multifunctional, high-performance materials; however, polymer blends often suffer from coarse phase morphology and poor interfacial adhesion due to the immiscibility of most polymer pairs (Fredi and Dorigato 2024). Thus, the microstructure and properties of polymer blends are highly dependent on the compatibility between the polymers. Mainly depending on the chemical structure of polymers and their interactions, polymer blends can be classified as: Homogeneous/miscible blend when the constituents exist in a single-phase after blending; heterogeneous/immiscible blend when the constituents exist as separate phases after processing; and compatible polymer blend, an immiscible polymer blend with strong interactions between the constituents, resulting in macroscopically uniform properties (Simunec et al. 2023). A completely miscible blend displays a homogeneous morphology with a single glass transition temperature (T_g), which falls between the T_g values of the two components and varies with composition. In a partially miscible blend, two homogeneous phases coexist, with each phase containing some amount of the other polymer dissolved within it. This results in two T_g values, each shifting from the T_g of one component toward that of the other. A fully immiscible blend, on the other hand, exhibits macrophase separation with a coarse interface boundary, large dispersed phase size, poor phase adhesion, and two distinct T_g values that remain unchanged by the blend composition. https://doi.org/10.1039/c5ra01655j. For instance, PCL, PVA, and PHAs, usually form immiscible blends with PLA, while PCC, PBS or PPC present low compatibility (Srivastava et al. 2024).

In immiscible polymer blends, material properties like mechanical strength are heavily influenced by the interaction between the phases. Due to poor thermodynamic phase stability, these blends tend to undergo phase separation, where the less abundant polymer forms coarse domains within the more abundant matrix. Weak interfacial adhesion between immiscible polymers often undermines the blend's mechanical properties. To improve strength, microstructural manipulation is required, either by reducing the size of the dispersed polymer or by enhancing adhesion between the phases. This can be achieved through compatibilization strategies that promote chemical or physical bonding between the phases. Non-reactive compatibilization uses graft or block copolymers, which localize at the interface and stabilize the blend, while reactive compatibilization generates these copolymers in situ, improving phase adhesion and reducing coalescence of the dispersed domains (Simunec et al. 2023). As example, the diblock copolymer PLA-b-PEO of "A–C" type was employed by Maglio et al. to compatibilize PLA/PCL blend and it exhibited similar behavior to PLA-b-PCL-b-PLA triblock copolymer (Maglio et al. 2004). In the same way, block copolymer of poly(ethylene glycol) and poly(propylene glycol) was used also to enhance the compatibility between PLA and PCL (Chavalitpanya and Phattanarudee 2013), demonstrating that PCL phase was better distributed and more miscible in the PLA matrix enhancing also the final mechanical properties. Also, Poly(ε-caprolactone) (PCL) and poly(lactic acid) (PLA) blends were compatibilized by reactive blending and by copolymers formed during reaction in the solution, using di-(2-tert-butyl-peroxyisopropyl)benzene (BIB) or dicumyl peroxide (DCP) as radical initiator (Przybysz-Romatowska et al. 2020). In the same way, maleic anhydride (MA) was used as reactive compatibilizer, bridging among the polymers in the blends, as PHB and PLA. MA acted like a bridging unit between PLA and PHB through covalent bonds by a transesterification reaction. The enhanced interaction among biopolymers led to improved miscibility, resulting in a transition from brittle fracture to ductile fracture following reactive blending (Jandas et al. 2014). Thus, depending on the chemical structure of polymers, different alternatives of reactive modifications for PLA blends can be found in the literature (Zeng et al. 2015).

Additionally, other strategies were explored in the literature to enhance the compatibility in PLA blends (Zeng et al. 2015). Adding micro- and nanoparticles to an immiscible blend offers an alternative approach to enhancing compatibility (Fredi and Dorigato 2024; Zeng et al. 2015). Micro–nanoparticles typically position themselves at the interface between the blend components, serving as interfacial modifiers that strengthen adhesion and improve the overall performance of the blend. They help prevent the coalescence of finely dispersed phase particles during melt processing, thereby stabilizing the morphology and preserving the blend's properties. Unlike polymer compatibilizers, nanoparticles are not specific to the nature of the immiscible components and can be easily incorporated through blending. Thus, Hoidy et al. (2010) compatibilized PLA/PCL blends using organoclay (OMMT) as a filler. Their findings revealed that OMMT not only improved the dispersion and interfacial adhesion within the polymer matrix but also enhanced the mechanical properties and thermal stability of the PLA/PCL blends. In the same way, organoclay (Cloisite® 30B) was used to compatibilize PLA blend of poly(butylene succinate) (PBS) (Risse et al. 2014) and poly[(butylene adipate)-co-terephthalate] (PBAT) (Nofar et al. 2016). The blending parameters also play a crucial role in polymers interactions and blend miscibility. In solvent blending, it has been demonstrated that the choice of solvent and composition significantly affects the miscibility and compatibility of the polymers (Marwat and Baloch 2015). In the same way, the optimization of melt-mixing parameters like temperature, shear rate, and time offers could synergistically complement the effect of chemical or physical compatibilization. During mixing, shear forces fragment the dispersed phase into smaller domains, with the ultimate size influenced by viscosity ratio, elasticity, and interfacial interactions. Increased shear and stronger interfacial adhesion yield finer dispersions, improving overall properties (Fredi and Dorigato 2024).

9.4 Characterization of Polylactic Acid-Based Blends

The performance and properties of PLA and their blends could not be easily analyzed by distinguishing or isolating every involved parameter or characteristic. Superposition and interdependence of characteristics and features have been largely reported (Khouri et al. 2024; Niaounakis 2014). These factors could be grouped as: (i) PLA characteristics, (ii) incorporated material, (iii) processing, and (iv) product features.

(i) *Polymer Characteristic*: Chemical composition and structure, stereochemistry, molecular weight, polymer chain orientation, branching, crystalline structure and phases, crystallinity degree, and purity are important factors influencing the final properties (Khouri et al. 2024; Niaounakis 2014; Sha et al. 2016; Murariu and Dubois 2016). PLA displays some relevant features as high tensile strength and elastic modulus, biocompatibility, biodegradability, processability, and gas barrier capabilities (Singhvi et al. 2019; Murariu and Dubois 2016; Hussain et al. 2024; Puthumana et al. 2020; Kubisa et al. 2021); however, some drawbacks largely reported are the brittle performance, hydrophobicity, and low degradation rate (Singhvi et al. 2019; Saini et al. 2016; Murariu and Dubois 2016; Hussain et al. 2024; Puthumana et al. 2020; Kubisa et al. 2021; Bayer 2017). As mentioned above, the presence of L and D isomers influences the thermal, mechanical, and degradation performance (Murariu and Dubois 2016). PLLA is used for higher mechanical requirements along degradation time (Murariu and Dubois 2016). PDLA (content higher to 10 wt% of D isomer) tends to be faster degraded due to the amorphous state (Singhvi et al. 2019; Sha et al. 2016; França et al. 2022), while stereocomplex PLA enhances the thermal and hydrolysis resistance (Murariu and Dubois 2016; Bayer 2017).

(ii) *Incorporated Material*: Blending, grafting, isomers, oligomers, catalysts, plasticizers, stabilizers, pigments, antioxidants, antistatic agents, chain extenders, and crosslinking agents also determine the final properties of the product (Khouri et al. 2024; Niaounakis 2014; Deng et al. 2016; Hamad et al. 2015; Ghalia and Dahman 2017). As mentioned above, polymers commonly blended with neat PLA could be: PCL, TPU, PBT, PHBV, PBAT, TPS, BF, PEG, PHB, PBS, chitosan, collagen, elastin, and hyaluronic acid (Saini et al. 2016; Ebrahimi and Ramezani Dana 2022). Compounding looks able to modify the crystalline structure, mechanical behavior, thermal performance, hydrophilicity, and degradation rate (Saini et al. 2016; Sha et al. 2016; Ebrahimi and Ramezani Dana 2022; Mokarizadeh Haghighi Shirazi et al. 2020). The blend morphology, phase dispersion, and processing conditions directly influence on the mechanical properties, thermal features, and degradation (Saini et al. 2016). Plasticizers with low molecular weight have been largely used (sorbitol, glycerol, glucose monoesters, PEG, and partial fatty acid esters) (Sha et al. 2016). Furthermore, the plasticizing effect, PEG content of 5 wt% (hydrophilic polymer), has been used to enhance the hydrophobicity of PLA (Khouri et al. 2024; Ebrahimi and Ramezani Dana 2022; Sin et al. 2012; Oh 2011; Tyler et al. 2016). In the similar way, lactide has been employed as plasticizer, but it could easily migrate to the surface becoming sticky (Sin et al. 2012).

(iii) *Processing*: Methods (extrusion, injection molding, internal mixing, blown film extrusion, injection blow molding, thermoforming, multilayer lamination, casting, additive manufacturing, spinning), parameters (time, temperature, speed, shear forces), post-processing, heat treatment (annealing, quenching), thermal history, coatings, and surface modifications (plasma treatment, attachment of functional groups, photografting) are some factors influencing the final properties (Taib et al. 2023; Niaounakis 2014; Sin et al. 2013b; Hurlburt 2015; Li et al. 2015). Extrusion of PLA requires a low shear processing condition (length/diameter ratio: 24–40) due to its structure and thermal stability in order to prevent the thermo-mechanical degradation (Hamad et al. 2015). Injection molding conditions could be optimized by using the pressure–volume–temperature diagram (PVT diagram) (Hamad et al. 2015). The injection conditions (temperature and pressure) are close related to the product volume and also to the chemical structure, molecular weight, morphology, among other parameters (Hamad et al. 2015). The low crystallization rate of PLA requires a long injection molding cycle (Murariu and Dubois 2016), when this property is desirable. PLLA/ PGA used as barrier materials should be obtained processed under elongation efforts inducing oriented films or under multilayer lamination conditions (Niaounakis 2014).

(iv) *Product Features*: Simple dimensions, geometry, threaded surfaces, thickness variations, porosity, internal structure, superficial condition (roughness, morphology, functional groups), and sterilization process. The in-service performance of PLA-based biomedical products could be assumed as a combination of the others group of involved factors (Niaounakis 2014; Puthumana et al. 2020). In this sense, PLA blends looks more appropriated for 3D printing scaffolds by tailoring the degradation rate, mechanical properties, and also manufacturing hierarchically porous internal structures favorable for cell differentiation (França et al. 2022).

9.4.1 Physicochemical Characterization

The structural characterization as well as the determination of the kind of polymer present in a blend, and its quantification, is frequently done by spectroscopic techniques. In this sense, Fourier transform infrared spectroscopy (FT-IR) gives information about the functional groups and also provides physical information related to physical interactions between the polymers or the phase segregation (Mazumder et al. 2018). Additionally, nuclear magnetic resonance spectroscopy (NMR) can also be used to determine the structural properties of polymer blends, while mass spectroscopy is applied to obtain information about the molecular weight and surface concentration of the polymer blends (Mazumder et al. 2018). Density is also a property investigated and is an indicative to the crystalline or amorphous state of PLA and its blends (Lasprilla et al. 2012). For instance, crystalline PLA presents a density 1.3 g/cm^3, while amorphous state has a density around 1.25 g/cm^3 (Lasprilla et al. 2012). Semi-crystalline version displays slightly higher density values (PLLA: 1.36 g/cm^3, meso-lactide PLA: 1.33 g/cm^3) affected by polymer chain orientation, thermal history, purity, composition, among other factors. Solubility is affected by thermal history, crystallinity, polarity, purity, and chemical composition (DeStefano et al. 2020; Hussain et al. 2024; Ebrahimi and Ramezani Dana 2022; Lasprilla et al. 2012). Amorphous PLA could be easier dissolved in organic solvents compared to crystalline or stereocomplex ones. PLA is soluble in dioxane, acetonitrile, chloroform, methylene chloride, and dichloroacetic acid (DeStefano et al. 2020; Ebrahimi and Ramezani Dana 2022; Lasprilla et al. 2012). Melted PLA is also soluble with ethyl benzene, toluene, acetone, and tetrahydrofuran (DeStefano et al. 2020; Ebrahimi and Ramezani Dana 2022; Lasprilla et al. 2012). PLA is not soluble in water, alcohols, methanol, ethanol, propylene glycol, or linear hydrocarbons (DeStefano et al. 2020; Hussain et al. 2024; Ebrahimi and Ramezani Dana 2022). Crystalline PLA is not soluble in acetone, ethyl acetate, or tetrahydrofuran (Ebrahimi and Ramezani Dana 2022; Lasprilla et al. 2012). The two different chiral configurations of PLA (L-lactic and D-lactic) allow variations of this biopolymer (DeStefano et al. 2020). The D-lactic cannot be assimilated by the human body due to the lack of appropriate enzymes, while the non-toxic L-lactic exists in the human body (Sin et al. 2012). PLA, as most hydrophobic aliphatic polyesters, could be electrically charged by friction facilitating the adhesion of dust and dirt. Fibers and films could be charged along production inducing electrical discharges and unstable processing conditions (sticking) (Niaounakis 2014). The permeability of PLA and its blends also is an usual characterization, particularly for packaging applications, including medical packaging uses. Different methods are available to determine the gas transport properties, including static permeation with pure penetrants, static or dynamic methods with concentration analysis, or dynamic permeation with specific sensor (Giacinti Baschetti and Minelli 2020). In general, permeability of PLA looks higher to PET but lower to PS (Hamad et al. 2015). Crystallinity and also annealing, related to recrystallization process, promote the barrier performance (Hamad et al. 2015). Injection molded PLA tends to exhibit restricted oxygen or carbon dioxide barrier performance (Niaounakis 2014). The blending and the resulting barrier properties will depend on polymer nature and the interactions between them, as well as of other final properties such as crystallinity. For instance, the incorporation of chitosan in a PLA matrix results in an increase of water vapor permeability (WVP) and water vapor transmission rate (WVTR), attributed to the hydrophilicity of chitosan (Shan Teo and Shyang Chow 2014). In the same of blend, the incorporation of PVOH improved compatibility and interfacial bonding of PLA/chitosan blends, resulting in a lower WVP and WVTR (Shan Teo and Shyang Chow 2014). On the other hand, the gas permeation (O_2, N_2, and CO_2) as well as their diffusivity decreased with the increase of PHB ratio, mainly because of the crystallinity growth in the blends (Siracusa et al. 2020).

The solubility parameter (δ) and interaction parameter (χ) are used to evaluate the compatibility between PLA and added polymers or plasticizers (Sin et al. 2012). For χ around 0.5, the mixture is miscible and no phase separation should be expected (Sin et al. 2012). This feature is close related to the thermal behavior of blends (Saini et al. 2016). Contact angle provides information about the hydrophobic properties of PLA and its blends and the wettability of the surfaces of the final products. For instance, the water contact angle was gradually reduced by increasing the concentration of PEG in the PLA/PEG blend, because of the high hydrophilicity of the PEG due to the presence of hydrophilic hydroxyl groups in its structure (Kumar et al. 2023). The wettability and water sorption is a key factor in degradation behavior.

9.4.2 Rheological Characterization

The flow characteristics of polymers are important for molten processing, such as extrusion or injection molding. Rheological tests can be used to characterize a variety of polymer properties, including viscosity of the melt, flow behavior, viscoelastic properties, temperature-dependent behavior, glass transition temperature, and aging behavior. The rheological features of PLA-based blends have been related to the properties of the components, molecular weight, branching, crosslinking, D isomer content, dispersed phase size, ratio of polymers viscosity, interfacial interactions, compatibilizer incorporation, crystalline structure, morphology stability, degradation, and processing conditions (temperature, residence time) (Nofar et al. 2016, 2019; Sin et al. 2012). Melted PLA normally display two distinct regions: Newtonian behavior at low shear rates followed by shear thinning at higher frequencies (Hamad et al. 2015; Sin et al. 2012). This feature depends on the molecular weight, low values (40,000 Dalton) display Newtonian performance with shear rate, while high ones (100,000–300,000 Dalton) shear thinning as a consequence of the gradual disentanglement of polymer chains (Nofar et al. 2019). Viscosity of PLA also displays shear thinning with temperature (Sin et al. 2012). Crystalline PLA exhibits larger shear viscosity than the amorphous one due to crystal lamellae are organized patterns with strong large resistance to flow (Sin et al. 2012). Higher temperatures tend to diminish this difference by higher vibrational movements of molecules allowing weaker polymer chain connections (Sin et al. 2012). Degradation of PLA along processing could be monitored by rheological measurements. Transient dynamic time sweep rheological tests displayed viscosity reductions related to hydrolysis reaction and thermal degradation (Nofar et al. 2019). Branched PLA, compared to linear one, showed higher viscosity at low frequencies, while the opposite trend was observed into the non-Newtonian region (Hamad et al. 2015).

As mentioned above, PLA blends could be miscible or immiscible, for the first ones both polymers are compatible with interfacial tension close to zero, while for the second ones, an interphase and interfacial interactions are present (Nofar et al. 2019). For content lower than 10 wt%, it has been reported a quite independent dynamic melt viscoelastic performance to the blending composition, morphology, and interfacial interactions (Nofar et al. 2019). This feature has been observed for immiscible blends of PLA with PBAT, PBSA, PCL, PHBV, or starch (Nofar et al. 2019). PLA/starch blends displayed droplet-like, rod-like, or co-continuous morphologies with highly similar dynamic viscoelastic performance (Nofar et al. 2019). Blends of compatible high and low molecular weight PLA displayed storage modulus almost independent of the composition into the analyzed frequency range, while the melt viscosity decreased at low frequencies (Yang and Du 2024). PLA blends (D isomer 5–18 wt%) with low D content exhibited higher viscosity in accordance with a crystalline structure influenced by the ratio of isomers, molecular weight and compounding method (Hamad et al. 2015; Nofar et al. 2019). Incorporation of PDLA with low molecular weight looks more effective to enhance the viscosity at low frequencies, through the formation of larger crystals acting as crosslinking points (Nofar et al. 2019).

Blends of PLA with PBAT (5–30 wt%) shows a larger viscosity with fluctuation along shear rates indicative of immiscibility (Sin et al. 2012). Plasticizers as PEG or acetyl triethyl citrate (ATC) (until 10 wt%) could accelerate the crystallization rate of PLA, related to the rheological performance, without degradation or phase separation (Nofar et al. 2019). Plasticized PLA displayed slightly reduced viscosity values and storage modulus with PEG content (Nofar et al. 2019). High molecular weight of PLA facilitates wall slip and melt fracture, the incorporation of PCL (0.5 wt%) shifted it to higher shear rate facilitating the extrusion process (Nofar et al. 2019).

9.4.3 Thermal Characterization

Thermal properties of PLA involve heat capacity, thermal transitions, thermal decomposition, and crystallization profile. Commonly used techniques are: Differential Scanning Calorimetry (DSC), Thermal Gravimetric Analysis (TGA), Dynamic Mechanical Analysis (DMA), X-ray Diffraction (XRD), and Polarized Optical Microscopy (POM). Thermal conductivity, heat capacity, and pressure–volume–temperature (PVT) should be also considered in accordance with their relevance in processing. Thermal profile has been related to: crystallinity degree and crystalline structure, polymer molecular weight, polymerization conditions, thermal history, purity, and stereochemical composition (combination of L and D isomers) (Sin et al. 2012). Crystallization represents a very complex thermal process related to the chemical composition, polymer chain orientation, branching, molecular weight, plasticizers, heating cycles, aging, annealing, and manufacturing conditions (Niaounakis 2014; Hussain et al. 2024; Bayer 2017; Ebrahimi and Ramezani Dana 2022). The crystallization temperature increases with molecular weight, higher energy is required to breakdown intermolecular bonds (Sin et al. 2012), while the heat and free energy of crystallization remains almost constant (Sin et al. 2012). Plasticizers incorporation influences the crystallization temperature, crystalline structure (shape, size, type, and density of crystals), and crystallization degree (Bayer 2017). In general, biopolymers display a quite low flame retardancy and a high sensitivity to thermal degradation along processing, not easily improved by blending nor plasticizer incorporation (Niaounakis 2014). The T_g of PLA related to the molecular weight, purity, and thermal history is close linked to the in-service temperature (Sin et al. 2012). Cooling conditions affect the crystalline structure (mainly formed by L isomer), crystallinity, and the presence of different crystalline phases (α, β, and γ) (Singhvi et al. 2019; DeStefano et al. 2020; Hussain et al. 2024; Ebrahimi and Ramezani Dana 2022; Sin et al. 2012). The most stable α phase shows a higher melting temperature compared to β and γ phases. PLA commonly exhibits low crystallinity mainly conformed by α-crystals (DeStefano et al. 2020). At temperatures lower than 100 °C just α-crystals could be formed. The presence of α crystals shows higher water vapor barrier and elastic modulus with lower strain at break (Ma et al. 2021). Alternative α′-crystals could be obtained at 100–120 °C with lower arrangement and performance (barrier properties, elastic modulus) (DeStefano et al. 2020). DSC could reveal α′-crystals by the presence of a small exotherm (transformation from α′ to α-crystals) close to the melting point (DeStefano et al. 2020; Tábi et al. 2010). The melting enthalpy of 100% crystalline PLLA (ΔH°_{m}) is commonly cited around 93 J/g, although higher values have been also reported (148 J/g) (Lasprilla et al. 2012).

As mentioned above, PLA, based on the L isomer, generally displays a crystallinity around 10–37% with glass transition of 55–65, 67, crystallization temperature 100, and melting temperature of 160–180, with also low thermal stability (DeStefano et al. 2020; Niaounakis 2014; Hussain et al. 2024; Ebrahimi and Ramezani Dana 2022; Ranakoti et al. 2022). The decomposition temperature of PLA (normally around 230–260 °C) occurs by hydrolysis reaction followed by lactide reformation, oxidative chain scission, and molecular transesterificacion (Sin et al. 2012). Normally, toxic substances are not released, but residues of plasticizer or oligomers should be taken into account (Sin et al. 2012). PLA is highly sensitive to heating process, and its response mainly depends on the molecular weight, crystalline structure, contaminants, functional end-groups, temperature, pH, and additives. Particularly, residual metals used along polymerization could activate a selective depolymerization process. For example, acetylated PLA combined with high stannous catalyst residues displays higher degradation temperatures and activation energy (Sin et al. 2012). Injection molding of crystalline PLA could display shrinkage, and thermal process helps to eliminate this feature (Niaounakis 2014). Intermediate cooling rate promotes more regular shape and size crystals. While, higher cooling rates, by injection molding or quenching, induce amorphous polymers (Niaounakis 2014). On the contrary, slower process enhanced the crystallinity with irregular shaped crystals (Sin et al. 2012). Thermal treatment influences the crystalline structure and crystallinity finally linked to the in-service performance (Agrawal et al. 2000). Annealing is a possible way to enhance the polymer chain arrangement of crystalline structure and also its performance (crystallinity, melting temperature, elastic modulus, and tensile strength) (Sin et al. 2012; Ma et al. 2021). Annealed injection molded PLA at temperature around T_c under slow cooling process could display higher crystallinity improving the heat resistance (DeStefano et al. 2020; Niaounakis 2014). It has been proposed a temperature of 75 °C lower to the melting point as a proper annealing condition (Sin et al. 2012). Recrystallization of PLA could promote higher thermal profile compared to amorphous products (Tábi et al. 2010). The shrinking or warping of finale products along recrystallization should be also taken into account (Tábi et al. 2010). Drawn PLA under high temperatures promotes the less stable β-crystals formation, detectable by a reduction of 10 °C in the melting point (DeStefano et al. 2020; Ebrahimi and Ramezani Dana 2022). Highly oriented PLLA fibers displayed superior elastic modulus (3–7 GPa) and strain at break (20–100%) related to the polymer chain orientation and crystallinity (Bergström and Hayman 2016).

Neat PLA could show different crystalline structure compared to the blends detected by X-ray diffractions (Sin et al. 2012). PLLA is a crystalline polymers exhibiting a melting point, while blends of PPLA/PDLA (more than 10% of D isomer) are amorphous polymer just displaying a glass transition temperature (Singhvi et al. 2019; DeStefano et al. 2020; Sin et al. 2012; Lasprilla et al. 2012). In addition, blending of PLLA with PDLA can enhance the melting point, thermal resistance, and hydrolytic degradation, by the development of a stereo complex structure, compared with neat PLA (Sin et al. 2012; Vayshbeyn et al. 2023). Blending of PLLA with PDLA (8–50%) represents a widely used technique to enhance the thermal stability, the melting temperature, or the heat deflection point (Khouri et al. 2024; DeStefano et al. 2020; Niaounakis 2014). The blend with 50% of both isomers displays the best thermal performance with a shift of T_m to 230 °C (Khouri et al. 2024). Blends of low and high molecular weight PLA displayed co-crystallization with lower crystallization rate, crystallinity, and crystal size with content of the low molecular weight (Yang and Du 2024). Incorporation of plasticizer (Poly(ethylene glycol) (PEG), Poly(1,3-butanediol) (PBOH), Acetyl glycerolmonolaurate (AGM), Dibutyl sebacate (DBS)) could improve the processability with simultaneous presence of irregularities shifting to lower T_m and T_g values. PEG displays a superior ability to promote the crystallization of PLA due to higher chain mobility and also miscibility. While, low molecular weight of PEG has been related to increased melting enthalpy (ΔH_m) and macroscopic phase separation (Sin et al. 2012).

9.4.4 Mechanical Characterization

A variety of standardized and specific methods could be employed to determine the mechanical performance of polymers and their biomedical products, such as: uniaxial tensile, compression, bending, shear, fracture, impact, fatigue test, and creep tests (Gleadall 2015a). All of these procedures allow to obtain experimental curves and characteristic parameter values (stiffness, strength, strain at break, toughness) to investigate the influence of different characteristics. A lack of experimental data and simulations could be detected in the literature due to the large quantity of involved parameters. Crystalline structure (crystals size and morphology), crystallinity degree, stereochemistry, molecular weight, plasticizers, additives, chemical composition, processing conditions, and method testing set-up (loading condition, loading rate, temperature) have been close related to the mechanical performance of PLA and its based blends (Ebrahimi and Ramezani Dana 2022; Sin et al. 2012; Ma et al. 2021; Gleadall 2015a). PLA displays similar mechanical properties (Table 9.2) to some industrially used thermoplastics as PP, PS or PET (Khouri et al. 2024; Saini et al. 2016; Marwat et al 2015, Riaz et al. 2018) with comparative high tensile strength and elastic modulus (DeStefano et al. 2020; Saini et al. 2016; Niaounakis 2014; Sha et al. 2016, Hussain et al. 2024). PLA also displays brittle features, as most biopolymers, (limited impact and fracture toughness, elongation at break and plastic flow capability) highly restrictive for severe loading conditions (Khouri et al. 2024; DeStefano et al. 2020; Saini et al. 2016; Hussain et al. 2024). The in-service temperature of PLA is normally placed below the T_g for biomedical products. In this stage, the polymer chain mobility is highly restricted and only small elastic deformation can be obtained. Above the T_g, polymer chains can slide and elongate promoting a strain hardening effect (Gleadall 2015a). Different strategies, as blends or processing methods, have been considered to improve the final performance of PLA (Sin et al. 2012). In a general way, large molecular weight could be linked to enhanced mechanical performance (DeStefano et al. 2020; Niaounakis 2014). The polymer chain arrangement, and also the crystalline structure, could be promoted by increasing the molecular weight of annealed PLLA conducting to an improvement of strain at break, plastic deformation, and toughness (DeStefano et al. 2020; Sin et al. 2012), while a marginal effect on the PLA strength has been detected. Crystalline PLLA is a hard transparent polymer with a tensile strength of 45–70 MPa (Tyler et al. 2016). The presence of D stereoisomer, amorphous internal structure, could suppress the profitable effect of molecular weight increment promoting an almost constant impact performance (Sin et al. 2012). The restricted wear resistance and comparative high friction coefficient of PLA could limited its final appliances or also be a drawback during processing. Sticking of the polymer facilitates thermomechanical degradation due to higher shear forces and friction heating (extrusion) or reduces the ability to flow through small cavities (injection molding) (Niaounakis 2014).

PLA blends could introduce enhancements superposed to some drawbacks mainly in accordance with the complex miscibility between phases in accordance with interfacial adhesion. Enhanced performances could be longer remained with the incorporation of PEG, PEO, EVA, PBOH, DBS, AGM, or FAEs. The effect depends on the percentage of plasticizer or polymer, molecular weight, end-groups, and polarity (DeStefano et al. 2020; Ebrahimi and Ramezani Dana 2022; Sin et al. 2012). Unfortunately, higher ductility generally corresponds to drastic lower elastic modulus and quite reduced tensile strength (Sin et al. 2012). The incorporated plasticizer should not act as a weak zone, due to a low interface adhesion between phases, promoting embrittlement and also crack propagation. In this situation, tenacity could be restricted as observed with FAEs (Sin et al. 2012). Lactide looks as an effective additive to improve the strain at break of PLA, but its high migration rate shortly promote a stiffener material. Compatibilizers are incorporated into blends in order to increase the interfacial adhesion between polymers (Saini et al. 2016). It could highlight, the improved of strain at break with: 10–20 wt% of PEG or 20–30 wt% of: PEO, AGM, PBOH, or DBS (Sin et al. 2012). PLA with 20–30 wt% of PCL (reactive blending), 5–20 wt% of PBAT (melt blending), 5–30 wt% PHAs (melt blending) could improve ductility and impact strength with decay of tensile strength and elastic modulus (Sin et al. 2012). In contrast, higher strain at break could be also obtained without detriment of elastic modulus (20 wt% of PPD) or tensile strength (25 wt% of PTAT) (Sin et al. 2012). Incorporation of poly (butylene succinate-co-L-lactate) (PBSL) reveals the importance of phase separation on the mechanical behavior. 1–10 wt% of PBSL exhibited increased ductility with almost constant values of elastic modulus and tensile strength related with a properly phases adhesion. On the contrary, 10–60 wt% of PBSL exhibited phase separation and a positive trend for strain at break with a clearly detriment of the other tensile parameters (Shibata et al. 2006). Blends of PLA/TPU (80/20 wt%) enhanced toughness and elongation at break and displayed shape memory behavior, attributed to the crystalline regions of PLA (Saini et al. 2016). PLA with PCL 3 wt% showed higher elastic modulus (54 MPa) and elongation at break (9%). While, the blend with PCL 9 wt% also displayed increased strain at break (10%) and lower elastic modulus (37 MPa) (Hussain et al. 2024).

Additionally, PCL incorporation could promote the brittle to ductile transition on the fracture behavior (Saini et al. 2016). PLA/starch (20 wt%) blends simultaneously increased tensile strength, elastic modulus, and strain at break, but processing conditions (temperature and time) should be optimized to do not diminish the tensile toughness (Saini et al. 2016). PLA grafted with oligomers and blended with starch improved ductility, melt flow features, gas barrier, and thermal stability (Hussain et al. 2024). Compatibilized (poly(2-ethyl-2-oxazoline)) blends of PLA/soy protein enhanced tensile strength (Saini et al. 2016). PLA/starch with MDI (1 wt%) exhibited larger tensile strength and modulus with additional increment of impact toughness (Saini et al. 2016).

9.4.5 Degradation Characterization

Biodegradable polymers should be biocompatible with non-toxic and easily metabolized degradation products (Puthumana et al. 2020; Ebrahimi and Ramezani Dana 2022). Biodegradation represents the progressive drop of any specific property along immersion time, such as: mechanical strength, toughness, transparency, molecular weight, and mass sample linked to physical, chemical or biological reactions (Niaounakis 2014; Elmowafy et al. 2019). Experimental testing or characterization techniques could be applied to analyze the degradation process, such as: DSC, TGA, high performance liquid chromatography (HPLC), FT-IR, NMR, Raman spectroscopy, or qualitative methods by image analysis (scanning electron microscopy (SEM), transmission electron microscopy (TEM)), and atomic force microscopy (AFM) (Elmowafy et al. 2019). The possible molecular weight reduction as a result of manufacturing or sterilization should be controlled in order to minimize its influence on the degradation profile of biodegradable products (DeStefano et al. 2020). Biodegradation of PLA is a complex chemical and physical-based process. Under in vivo or in vitro conditions, a long immersion time is normally requires to initiate the hydrolysis process mainly affected by the stereochemistry, crystalline structure, crystallinity, molecular weight, chain flexibility, chemical composition and structure, polymer reactivity, additives, residual lactic acid or catalysts, functional groups, purity, hydrophilicity, morphology, processing, sterilization, size, shape and geometry of the biomedical product and surrounding conditions (acidity, humidity, enzymes, human microbiota, body fluids) (Khouri et al. 2024; DeStefano et al. 2020; Niaounakis 2014; Hussain et al. 2024; Ebrahimi and Ramezani Dana 2022; Lasprilla et al. 2012; Gleadall 2015b). As drawback, it could be mentioned the normally long degradation period (from 6 months to 2 years) and low hydrophilicity of PLA for some biomedical applications (DeStefano et al. 2020; Hussain et al. 2024). The prevailing degradation mechanism of PLA (hydrolysis) produces lactic acid no potentially toxic or harmful (Singhvi et al. 2019; Puthumana et al. 2020; Ebrahimi and Ramezani Dana 2022; Lasprilla et al. 2012). The whole process is performed by sequentially steps: initial hydrolysis and oligomers formation, continuous molecular weight reduction, losing of plasticizers, and progressive drop of performance and mass (Khouri et al.

2024; Niaounakis 2014; Hussain et al. 2024; Ebrahimi and Ramezani Dana 2022; Lasprilla et al. 2012). The reaction with adsorbed water by hydrolysis promotes shortened of long chains and also larger superficial area, followed by further enzymatic or microorganism-based reactions (Lasprilla et al. 2012; Gleadall 2015b). Water diffusion easily occurs through incorporated plasticizer increasing the degradation rate (DeStefano et al. 2020). The absorbed water could also act as a plasticizer affecting the mechanical performance (Gleadall 2015a). The microorganisms placed on the surface of immersed products secrete enzymes helping the long polymer chains scission to oligomers or monomers (Niaounakis 2014; Hussain et al. 2024).

Crystalline polymers could enhance its crystallinity along degradation. A single polymer chain can take part of many crystals also forming a hierarchically structure called spherulites. Along degradation, the cleavaged amorphous region displays enough mobility to act as nucleation site promoting recrystallization (Hussain et al. 2024; Gleadall 2015b). The early stage of degradation is characterized by sample mass loss and almost constant molecular weight. At prolonged immersion times, both parameters display drastic reductions (Gleadall 2015b). Thick samples immersed in a buffer solution (37 °C, pH 7.4) exhibited larger bulk hydrolysis rate than superficial one (Ebrahimi and Ramezani Dana 2022; Gleadall 2015b). In this case, oligomers and acid lactic at the surface lay in contact with surrounding media could be easier diffuse in comparison with bulk products remaining longer accumulated (Gleadall 2015b). Incorporation of enzymes could fastener the superficial hydrolysis process (Gleadall 2015b). Autocatalytic hydrolysis of PLA could be regulated by the presence of oligomers, lactic acid or functionalized end-groups (Khouri et al. 2024; Ebrahimi and Ramezani Dana 2022; Gleadall 2015b). PLLA, based on the L stereoisomer, exhibits high degradation and resorption times through the activity of enzymes (Gleadall 2015b). Degradation of crystalline PLLA is activated into the amorphous regions due to the greater water adsorption, while crystals display higher viscosity and slower degradation rates (Khouri et al. 2024; Niaounakis 2014; Gleadall 2015b). Amorphous zones surrounded by crystals shown higher hydrolysis resistance compared to fully amorphous PLA (Gleadall 2015b). Large molecular weights decline the degradation and resorption rates by the presence of entanglements, higher hydrolysis resistance, hindering the polymer chain cleavage (Gleadall 2015b). Low molecular weight PLA (60.000 g/mol) exhibits a shorter degradation period favorable to reduced infection and inflammation probability (Khouri et al. 2024; Hussain et al. 2024).

The blend of PLLA with PDLA could exhibit a higher hydrolysis resistance, compared with both neat polymers, linked to a complex amorphous region (Sin et al. 2012; Gleadall 2015b). The D stereoisomer added in small quantity, to improve and longer retain mechanical properties, cannot be degraded by enzymes and a slower hydrolysis reaction is required (Sin et al. 2012). Blending of PLA with other polymers (PGA, PEG, PCL, PHB, PHBV) tailored the in-service performance mainly related to degradation profile, mass and strength loss along the bone healing process (Singhvi et al. 2019; Saini et al. 2016; Niaounakis 2014; Gleadall 2015b). The initial properties of processed blend could be restricted respect to neat PLA (Sin et al. 2012). PLA is mainly degraded by hydrolytically mechanism, while PCL or PHB are primarily attacked by enzymes (Niaounakis 2014). The products of degradation remain invariable, while the degradation rate is defined by composition, morphology, phase dispersion, processing conditions, and additives (Saini et al. 2016). PGA incorporation has been linked to easier polymer chain cleavage and limited crystallinity allowing a faster degradation. This feature is especially observed under the combination of high percentage of PGA and low molecular weight (Gleadall 2015b). PCL cannot be degraded by enzymes, this characteristic promotes longer degradation rates (Gleadall 2015b). PLA/PCL (12 wt%) blends with glycerin displayed longer retain of mechanical performance (Saini et al. 2016), while PLA/starch/chitosan blends showed higher hydrophilicity (related to starch) facilitate water diffusion (related to chitosan) with slow degradation profile (related to PLA) (Saini et al. 2016).

The mechanical performance of PLLA, PDLA, or PLLA/PDLA blends along in vitro degradation (phosphate buffer solution, pH 7.4 at 37 °C) exhibited a no monotonic decrease with immersed time. An initial stage characterized by erratic fluctuation of parameter values is normally followed by a sudden drop to zero (Gleadall 2015a). This kind of features is in accordance with simultaneous changes of a large number of involved factors, as: crystalline structure, crystallinity degree, or molecular weight (Gleadall 2015a). The variation of strength and elastic modulus have been linked to crystalline structure and molecular weight degradation; decreasing trends could be observed since outset or retarded, and also increased values have been reported (Gleadall 2015a). Polymer scission helps to increase the crystallinity degree (higher elastic modulus and strength) and to obtain shorter chains which can be easily untangled (lower values of both parameters) (Gleadall 2015a). In general, crystallinity prevails at short immersion time, while the lower molecular weight becomes prevalent along a second stage. Molecular weight degradation has been related to initial mechanical properties of the polymer and its distribution of molecular weight, residual monomer, crystalline structure, morphology, blends composition and plasticizer, processing, temperature, pH, polymer to volume ratio, sample size, and geometry (Gleadall 2015a).

Amorphous PLA could larger retain the initial behavior, even under a greater reduction of molecular weight, by a large increment of crystallinity degree and a stabilization of amorphous region. The presence of water helps to closer packed chains although they do not crystalized (Gleadall 2015a). Strain at break and ductility looked more erratic and sensitive than the other parameters. Its performance is defined by the local presence of defects and characteristics of degraded polymer, so no average or macroscopy criteria can be considered for its evaluation (Gleadall 2015a). Amorphous PLA could show an initial larger strain at break. The whole degradation process could also display an initial reduction followed by a longer retaining non-zero value. Initially, the adsorbed water allows the reorientation of polymer chains restricting their mobility. This phenomena also drops at long immersion time the molecular weight degradation rate (Gleadall 2015a). The non-linear relationship between characteristics and mechanical parameters suggests different trigger values of molecular weight for each situation. Beyond this point, the retaining capability is deteriorated in a relatively short period of time. Blend of L and D isomers could longer retain (around 15 months) the initial mechanical behavior in accordance with slower molecular weight reduction. Easily untangle of shortened polymer chain diminished the tensile parameters (Gleadall 2015a).

9.5 Application of Polylactic Acid-Based Blends

Along the last decades, PLA and its blends have been largely incorporated into the biomedical and pharmaceutical fields due to the high biocompatibility and biodegradability into the human body. Bioresorbable fixation devices, auricular, cranial and ocular implants, maxillofacial systems, interference screws, cardiovascular devices, tracheal replacements, artificial bones, scaffolds, stents, anchors, chest or external walls supports, pins, sutures, drug delivery systems, membranes, meshes, tendon and skin regeneration, hydrogels, films, foams, surgical tools, packaging and equipment represent some of the current appliances (Khouri et al. 2024; Singhvi et al. 2019; DeStefano et al. 2020; Saini et al. 2016; Hussain et al. 2024; Tyler et al. 2016; Gleadall 2015b; Mohammed-Sadhakathullah et al. 2024; Perry 2008; Kinoshita et al. 2008; Ferretti 2008). In Table 9.3, some relevant commercial biomedical products of PLA-based blends are summarized. Implantable biomedical devices should be biocompatible as they are foreign products, probably rejected by the immune system (Khouri et al. 2024; DeStefano et al. 2020; Tyler et al. 2016; Lasprilla et al. 2012). Their degradation products should not be cytotoxic nor induce inflammatory reaction and the degradation rate have to fit to tissue regeneration (Pawar et al. 2014; Tyler et al. 2016). PLA represents a biocompatible and biodegradable polymer not requiring implant removal, where the lactic acid is the non-toxic degradation product (Khouri et al. 2024; Hamad et al. 2015; Ebrahimi and Ramezani Dana 2022; Tyler et al. 2016; Lasprilla et al. 2012). Different types of PLA could be fit to a wide range of biomedical applications and products, while its brittle performance and crystallinity could be restrictive (Puthumana et al. 2020; Lasprilla et al. 2012). Blends of PLA with PGA, PEG, or PHB allow to enhance wettability or fastener biodegradability in accordance with required time of tissue growth or recovery (Singhvi et al. 2019; Saini et al. 2016). Superficial features play a critical role for biomedical products. Chemical and physical modifications, plasma and radiation treatments, gelatin or protein incorporation, functionalization, and coating could enhance the cell adhesion, hydrophilicity, antifouling activity, or hemocompatibility (Saini et al. 2016; Tyler et al. 2016; Lasprilla et al. 2012; Baran and Yildirim Erbil 2019; Lee et al. 2016).

Firstly, the production of PLA biomedical grade exhibited high costs and restricted technologies (Hamad et al. 2015; Gleadall 2015b). Nowadays, injection molding is used for implants, fiber spinning for sutures or artificial blood vessels and casting for scaffolds production (Khouri et al. 2024; Gleadall 2015b; Maleki et al. 2022). Recently incorporated, additive manufacturing allows cheap and fast production of customized implants for cartilage reconstruction, aortic valves, functionalized scaffolds, capsules for drug delivery, surgery tools (surgical retractors, needle drivers, hemostats, forceps, scalpel handles), and organ models or medical training devices

Table 9.3 Commercial biomedical applications of PLA-based blends

Material	Intended use	Trade mark
PLLA, PDLA, PLA/PGA, PLA/PEG	Absorbable medical devices	Purasorb
PLLA, PDLA, PLA/PGA	Release drug carrier, scaffolds, absorbable screws and plates, 3D printing, microcapsules	Esunmed
PLLA, PDLA, PLA/PGA	3D printing, release drug carrier, scaffolds, absorbable implants	Resomer and Lactel (Evonik)
PLLA, PDLA	Release drug carriers, absorbable implants	Merck
PLLA, PDLA	Drug delivery systems, absorbable implants	Musashino
PLLA/PDLA	Miniplate, miniscrew, mesh, bone graft	Inion
PLLA	Miniplate, miniscrew, mesh	Fixsorb
PLLA/PDLA	Interference screw, meniscal arrow	Linvatec
PLA	Interference screw	Euroscrew
PDLA	Cranial and dental implant	KLS Martin
PLA/PGA	Cancer treatment	Zoladex
PLA/PGA	Ocular treatment, cartilage regeneration, scaffolds	BioTissue Technologies AG
PLA/PGA	Absorbable stent	Synergy, Boston Scientific
PLA/PGA	Cancer treatment, acromegaly	Ipsen
PLA/PGA	Wound healing	Medisorb
PLA/PGA	Suture	Zimmer

obtained in combination with image analysis (Khouri et al. 2024; DeStefano et al. 2020; Hussain et al. 2024; Ebrahimi and Ramezani Dana 2022). 3D printing of PLA also demonstrated its capability to massive and short time production of personal protective and medical equipment (ventilators and circuit splitter) along the pandemic of Coronavirus Disease of 2019 (Covid-19) (DeStefano et al. 2020).

Sutures were the first biomedical application of PLA, it should fast degrade on the contrary to fixation devices (Saini et al. 2016; Puthumana et al. 2020). Incorporation of PGA or PHBV allowed to warrant the in-service requirements (tensile modulus and strength, processability, biocompatibility, and degradation rate) (DeStefano et al. 2020; Saini et al. 2016; Hussain et al. 2024; Gleadall 2015b). Neat PLLA looks not suitable for suture (slow degradation), but favorable for ligament and tendon reconstruction and stents for vascular or urological surgery due to the long-term retention strength (similar to bone) (Singhvi et al. 2019; Lasprilla et al. 2012). Oriented PLA also displays strengthener performance similar to the under healing bone (Khouri et al. 2024). PLA fibers have been also successfully used for cardiac stent (composed by a zigzag helical structure), scaffolds, biosensors, and wound dressing (Khouri et al. 2024; Saini et al. 2016). Biodegradable fixation systems (mini plate and screws), pins, and wires have been firstly focused to no load bearing situations due to the low fatigue and fracture resistance (DeStefano et al. 2020; Lasprilla et al. 2012). Improved blends of PLLA (85 wt%) with PDLA (15 wt%) let the progressive introduction to maxillofacial fractures under limited loading conditions (Singhvi et al. 2019; Ebrahimi and Ramezani Dana 2022; Cheung et al. 2008). The relative large complication rates (around 25%), compared to metallic implants (around 10%), have been mostly related to the low adhesion of post-surgical protocol (Ferretti 2008; Cheung et al. 2008). Independently of this fact, both kind of implants displayed long-term success for similar surgical cases (Cheung et al. 2008). The degradation rate should be fit to the recovering process of load bearing capability by damaged tissue, minimizing the exposure time and consequently the inflammatory response (Khouri et al. 2024). Molecular weight should be as low as possible to shorter the degradation time, PLLA of 100 kDa ensured the required strength for orthopedic devices (rods, screws, ligament reconstruction) (DeStefano et al. 2020; Ebrahimi and Ramezani Dana 2022). Biodegradable fixation screws promote bone healing across a slow transfer of load from the biomedical device to the partially healed tissue. This performance involves the screw geometry as, external and internal diameter, length, and pitch screw. A smaller internal hole, related to thicker cross-section, exhibited faster degradation with larger accumulation of degraded polymer chains (Gleadall 2015b). In a similar way, stents faster degraded at the union of branch, where thicker size is observed (Gleadall 2015b). Crystallinity could delay the degradation while amorphous blend of L and D isomers become faster this process (Khouri et al. 2024). Incorporation of bioactive substances or superficial modifications could enhance the cell interaction, the hydrophilicity or introduce antibacterial activity (DeStefano et al. 2020). Craniofacial bone screw of PLLA and PDLA (70/30 wt%) retained, in vitro evaluations, 100% of the pull-out strength until 2 months (Agrawal et al. 2000). Blends of PLLA/PDLA (85/15 wt%) with additional incorporation of surfactants showed higher toughness indicative of its potential use for dental or orthopedic implants (Singhvi et al. 2019). Meniscal arrows of PLLA shows a large rate of clinical success (95%) retaining the initial strength for one year (Ebrahimi and Ramezani Dana 2022). Membranes of PLA and PAV (1–10 v/v%) blends exhibited a higher cell viability, bacteriostatic activity, and antioxidant property by functionalization with PEG-cholesterol molecules (Mohammed-Sadhakathullah et al. 2024). 4D printed biomedical products look as a promising way to minimize the invasiveness of surgery. Blend of PLA with PCL (5 wt%) enhanced the shape memory and toughness with tailored degradation rate (Hussain et al. 2024; Song et al. 2023; Kabirian et al. 2022). Minipins also displayed enough fixing capacity for bioresorbable membranes (Agrawal et al. 2000).

Scaffolds have been introduced as artificial skin grafts or to guide the formation of new bone while supporting compressive or occlusion forces, used for muscle, bone and cartilages tissue regeneration, and cardiovascular or neurological treatments (DeStefano et al. 2020; Tyler et al. 2016; Lasprilla et al. 2012; Kinoshita et al. 2008). Scaffolds are porous and flexible tridimensional structures allowing fixation, migration, proliferation and differentiation of living cells, and also providing support for dual systems with long-term drug release (Khouri et al. 2024; DeStefano et al. 2020). This kind of products should retain mechanical, geometrical, and permeability features along healing process (Tyler et al. 2016). The in-service performance has been linked to the wall thickness, pore size distribution, and porosity designed for each clinical situation. Thin internal walls facilitated the degradation rate, independently of the other parameters (Gleadall 2015b). Neat PLA and blends with PBS (50/50 wt%) or PCL displayed favorable bioactivity and mechanical performance for short bowel syndrome, vascularization, cardiac patches, and skin or tendon regeneration (DeStefano et al. 2020; Hussain et al. 2024; Ebrahimi and Ramezani Dana 2022; Pereira-Lobato et al. 2024). Scaffolds of PLA/PGA blends successfully assisted the meniscal regeneration and were fully degraded just after 2 months (Ebrahimi and Ramezani Dana 2022; Tyler et al. 2016). PLA/PGS blends showed tailored porosity for cell penetration and for specific tissue requirements with higher hydrophilicity and mechanical properties (Saini et al. 2016). 3D printed scaffolds permits the design of complex porous structures (pore shape, size distribution, porosity, and interconnections) with proper rigidity, strength, hydrophilicity, and degradation rates (DeStefano et al. 2020; Tyler et al. 2016; Myers et al. 2022). Printing of PLA/PCL blends represents a promising way to improve and extend the appliances of biodegradable scaffolds (Khouri et al. 2024). In addition, 3D printed scaffolds with honeycomb infill displayed mechanical parameter values close similar to the cancellous bone (Agrawal et al. 2000). While, schoen gyroid structure exhibited a superior performance (ductile failure of unit cells) compared with others internal architectures (DeStefano et al. 2020; Myers et al. 2022). 4D printing could produce tailored biomedical products with adaptable shapes, functions and structures for stents, cardiac valves, vascular grafts, or prostheses (Kabirian et al. 2022). Their reactions to different stimuli allow to minimize the implantation and to customize the tissue interaction (Kabirian et al. 2022). Bioprinting looks as another promising technology to produce organs. For example, PLA has been used as the matrix combined with alginate hydrogel and human stem cells for soft tissue regeneration (Tümer and Erbil 2021). Incorporation of PGA fibers (20 wt%) enhanced the retaining ability of compressive behavior (modulus and strength) (Agrawal et al. 2000). Multilayer scaffolds with locally variable properties mimic the biological and mechanical functions, considering the morphology and topography of bones and the specific patient requirements (Scaffaro et al. 2017; Chen et al. 2023). Electrospun fibrous scaffolds of PLA adjust the performance by control of the diameter and alignment of fibers, and also the sample thickness (Khouri et al. 2024). Electrospun scaffolds of PLA could mimic the extracellular matrix providing a favorable substrate for bone formation with additional advantage as high superficial area and porosity, small pore size, and tunable mechanical properties (Singhvi et al. 2019; Saini et al. 2016; Maleki et al. 2022). PLA fibers doped with ibuprofen displayed higher efficiency against inflammation (DeStefano et al. 2020).

Drug delivery systems exhibit improved absorption without premature degradation, targeting of reactions, controlled substances concentration and reduction of side effects (Khouri et al. 2024; Tyler et al. 2016). Blends of biodegradable polymers

(PLA, PGA, PCL, EVA) loaded with therapeutic substances look more suitable to enhance delivery efficiency than to develop new medication routes (Singhvi et al. 2019; Saini et al. 2016; Tyler et al. 2016). Adaptable degradation rates into the same delivery system helps short (anti-inflammatory) and long-term (anti-tumor) therapies. PLA combined with proteins promotes the interaction between with tissues or cells (DeStefano et al. 2020). Non-traditional applications, such as dental implantology, could be assisted through drug delivery systems based on blends of PLA with PGA or PVA (DeStefano et al. 2020; Hussain et al. 2024). Microspheres of neat PLA and blends with PGA or PEG filled with bioactive substances have been successfully applied for local delivery, prolonged drug release (contraceptives, narcotic antagonists, local anesthetics, cancer, or diabetes treatments), vaccines or wound dressing applications (Singhvi et al. 2019; Saini et al. 2016; Ebrahimi and Ramezani Dana 2022; Lasprilla et al. 2012). The drug release rate and period could be tailored by varying the molecular weight, crystallinity, mechanical stability, chemical properties, size and shape of the spheres (Ebrahimi and Ramezani Dana 2022; Lee et al. 2016). PLA/PEG micelles were applied for anticancer treatment with promising results (Ebrahimi and Ramezani Dana 2022). Porous matrix of this blend enhanced the performance (porosity, pore size, degradation rate, and drug release) of wound dressing with drug delivery (Saini et al. 2016). Nanodelivery systems, for periodontal pocket treatment, hydrogels or nanoemulsions, display larger volume to surface ratio, adaptable external shell, biodegradability, biocompatibility, low cytotoxicity, and also tailored response capabilities (Saini et al. 2016; Tyler et al. 2016).

References

Abdelwahab MA, Flynn A, Sen Chiou B, Imam S, Orts W, Chiellini E (2012) Thermal, mechanical and morphological characterization of plasticized PLA-PHB blends. Polym Degrad Stab 97:1822–1828. https://doi.org/10.1016/j.polymdegradstab.2012.05.036

Adnan M, Siddiqui AJ, Ashraf SA, Snoussi M, Badraoui R, Alreshidi M, Elasbali AM, Al-Soud WA, Alharethi SH, Sachidanandan M, Patel M (2022) Polyhydroxybutyrate (PHB)-based biodegradable polymer from agromyces indicus: enhanced production, characterization, and optimization. Polymers (Basel) 14. https://doi.org/10.3390/polym14193982

Agrawal CM, Parr JE, Lin ST (2000) Synthetic bioabsorbable polymers for implants.https://doi.org/10.1520/stp1396-eb

Al Nakib R, Toncheva A, Fontaine V, Vanheuverzwijn J, Raquez JM, Meyer F (2022) Thermoplastic polyurethanes for biomedical application: a synthetic, mechanical, antibacterial, and cytotoxic study. J Appl Polym Sci 139:1–17. https://doi.org/10.1002/app.51666

Al-Itry R, Lamnawar K, Maazouz A (2014) Rheological, morphological, and interfacial properties of compatibilized PLA/PBAT blends. Rheol Acta 53:501–517. https://doi.org/10.1007/S00397-014-0774-2/METRICS

Allami T, Alamiery A, Nassir MH, Kadhum AH (2021) Investigating physio-thermo-mechanical properties of polyurethane and thermoplastics nanocomposite in various applications. Polymers (Basel) 13:1–24. https://doi.org/10.3390/polym13152467

Álvarez I, Gutiérrez C, Rodríguez JF, De Lucas A, García MT (2020) Production of drug-releasing biodegradable microporous scaffold impregnated with gemcitabine using a CO_2 foaming process. J CO_2 Util 41:101227. https://doi.org/10.1016/j.jcou.2020.101227

Aranaz I, Alcántara AR, Civera MC, Arias C, Elorza B, Caballero AH, Acosta N (2021) Chitosan: an overview of its properties and applications. Polymers (Basel) 13. https://doi.org/10.3390/polym13193256

Astner AF, Hayes DG, O'Neill H, Evans BR, Pingali SV, Urban VS, Schaeffer SM, Young TM (2023) Assessment of cryogenic pretreatment for simulating environmental weathering in the formation of surrogate micro- and nanoplastics from agricultural mulch film. Sci Total Environ 870:161867. https://doi.org/10.1016/J.SCITOTENV.2023.161867

Avinc O, Khoddami A (2009) Overview of poly(lactic acid) (PLA) fibre. Fibre Chem 41:391–401. https://doi.org/10.1007/s10692-010-9213-z

Balla E, Daniilidis V, Karlioti G, Kalamas T, Stefanidou M, Bikiaris ND, Vlachopoulos A, Koumentakou I, Bikiaris DN (2021) Poly(lactic acid): a versatile biobased polymer for the future with multifunctional properties applications. Polymers (Basel) 13:1822

Baran EH, Yildirim Erbil H (2019) Surface modification of 3d printed pla objects by fused deposition modeling: a review. Colloids Interfaces 3. https://doi.org/10.3390/colloids3020043

Bayer IS (2017) Thermomechanical properties of polylactic acid-graphene composites: a state-of-the-art review for biomedical applications. Materials (Basel) 10. https://doi.org/10.3390/ma10070748

Bergström JS, Hayman D (2016) An overview of mechanical properties and material modeling of polylactide (PLA) for medical applications. Ann Biomed Eng 44:330–340. https://doi.org/10.1007/s10439-015-1455-8

Blend P (2011) Introduction to polymer blends. In: SpringerReference, pp 1–7. https://doi.org/10.1007/springerreference_1

Brown CD, Stayton PS, Hoffman AS (2005) Semi-interpenetrating network of poly(ethylene glycol) and poly(D, L-lactide) for the controlled delivery of protein drugs. J Biomater Sci Polym Ed 16:189–201. https://doi.org/10.1163/1568562053115471

Budak K, Sogut O, Aydemir Sezer U (2020) A review on synthesis and biomedical applications of polyglycolic acid. J Polym Res 27:1–19. https://doi.org/10.1007/s10965-020-02187-1

Cantón I, Mckean R, Charnley M, Blackwood KA, Fiorica C, Ryan AJ, MacNeil S (2010) Development of an Ibuprofen-releasing biodegradable PLA/PGA electrospun scaffold for tissue regeneration. Biotechnol Bioeng 105:396–408. https://doi.org/10.1002/BIT.22530

Cao A, Wan D, Gao C, Elverum CW (2024) A novel method of fabricating designable polylactic acid (PLA)/thermoplastic polyurethane (TPU) composite filaments and structures by material extrusion additive manufacturing. J Manuf Process 118:432–447. https://doi.org/10.1016/J.JMAPRO.2024.03.015

Casalini T, Rossi F, Castrovinci A, Perale G (2019) A perspective on polylactic acid-based polymers use for nanoparticles synthesis and applications. Front Bioeng Biotechnol 7:483145. https://doi.org/10.3389/FBIOE.2019.00259/BIBTEX

Chavalitpanya K, Phattanarudee S (2013) Poly(lactic acid)/polycaprolactone blends compatibilized with block copolymer. Energy Procedia 34:542–548. https://doi.org/10.1016/j.egypro.2013.06.783

Cheerarot O, Saikrasun S (2023) Effects of different preparation methods (solvent casting and melt blending) on properties of chitosan-filled polylactic acid biocomposite films. J Elastomers Plast 55:1173–1198. https://doi.org/10.1177/00952443231198468

Chen Y, Geever LM, Higginbotham CL, Devine DM (2014) Analysis of the mechanical properties of solvent cast blends of PLA/PCL. Appl Mech Mater 679:50–56. https://doi.org/10.4028/www.scientific.net/AMM.679.50

Chen T, Zhao X, Weng Y (2023) Self-assembled polylactic acid (PLA): synthesis, properties and biomedical applications. Front Chem 10:1–8. https://doi.org/10.3389/fchem.2022.1107620

Cheung LK, Yip IHS, Chow RLK (2008) Stability and morbidity of Le Fort I osteotomy with bioresorbable fixation: a randomized controlled trial. Int J Oral Maxillofac Surg 37:232–241. https://doi.org/10.1016/j.ijom.2007.09.169

Choi RS, Riegler M, Pothoulakis C, Kim BS, Mooney D, Vacanti M, Vacanti JP (1998) Studies of brush border enzymes, basement membrane components, and electrophysiology of tissue-engineered neointestine. J Pediatr Surg 33:991–997. https://doi.org/10.1016/S0022-3468(98)90520-6

Chuaponpat N, Ueda T, Ishigami A, Kurose T, Ito H (2020) Morphology, thermal and mechanical properties of co-continuous porous structure of PLA/PVA blends by phase separation. Polymers 12:1083. https://doi.org/10.3390/POLYM12051083

D'Amico DA, Iglesias Montes ML, Manfredi LB, Cyras VP (2016) Fully bio-based and biodegradable polylactic acid/poly(3-hydroxybutirate) blends: use of a common plasticizer as performance improvement strategy. Polym Test 49:22–28. https://doi.org/10.1016/j.polymertesting.2015.11.004

D'souza AA, Shegokar R (2016) Polyethylene glycol (PEG): a versatile polymer for pharmaceutical applications. Expert Opin Drug Deliv 13:1257–1275. https://doi.org/10.1080/17425247.2016.1182485

da Silva D, Kaduri M, Poley M, Adir O, Krinsky N, Shainsky-Roitman J, Schroeder A (2018) Biocompatibility, biodegradation and excretion of polylactic acid (PLA) in medical implants and theranostic systems. Chem Eng J 340:9–14. https://doi.org/10.1016/j.cej.2018.01.010

de França JOC, da Silva Valadares D, Paiva MF, Dias SCL, Dias JA (2022) Polymers based on PLA from synthesis using D,L-lactic acid (or racemic lactide) and some biomedical applications: a short review. https://doi.org/10.3390/polym14122317

Degradable PHA/PLA mixture film and preparation method thereof (2014)

Deng S, Huang R, Zhou M, Chen F, Fu Q (2016) Hydrophobic cellulose films with excellent strength and toughness via ball milling activated acylation of microfibrillated cellulose. Carbohydr Polym 154:129–138. https://doi.org/10.1016/J.CARBPOL.2016.07.101

DeStefano V, Khan S, Tabada A (2020) Applications of PLA in modern medicine. Eng Regen 1:76–87. https://doi.org/10.1016/j.engreg.2020.08.002

Dogan SK, Boyacioglu S, Kodal M, Gokce O, Ozkoc G (2017) Thermally induced shape memory behavior, enzymatic degradation and biocompatibility of PLA/TPU blends: "effects of compatibilization." J Mech Behav Biomed Mater 71:349–361. https://doi.org/10.1016/J.JMBBM.2017.04.001

Dominguez-Candela I, Sempere-Jos L, Sandoval-Perez I (2024) Evaluation of additives on the cell metabolic activity of new PHB/PLA-based formulations by means of material extrusion 3D printing for scaffold applications

Dong Y, Ruan Y, Wang H, Zhao Y, Bi D (2004) Studies on glass transition temperature of chitosan with four techniques. J Appl Polym Sci 93:1553–1558. https://doi.org/10.1002/APP.20630

Dwivedi R, Kumar S, Pandey R, Mahajan A, Nandana D, Katti DS, Mehrotra D (2020) Polycaprolactone as biomaterial for bone scaffolds: review of literature. Craniofac Res Found. https://doi.org/10.1016/j.jobcr.2019.10.003

Ebrahimi F, Ramezani Dana H (2022) Poly lactic acid (PLA) polymers: from properties to biomedical applications. Int J Polym Mater Polym Biomater 71:1117–1130. https://doi.org/10.1080/00914037.2021.1944140

Elmowafy EM, Tiboni M, Soliman ME (2019) Biocompatibility, biodegradation and biomedical applications of poly(lactic acid)/poly(lactic-co-glycolic acid) micro and nanoparticles. Springer Singapore.https://doi.org/10.1007/s40005-019-00439-x

Fang Y, Zhu X, Wang N, Zhang X, Yang D, Nie J, Ma G (2019) Biodegradable core-shell electrospun nanofibers based on PLA and γ-PGA for wound healing. Eur Polym J 116:30–37. https://doi.org/10.1016/j.eurpolymj.2019.03.050

Farah S, Anderson DG, Langer R (2016) Physical and mechanical properties of PLA, and their functions in widespread applications—a comprehensive review. Adv Drug Deliv Rev 107:367–392. https://doi.org/10.1016/j.addr.2016.06.012

Fernández-Tena A, Guerrica-Echevarria G, Aranburu N, Wang Z, Cavallo D, Müller AJ (2024) How to achieve a highly toughened 70/30 PLA/PCL blend by using nucleating agents and tailoring processing conditions. ACS Appl Polym Mater 6:583–595. https://doi.org/10.1021/ACSAPM.3C02220/ASSET/IMAGES/LARGE/AP3C02220_0009.JPEG

Ferretti C (2008) A prospective trial of poly-L-lactic/polyglycolic acid co-polymer plates and screws for internal fixation of mandibular fractures. Int J Oral Maxillofac Surg 37:242–248. https://doi.org/10.1016/j.ijom.2007.11.015

Fossion E (1993) Advances in maxillofacial surgery. Acta Chir Belg 83:XXXII–XXXVI. http://www.ncbi.nlm.nih.gov/pubmed/6858531

Fredi G, Dorigato A (2024) Compatibilization of biopolymer blends: a review. Adv Ind Eng Polym Res 7:373–404. https://doi.org/10.1016/j.aiepr.2023.11.002

Gadomska-Gajadhur A, Ruśkowski P (2020) Biocompatible catalysts for lactide polymerization—catalyst activity, racemization effect, and optimization of the polymerization based on design of experiments. Org Process Res Dev 24:1435–1442. https://doi.org/10.1021/acs.oprd.0c00149

Gao L, Drozdov AD (2024) Exploring the performance of bio-based PLA/PHB blends: a comprehensive analysis. 15:358–374. https://doi.org/10.1177/20412479241266954

Garlotta D (2001) A literature review of poly(lactic acid). J Polym Environ 9:63–84

Ghalia MA, Dahman Y (2017) Biodegradable poly(lactic acid)-based scaffolds: synthesis and biomedical applications. J Polym Res 24. https://doi.org/10.1007/s10965-017-1227-2

Giacinti Baschetti M, Minelli M (2020) Test methods for the characterization of gas and vapor permeability in polymers for food packaging application: a review. Polym Test 89:106606. https://doi.org/10.1016/J.POLYMERTESTING.2020.106606

Gleadall A (2015a) Mechanical properties of biodegradable polymers for medical applications. Woodhead Publishing Limited.https://doi.org/10.1533/9781782420255.2.163

Gleadall A (2015b) Modelling degradation of bioresorbable polymeric medical devices. http://www.sciencedirect.com:5070/book/9781782420163/modelling-degradation-of-bioresorbable-polymeric-medical-devices. Accessed 27 Nov 2024

Gorrasi G, Pantani R (2018) Hydrolysis and biodegradation of poly(lactic acid). Adv Polym Sci 279:119–151. https://doi.org/10.1007/12_2016_12

Hamad K, Kaseem M, Yang HW, Deri F, Ko YG (2015) Properties and medical applications of polylactic acid: a review. Express Polym Lett 9:435–455. https://doi.org/10.3144/expresspolymlett.2015.42

Hamad K, Kaseem M, Ayyoob M, Joo J, Deri F (2018) Polylactic acid blends: the future of green, light and tough. Prog Polym Sci 85:83–127. https://doi.org/10.1016/j.progpolymsci.2018.07.001

Hoidy WH, Al-Mulla EAJ, Al-Janabi KW (2010) Mechanical and thermal properties of PLLA/PCL modified clay nanocomposites. J Polym Environ 18:608–616. https://doi.org/10.1007/S10924-010-0240-X/METRICS

Hong H, Yang L, Yuan Y, Qu X, Chen F, Wei J, Liu C (2016) Preparation, rheological properties and primary cytocompatibility of TPU/PLA blends as biomedical materials. J Wuhan Univ Technol Mater Sci Ed 31:211–218. https://doi.org/10.1007/S11595-016-1354-3/METRICS

https://www.marketsandmarkets.com/Market-Reports/polylactic-acid-pla-market-29418964.html (n.d.)

https://www.matweb.com/search/datasheet.aspx?matguid=1932586b674346e2a9d5cb4c7462dd33

Hu Y, Rogunova M, Topolkaraev V, Hiltner A, Baer E (2003) Aging of poly(lactide)/poly(ethylene glycol) blends. Part 1. Poly(lactide) with low stereoregularity. Polymer (Guildf) 44:5701–5710. https://doi.org/10.1016/S0032-3861(03)00614-1

Hui I, Pasquier E, Solberg A, Agrenius K, Håkansson J, Chinga-Carrasco G (2023) Biocomposites containing poly(lactic acid) and chitosan for 3D printing—assessment of mechanical, antibacterial and in vitro biodegradability properties. J Mech Behav Biomed Mater 147:106136. https://doi.org/10.1016/J.JMBBM.2023.106136

Hurlburt RT (2015) Readiness. https://doi.org/10.1016/B978-0-323-26698-7.00002-7

Hussain M, Khan SM, Shafiq M, Abbas N (2024) A review on PLA-based biodegradable materials for biomedical applications. Giant 18. https://doi.org/10.1016/j.giant.2024.100261

Jandas PJ, Mohanty S, Nayak SK (2014) Morphology and thermal properties of renewable resource-based polymer blend nanocomposites influenced by a reactive compatibilizer. ACS Sustain

Chem Eng 2:377–386. https://doi.org/10.1021/SC400395S/ASSET/IMAGES/MEDIUM/SC-2013-00395S_0010.GIF
Jem KJ, Tan B (2020) The development and challenges of poly (lactic acid) and poly (glycolic acid). Adv Ind Eng Polym Res 3:60–70. https://doi.org/10.1016/j.aiepr.2020.01.002
Jung J, Kasi G, Seo J (2018) Development of functional antimicrobial papers using chitosan/starch-silver nanoparticles. Int J Biol Macromol 112:530–536. https://doi.org/10.1016/j.ijbiomac.2018.01.155
Jung YS, Woo J, Lee E, Lee S, Shin EJ (2022) Synthesis and properties of bio-based thermoplastic poly(ether urethane) for soft actuators. J Polym Res 29. https://doi.org/10.1007/s10965-022-03375-x
Kabirian F, Mela P, Heying R (2022) 4D printing applications in the development of smart cardiovascular implants, front. Bioeng Biotechnol 10:1–11. https://doi.org/10.3389/fbioe.2022.873453
Karst D, Yang Y (2006) Molecular modeling study of the resistance of PLA to hydrolysis based on the blending of PLLA and PDLA. Polymer (Guildf) 47:4845–4850. https://doi.org/10.1016/j.polymer.2006.05.002
Khouri NG, Bahú JO, Blanco-Llamero C, Severino P, Concha VOC, Souto EB (2024) Polylactic acid (PLA): properties, synthesis, and biomedical applications—a review of the literature. J Mol Struct 1309:138243. https://doi.org/10.1016/j.molstruc.2024.138243
Kinoshita Y, Matsuo M, Todoki K, Ozono S, Fukuoka S, Tsuzuki H, Nakamura M, Tomihata K, Shimamoto T, Ikada Y (2008) Alveolar bone regeneration using absorbable poly(L-lactide-co-ε-caprolactone)/β-tricalcium phosphate membrane and gelatin sponge incorporating basic fibroblast growth factor. Int J Oral Maxillofac Surg 37:275–281. https://doi.org/10.1016/j.ijom.2007.11.010
Kossentini Kallel T, Houichi H, Sayari A, Kammoun M (2015) Assessment of biodegradation of PLA/PCL and PLA/PEG biopolymers under aerobic and anaerobic conditions. J Mater Environ Sci 6:2542–2549
Kou SG, Peters LM, Mucalo MR (2021) Chitosan: a review of sources and preparation methods. Int J Biol Macromol 169:85–94. https://doi.org/10.1016/j.ijbiomac.2020.12.005
Król P (2007) Synthesis methods, chemical structures and phase structures of linear polyurethanes. Properties and applications of linear polyurethanes in polyurethane elastomers, copolymers and ionomers. Prog Mater Sci 52:915–1015. https://doi.org/10.1016/j.pmatsci.2006.11.001
Kubisa P, Lapienis G, Biela T (2021) Star-shaped copolymers with PLA–PEG arms and their potential applications as biomedical materials. Polym Adv Technol 32:3857–3866. https://doi.org/10.1002/pat.5297
Kumar M, Mohanty S, Nayak SK, Rahail Parvaiz M (2010) Effect of glycidyl methacrylate (GMA) on the thermal, mechanical and morphological property of biodegradable PLA/PBAT blend and its nanocomposites. Bioresour Technol 101:8406–8415. https://doi.org/10.1016/J.BIORTECH.2010.05.075
Kumar R, Alex Y, Nayak B, Mohanty S (2023) Effect of poly (ethylene glycol) on 3D printed PLA/PEG blend: a study of physical, mechanical characterization and printability assessment. J Mech Behav Biomed Mater 141:105813. https://doi.org/10.1016/J.JMBBM.2023.105813
Kun H, Wei Z, Xuan L, Xiubin Y (2012) Biocompatibility of a novel poly(butyl succinate) and polylactic acid blend. ASAIO J 58:262–267. https://doi.org/10.1097/MAT.0b013e31824709ee
Labet M, Thielemans W (2009) Synthesis of polycaprolactone: a review. Chem Soc Rev 38:3484–3504. https://doi.org/10.1039/b820162p
Lasprilla AJR, Martinez GAR, Lunelli BH, Jardini AL, Filho RM (2012) Poly-lactic acid synthesis for application in biomedical devices—a review. Biotechnol Adv 30:321–328. https://doi.org/10.1016/j.biotechadv.2011.06.019
Lee BK, Yun Y, Park K (2016) PLA micro- and nano-particles. Adv Drug Deliv Rev 107:176–191. https://doi.org/10.1016/j.addr.2016.05.020
Li J, Xiao P, Li H, Zhang Y, Xue F, Luo B, Huang S, Shang Y, Wen H, De Claville Christiansen J, Yu D, Jiang S (2015) Crystalline structures and crystallization behaviors of poly(L-lactide) in

poly(L-lactide)/graphene nanosheet composites. Polym Chem 6:3988–4002. https://doi.org/10.1039/C5PY00254K

Li X, Shang X, Lyu J, Tong Y, Situ W, Yu L, Wu T, Xie H, Qu J (2023) Efficient fabrication of PLA/PHB composites with enhanced mechanical properties, excellent thermal stability, fast crystallization ability, and degradation rate via the synergistic of weak shear field and melt quenching technique. Ind Crops Prod 196:116516. https://doi.org/10.1016/J.INDCROP.2023.116516

Lohmeijer BGG, Pratt RC, Leibfarth F, Logan JW, Long DA, Dove AP, Nederberg F, Choi J, Wade C, Waymouth RM, Hedrick JL (2006) Guanidine and amidine organocatalysts for ring-opening polymerization of cyclic esters. Macromolecules 39:8574–8583. https://doi.org/10.1021/MA0619381

Ma B, Wang X, He Y, Dong Z, Zhang X, Chen X, Liu T (2021) Effect of poly(lactic acid) crystallization on its mechanical and heat resistance performances. Polymer (Guildf) 212:123280. https://doi.org/10.1016/j.polymer.2020.123280

Maglio G, Malinconico M, Migliozzi A, Groeninckx G (2004) Immiscible poly(L-lactide)/poly(ε-caprolactone) blends: influence of the addition of a poly(L-lactide)-poly(oxyethylene) block copolymer on thermal behavior and morphology. Macromol Chem Phys 205:946–950. https://doi.org/10.1002/MACP.200300150

Maleki H, Azimi B, Ismaeilimoghadam S, Danti S (2022) Poly(lactic acid)-based electrospun fibrous structures for biomedical applications. Appl Sci 12. https://doi.org/10.3390/app12063192

Marwat ZK, Baloch MK (2015) Solvent induced miscibility between polymers, and its influence on the morphology, and mechanical properties of their blends. Eur Polym J 66:520–527. https://doi.org/10.1016/j.eurpolymj.2015.02.039

Mazumder MAJ, Sheardown H, Al-Ahmed A (eds) (2018) Polymer blends. https://doi.org/10.1007/978-3-319-95987-0

Meereboer KW, Misra M, Mohanty AK (2020) Review of recent advances in the biodegradability of polyhydroxyalkanoate (PHA) bioplastics and their composites. Green Chem 22:5519–5558. https://doi.org/10.1039/d0gc01647k

Mi HY, Salick MR, Jing X, Jacques BR, Crone WC, Peng XF, Turng LS (2013) Characterization of thermoplastic polyurethane/polylactic acid (TPU/PLA) tissue engineering scaffolds fabricated by microcellular injection molding. Mater Sci Eng C 33:4767–4776. https://doi.org/10.1016/j.msec.2013.07.037

Mohammed-Sadhakathullah AHM, Paulo-Mirasol S, Molina BG, Torras J, Armelin E (2024) PLA-PEG-cholesterol biomimetic membrane for electrochemical sensing of antioxidants. Electrochim Acta 476:143716. https://doi.org/10.1016/j.electacta.2023.143716

Mokarizadeh Haghighi Shirazi M, Khajouei-Nezhad M, Zebarjad SM, Ebrahimi R (2020) Evolution of the crystalline and amorphous phases of high-density polyethylene subjected to equal-channel angular pressing. Polym Bull 77:1681–1694. https://doi.org/10.1007/s00289-019-02827-7

Murariu M, Dubois P (2016) PLA composites: from production to properties. Adv Drug Deliv Rev 107:17–46. https://doi.org/10.1016/j.addr.2016.04.003

Myers D, Abdel-Wahab A, Hafeez F, Kovacev N, Essa K (2022) Optimisation of the additive manufacturing parameters of polylactic acid (PLA) cellular structures for biomedical applications. J Mech Behav Biomed Mater 136:105447. https://doi.org/10.1016/j.jmbbm.2022.105447

Navarro-Baena I, Sessini V, Dominici F, Torre L, Kenny JM, Peponi L (2016) Design of biodegradable blends based on PLA and PCL: from morphological, thermal and mechanical studies to shape memory behavior. Polym Degrad Stab 132:97–108. https://doi.org/10.1016/j.polymdegradstab.2016.03.037

Niaounakis M (2014) Biopolymers: processing and products. Elsevier, pp 1–601. https://doi.org/10.1016/C2013-0-09982-3

Nofar M, Heuzey MC, Carreau PJ, Kamal MR (2016) Effects of nanoclay and its localization on the morphology stabilization of PLA/PBAT blends under shear flow. Polymer (Guildf) 98:353–364. https://doi.org/10.1016/j.polymer.2016.06.044

Nofar M, Salehiyan R, Sinha Ray S (2019) Rheology of poly (lactic acid)-based systems. Polym Rev 59:465–509. https://doi.org/10.1080/15583724.2019.1572185

Oh JK (2011) Polylactide (PLA)-based amphiphilic block copolymers: synthesis, self-assembly, and biomedical applications. Soft Matter 7:5096–5108. https://doi.org/10.1039/c0sm01539c

Olaiya NG, Nuryawan A, Oke PK, Khalil HPSA, Rizal S, Mogaji PB, Sadiku ER, Suprakas SR, Farayibi PK, Ojijo V, Paridah MT (2020) The role of two-step blending in the properties of starch/chitin/polylactic acid biodegradable composites for biomedical applications. Polymers (Basel) 12. https://doi.org/10.3390/polym12030592

Patrício T, Bártolo P (2013) Thermal stability of PCL/PLA blends produced by physical blending process. Procedia Eng. 59:292–297. https://doi.org/10.1016/J.PROENG.2013.05.124

Patrício T, Domingos M, Gloria A, Bártolo P (2013) Characterisation of PCL and PCL/PLA scaffolds for tissue engineering. Procedia CIRP 5:110–114. https://doi.org/10.1016/j.procir.2013.01.022

Patrício T, Domingos M, Gloria A, D'Amora U, Coelho JF, Bártolo PJ (2014) Fabrication and characterisation of PCL and PCL/PLA scaffolds for tissue engineering. Rapid Prototyp J 20:145–156. https://doi.org/10.1108/RPJ-04-2012-0037

Pawar RP, Tekale SU, Shisodia SU, Totre JT, Domb AJ (2014) Biomedical applications of poly(lactic acid). Rec Pat Regen Med 4:40–51. https://doi.org/10.2174/2210296504666140402235024

Peesan M, Rujiravanit R, Supaphol P (2006) Electrospinning of hexanoyl chitosan/polylactide blends. J Biomater Sci Polym Ed 17:547–565. https://doi.org/10.1163/156856206776986251

Pereira-Lobato C, Echeverry-Rendón M, Fernández-Blázquez JP, González C, LLorca J (2024) Mechanical properties, in vitro degradation and cytocompatibility of woven textiles manufactured from PLA/PCL commingled yarns. J Mech Behav Biomed Mater 150:106340. https://doi.org/10.1016/j.jmbbm.2023.106340

Perry M (2008) Advanced trauma life support (ATLS) and facial trauma: can one size fit all? Part 1: dilemmas in the management of the multiply injured patient with coexisting facial injuries. Int J Oral Maxillofac Surg 37:209–214. https://doi.org/10.1016/j.ijom.2007.11.003

Prabhakar MN, Sajankumarji U, Kumara Babu RP, Subha MCS (2013) Interpenetrating polymer network hydrogel membranes of PLA and SA for control release of penicillamine drug. Indian J Adv Chem Sci 1:240–249

Przybysz-Romatowska M, Haponiuk J, Formela K (2020) Poly(ε-caprolactone)/poly(lactic acid) blends compatibilized by peroxide initiators: comparison of two strategies. Polymers 12:228. https://doi.org/10.3390/POLYM12010228

Puthumana M, Santhana Gopala Krishnan P, Nayak SK (2020) Chemical modifications of PLA through copolymerization. Int J Polym Anal Charact 25:634–648. https://doi.org/10.1080/1023666X.2020.1830650

Ramot Y, Haim-Zada M, Domb AJ, Nyska A (2016) Biocompatibility and safety of PLA and its copolymers. Adv Drug Deliv Rev 107:153–162. https://doi.org/10.1016/j.addr.2016.03.012

Ranakoti L, Gangil B, Misha SK, Singh T, Sharma S, Ilyas RA, El-Khatib S (2022) Critical review on polylactic acid: properties, structure. Materials (Basel) 15:1–29

Relinque JJ, de León AS, Hernández-Saz J, García-Romero MG, Navas-Martos FJ, Morales-Cid G, Molina SI (2019) Development of surface-coated polylactic acid/polyhydroxyalkanoate (PLA/PHA) nanocomposites. Polymers 11:400. https://doi.org/10.3390/POLYM11030400

Restrepo I, Medina C, Meruane V, Akbari-Fakhrabadi A, Flores P, Rodríguez-Llamazares S (2018) The effect of molecular weight and hydrolysis degree of poly(vinyl alcohol)(PVA) on the thermal and mechanical properties of poly(lactic acid)/PVA blends. Polímeros 28:169–177. https://doi.org/10.1590/0104-1428.03117

Riaz S, Fatima N, Rasheed A, Riaz M, Anwar F, Khatoon Y (2018) Metabolic engineered biocatalyst: a solution for PLA based problems. Int J Biomater 2018. https://doi.org/10.1155/2018/1963024

Risse S, Tighzert L, Berzin F, Vergnes B (2014) Microstructure, rheological behavior, and properties of poly(lactic acid)/poly(butylene succinate)/organoclay nanocomposites. J Appl Polym Sci 131:40364. https://doi.org/10.1002/APP.40364

Saini P, Arora M, Kumar MNVR (2016) Poly(lactic acid) blends in biomedical applications. Adv Drug Deliv Rev 107:47–59. https://doi.org/10.1016/j.addr.2016.06.014

Scaffaro R, Lopresti F, Maio A, Sutera F, Botta L (2017) Development of polymeric functionally graded scaffolds: a brief review. J Appl Biomater Funct Mater 15:e107–e121. https://doi.org/10.5301/jabfm.5000332

Serra T, Ortiz-Hernandez M, Engel E, Planell JA, Navarro M (2014) Relevance of PEG in PLA-based blends for tissue engineering 3D-printed scaffolds. Mater Sci Eng C 38:55–62. https://doi.org/10.1016/J.MSEC.2014.01.003

Sha L, Chen Z, Chen Z, Zhang A, Yang Z (2016) Polylactic acid based nanocomposites: promising safe and biodegradable materials in biomedical field. Int J Polym Sci 2016. https://doi.org/10.1155/2016/6869154

Shan Teo P, Shyang Chow W (2014) Water Vapour permeability of poly(lactic acid)/chitosan binary and ternary blends. King Mongkut's Univ Technol North Bangkok Int J Appl Sci Technol 7:23–27. https://doi.org/10.14416/J.IJAST.2014.01.001

Shasteen C, Bin Choy Y (2011) Controlling degradation rate of poly(lactic acid) for its biomedical applications. Biomed Eng Lett 1:163–167. https://doi.org/10.1007/s13534-011-0025-8

Sheth M, Kumar RA, Dave V (2008) Biodegradable polymer blends of poly (lactic acid) and poly (ethylene glycol) PL 1495–1505

Shibata M, Inoue Y, Miyoshi M (2006) Mechanical properties, morphology, and crystallization behavior of blends of poly(l-lactide) with poly(butylene succinate-co-l-lactate) and poly(butylene succinate). Polymer (Guildf) 47:3557–3564. https://doi.org/10.1016/j.polymer.2006.03.065

Simunec DP, Jacob J, Kandjani AEZ, Trinchi A, Sola A (2023) Facilitating the additive manufacture of high-performance polymers through polymer blending: a review. Eur Polym J 201:112553. https://doi.org/10.1016/j.eurpolymj.2023.112553

Sin LT, Rahmat RR, Rahman WAWA (2012) Polylactic acid: PLA biopolymer technology and applications. In: Polylactic acid PLA, pp 1–341. https://doi.org/10.1016/C2010-0-65966-9

Sin LT, Rahmat AR, Rahman WAWA (2013a) Overview of poly(lactic acid). https://doi.org/10.1016/b978-1-4377-4459-0.00001-9

Sin LT, Rahmat AR, Rahman WAWA (2013b) Mechanical properties of poly(lactic acid). https://doi.org/10.1016/b978-1-4377-4459-0.00005-6

Singhal AR, Agrawal CM, Athanasiou KA (2007) Salient degradation features of a 50:50 PLA/PGA scaffold for tissue engineering. 2:197–207. https://doi.org/10.1089/TEN.1996.2.197

Singhvi MS, Zinjarde SS, Gokhale DV (2019) Polylactic acid: synthesis and biomedical applications. J Appl Microbiol 127:1612–1626. https://doi.org/10.1111/jam.14290

Siracusa V, Karpova S, Olkhov A, Zhulkina A, Kosenko R, Iordanskii A (2020) Gas transport phenomena and polymer dynamics in PHB/PLA blend films as potential packaging materials. Polymers (Basel) 12. https://doi.org/10.3390/POLYM12030647

Song JJ, Srivastava I, Kowalski J, Naguib HE (2014) Fabrication and characterization of a foamed polylactic acid (PLA)/thermoplastic polyurethane (TPU) shape memory polymer (SMP) blend for biomedical and clinical applications. Behav Mech Multifunct Mater Compos 2014(9058):90580B. https://doi.org/10.1117/12.2046494

Song M, Li S, Zhu G, Guo J (2023) Compatibilised and toughened of PLA/PCL blends via modified-chitosan linking amorphous regions: 4D printing and shape memory processes. Polym Test 125:108105. https://doi.org/10.1016/j.polymertesting.2023.108105

Srivastava A, Bhati P, Singh S, Agrawal M, Kumari N, Vashisth P, Chauhan P, Bhatnagar N (2024) A review on polylactic acid-based blends/composites and the role of compatibilizers in biomedical engineering applications. Polym Eng Sci 64:1003–1044. https://doi.org/10.1002/pen.26626

Stjerndahl A, Finne-Wistrand A, Albertsson AC, Bäckesjö CM, Lindgren U (2008) Minimization of residual tin in the controlled Sn(II)octoate-catalyzed polymerization of ε-caprolactone. J Biomed Mater Res Part A 87:1086–1091. https://doi.org/10.1002/jbm.a.31733

Tábi T, Sajó IE, Szabó F, Luyt AS, Kovács JG (2010) Crystalline structure of annealed polylactic acid and its relation to processing. Express Polym Lett 4:659–668. https://doi.org/10.3144/expresspolymlett.2010.80

Taib NAAB, Rahman MR, Huda D, Kuok KK, Hamdan S, Bin Bakri MK, Bin Julaihi MRM, Khan A (2023) A review on poly lactic acid (PLA) as a biodegradable polymer. Springer Berlin Heidelberg. https://doi.org/10.1007/s00289-022-04160-y

Takayama T, Daigaku Y, Ito H, Takamori H (2014) Mechanical properties of bio-absorbable PLA/PGA fiber-reinforced composites. J Mech Sci Technol 28:4151–4154. https://doi.org/10.1007/S12206-014-0927-3/METRICS

Tan GYA, Chen CL, Li L, Ge L, Wang L, Razaad IMN, Li Y, Zhao L, Mo Y, Wang JY (2014) Start a research on biopolymer polyhydroxyalkanoate (PHA): a review. Polymers (Basel) 6:706–754. https://doi.org/10.3390/polym6030706

Tang X, Thankappan SK, Lee P, Fard SE, Harmon MD, Tran K, Yu X (2014) Polymeric biomaterials in tissue engineering and regenerative medicine. Elsevier Inc. https://doi.org/10.1016/B978-0-12-396983-5.00022-3

Torres-Hernández YG, Ortega-Díaz GM, Téllez-Jurado L, Castrejón-Jiménez NS, Altamirano-Torres A, García-Pérez BE, Balmori-Ramírez H (2018) Biological compatibility of a polylactic acid composite reinforced with natural chitosan obtained from shrimp waste. Materials (Basel) 11. https://doi.org/10.3390/ma11081465

Tsuji H, Tsuruno T (2010) Accelerated hydrolytic degradation of poly(l-lactide)/poly(d-lactide) stereocomplex up to late stage. Polym Degrad Stab 95:477–484. https://doi.org/10.1016/j.polymdegradstab.2010.01.008

Tümer EH, Erbil HY (2021) Extrusion-based 3D printing applications of PLA composites: a review. Coatings 11:390. https://doi.org/10.3390/COATINGS11040390

Tyler B, Gullotti D, Mangraviti A, Utsuki T, Brem H (2016) Polylactic acid (PLA) controlled delivery carriers for biomedical applications. Adv Drug Deliv Rev 107:163–175. https://doi.org/10.1016/j.addr.2016.06.018

Vanovcanova Z, Alexy P, Feranc J, Plavec R, Bockaj J, Kalinakova L, Tomanova K, Perdochova D, Sarisky D, Galisova I (2016) Effect of PHB on the properties of biodegradable PLA blends. Chem Pap 70:1408–1415. https://doi.org/10.1515/CHEMPAP-2016-0075/METRICS

Vayshbeyn LI, Mastalygina EE, Olkhov AA, Podzorova MV (2023) Poly(lactic acid)-based blends: a comprehensive review. Appl Sci 13. https://doi.org/10.3390/app13085148

Vogel C, Hoffmann GG, Siesler HW (2009) Rheo-optical FT-IR spectroscopy of poly(3-hydroxybutyrate)/poly(lactic acid) blend films. Vib Spectrosc 49:284–287. https://doi.org/10.1016/J.VIBSPEC.2008.10.014

Woodard LN, Kmetz KT, Roth AA, Page VM, Grunlan MA (2017) Porous poly(ϵ-caprolactone)-poly(l-lactic acid) semi-interpenetrating networks as superior, defect-specific scaffolds with potential for cranial bone defect repair. Biomacromolecules 18:4075–4083. https://doi.org/10.1021/ACS.BIOMAC.7B01155/SUPPL_FILE/BM7B01155_SI_001.PDF

Wu CS, Liao HT (2005) A new biodegradable blends prepared from polylactide and hyaluronic acid. Polymer (Guildf) 46:10017–10026. https://doi.org/10.1016/J.POLYMER.2005.08.056

Xie Y, Zhang F, Akkus O, King MW (2022) A collagen/PLA hybrid scaffold supports tendon-derived cell growth for tendon repair and regeneration. J Biomed Mater Res—Part B Appl Biomater 110:2624–2635. https://doi.org/10.1002/jbm.b.35116

Xu J, Song J (2015) Polylactic acid (PLA)-based shape-memory materials for biomedical applications. In: Shape memory polymers for biomedical applications. pp 197–217. https://doi.org/10.1016/B978-0-85709-698-2.00010-6

Yan C, Hou DF, Zhang K, Yang MB (2023) Effects of PDLA molecular weight on the crystallization behaviors and rheological properties of asymmetric PDLA/PLLA blends. Polymer (Guildf) 270:125764. https://doi.org/10.1016/j.polymer.2023.125764

Yang H, Du J (2024) Crystallinity, rheology, and mechanical properties of low-/high-molecular-weight PLA blended systems. Molecules 29. https://doi.org/10.3390/molecules29010169

Yang Z, Chen L, McClements DJ, Qiu C, Li C, Zhang Z, Miao M, Tian Y, Zhu K, Jin Z (2022) Stimulus-responsive hydrogels in food science: a review. Food Hydrocoll 124:107218. https://doi.org/10.1016/J.FOODHYD.2021.107218

Yao Q, Cosme JGL, Xu T, Miszuk JM, Picciani PHS, Fong H, Sun H (2017) Three dimensional electrospun PCL/PLA blend nanofibrous scaffolds with significantly improved stem cells osteogenic differentiation and cranial bone formation. Biomaterials 115:115–127. https://doi.org/10.1016/J.BIOMATERIALS.2016.11.018

Zaaba NF, Jaafar M (2020) A review on degradation mechanisms of polylactic acid: hydrolytic, photodegradative, microbial, and enzymatic degradation. Polym Eng Sci 60:2061–2075. https://doi.org/10.1002/PEN.25511

Zainuddin MZ, Abu Bakar AA, Adam AN, Abdullah SM, Tamchek N, Alauddin MS, Mahat MM, Wiwatcharagoses N, Alforidi A, Ghazali MIM (2023) Mechanical and structural properties of polyhydroxybutyrate as additive in blend material in additive manufacturing for medical applications. Polymers (Basel) 15. https://doi.org/10.3390/polym15081849

Zeng JB, Li KA, Du AK (2015) Compatibilization strategies in poly(lactic acid)-based blends. RSC Adv 5:32546–32565. https://doi.org/10.1039/c5ra01655j

Zhang M, Thomas NL (2011) Blending polylactic acid with polyhydroxybutyrate: the effect on thermal, mechanical, and biodegradation properties. Adv Polym Technol 30:67–79. https://doi.org/10.1002/ADV.20235

Zhao X, Hu H, Wang X, Yu X, Zhou W, Peng S (2020) Super tough poly(lactic acid) blends: a comprehensive review. RSC Adv 10:13316–13368. https://doi.org/10.1039/d0ra01801e

Chapter 10
Therapeutic Applications of Sustainable Chitin and Chitosan Derivatives and Its Nanocomposites

Varsha Thadiyan, Vibhuti Sharma, Nivedita Sharma, and Reena Gupta

Abstract Chitin and chitosan are natural biopolymers derived from crustacean shells, fungi and insects and have emerged as versatile materials for therapeutic applications due to their biodegradability, biocompatibility and non-toxic nature. These polysaccharides, their derivatives and nanocomposites offer significant potential to address global health challenges sustainably. The functional modifications of chitin and chitosan enhance their physicochemical properties and enable their utilization in therapeutics. Chitosan-based nanocomposites often functionalized with metallic or polymeric nanoparticles, provide improved bioavailability and targeted therapeutic efficacy for various diseases, including cancer and microbial infections. Their antimicrobial, anti-inflammatory and antioxidant properties have also made them invaluable in wound care management, promoting faster healing and reducing scarring. Emerging therapeutic and medical applications include bioactive scaffolds for tissue regeneration and carriers for genetic material in gene therapy. The eco-friendly production methods and abundant availability of raw materials underline their sustainability and global applicability. This chapter emphasizes recent advancements in the multifunctional applications of chitin, chitosan, its derivatives and nanocomposites. It highlights their pivotal role in sustainable therapeutic interventions and the promise they hold for next-generation biomedical solutions.

Keywords Chitin · Chitosan · Biocompatibility · Non-toxic · Nanocomposites and biomedical

10.1 Introduction

Chitin a natural biopolymer is the second most abundant organic material on Earth, derived primarily from the exoskeletons of arthropods (e.g., shrimp, crabs and insects), fungal cell walls and mollusk shells (Pal and Kahawas 2024). Structurally, it

V. Thadiyan · V. Sharma · N. Sharma · R. Gupta (✉)
Department of Biotechnology, Himachal Pradesh University, Summerhill, Shimla, India
e-mail: reenagupta_2001@yahoo.com

P. M. Visakh (ed.), *Engineering Applications of Polymer based Nano Blends*, Advanced Structured Materials 235, https://doi.org/10.1007/978-3-031-90979-5_10

is a polysaccharide composed of *N*-acetylglucosamine units, providing strength and flexibility. Chitosan, a deacetylated derivative of chitin, is obtained through alkali treatment. Unlike chitin, chitosan is water-soluble under acidic conditions, making it versatile for various applications. Both compounds are widely used in agriculture, biomedicine and industry (Zargar et al. 2015). Chitin excels in wound healing, as a biodegradable matrix and in filtration technologies, while chitosan finds use as an antimicrobial agent, drug delivery vehicle and bioplastic (Rai et al. 2021). The biocompatibility, renewability and functional versatility of these biomaterials have positioned them as key resources in sustainable technology and advanced scientific research. The exploration of sustainable biopolymers for therapeutic applications has been significantly enhanced by advancements in chitin and chitosan derivatives along with their nanocomposites. Chitin and chitosan exhibit remarkable biodegradability and unique physicochemical properties making them highly promising candidates for various medical and pharmaceutical applications (Baharlouei and Rahman 2022). Their ability of chemical modifications into diverse derivatives extends their functionality enabling targeted and effective therapeutic interventions. The developments in nanotechnology have propelled the utility of chitin and chitosan derivatives into the nanoscopic realm offering improved drug delivery systems, enhanced therapeutic efficacy and controlled drug release mechanisms (Lanka 2020). One of the most prominent therapeutic applications of chitin, chitosan and its derivatives is in the domain of drug delivery systems. Their inherent mucoadhesive properties and the ability to open tight junctions in epithelial cells facilitate the transport of therapeutic molecules across biological membranes (Mura et al. 2022). For instance, chitosan nanoparticles are increasingly utilized to encapsulate drugs, ensuring their stability and providing controlled release, critical in diseases requiring precise dosages over extended periods (Herdiana et al. 2022).

Chemical modifications like carboxymethylation or grafting with functional molecules have augmented chitosan solubility and target specificity, paving the way for its use in targeted cancer therapy (Begum et al. 2024). Chitosan-based nanocarriers have also shown promise in gene delivery, addressing the critical challenges of efficient transfection and minimal cytotoxicity (Sharma et al. 2019). By forming stable polyplexes with nucleic acids, chitosan derivatives offer a non-viral alternative for gene therapy, reducing the risk of immunogenicity which is commonly associated with viral vectors (Kritchenkov et al. 2017). Chitin and chitosan derivatives exhibit inherent antimicrobial, antioxidant and wound-healing properties, which have been leveraged in wound management and tissue engineering (Mawazi et al. 2024). The hemostatic and bacteriostatic capabilities of chitosan support rapid clot formation and infection prevention in wound dressings (Wang et al. 2019). Modified chitosan-based hydrogels and scaffolds have also been employed in tissue regeneration, particularly in skin, cartilage and bone repair, owing to their ability to promote cell adhesion, proliferation and differentiation (Haider et al. 2024). These biomaterials can be tailored with bioactive molecules such as growth factors, enhancing their regenerative efficacy. The therapeutic potential of chitosan nanocomposites has extended into oncology, where they play a vital role in anticancer strategies (Herdiana et al. 2021). Chitosan nanoparticles conjugated with chemotherapeutic agents, siRNA, or photodynamic therapy agents have demonstrated significant anti-cancer activity, with minimal off-target effects (Fathi et al. 2018). The versatility of chitosan derivatives in encapsulating hydrophobic drugs and protecting them from premature degradation makes them ideal for cancer treatments where drug stability is a major concern (Narmani and Jafari 2021). Nanocomposites integrating chitosan with inorganic nanoparticles, such as gold or magnetic nanoparticles, have further amplified their effectiveness in imaging-guided therapies and hyperthermia (Andoh et al. 2024). Chitin and chitosan derivatives have also carved a niche in regenerative medicine, specifically in the development of biomaterials for bone and cartilage engineering. The structural similarity of chitin to glycosaminoglycans in the extracellular matrix makes it a suitable scaffold for cell growth and tissue regeneration (Islam et al. 2017). Composite materials incorporating chitin or chitosan with hydroxyapatite, graphene oxide or bio-ceramics have shown remarkable success in mimicking the natural bone matrix, promoting osteogenesis and supporting the sustained release of growth factors for effective bone repair (Venkatesan et al. 2023).

The versatility of chitin and chitosan derivatives extends further into cardiovascular and neural therapeutics. In cardiovascular medicine, chitosan hydrogels and nanoparticles have been utilized to deliver drugs and growth factors, aiding in myocardial repair and vascularization (Cao et al. 2024). In neural applications, chitosan-based scaffolds and drug delivery systems are being explored for repairing nerve injuries and delivering neuroprotective agents, offering hope for treatment (Khodaverdi et al. 2024). Chitin and chitosan derivatives and their nanocomposites exemplify the tremendous potential of sustainable biomaterials in advancing therapeutic solutions. With their wide array of functionalities, ranging from drug delivery to tissue engineering and antimicrobial applications, these biopolymers are poised to revolutionize modern medicine (Desai et al. 2023). Continued research into their properties and interactions, along with advancements in nanotechnology, will undoubtedly unlock new dimensions for their therapeutic applications, ensuring their prominent role in the future of healthcare and biomedicine.

10.2 Therapeutic Applications of Sustainable Chitin Derivatives and Its Nanocomposites

Chitin is derived primarily from the exoskeletons of crustaceans and certain fungal species, and the chemical structure of chitin is shown in Fig. 10.1. Its structural and functional versatility makes it an essential component in various biomedical applications (Ahmad et al. 2020). Its unique mechanical strength, natural biocompatibility and biodegradability allow it to serve as a foundation for creating innovative nanocomposites designed for precise medical interventions. The modifications of chitin, such as deacetylation or chemical grafting, yield derivatives with improved solubility, enhanced biological interactions and functional groups tailored for specific therapeutic purposes (Silva et al. 2017) (Fig. 10.2).

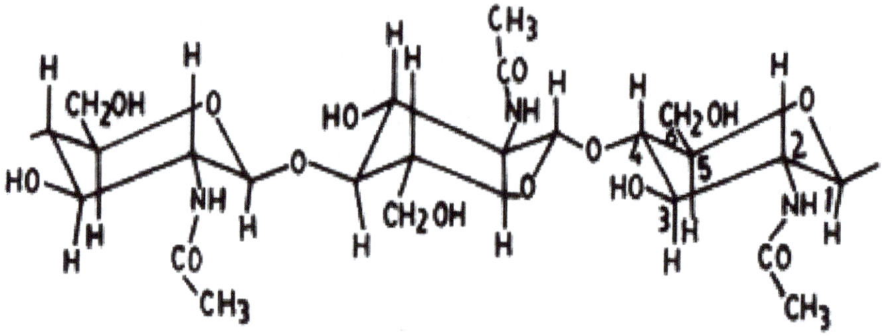

Fig. 10.1 Chemical structure of chitin

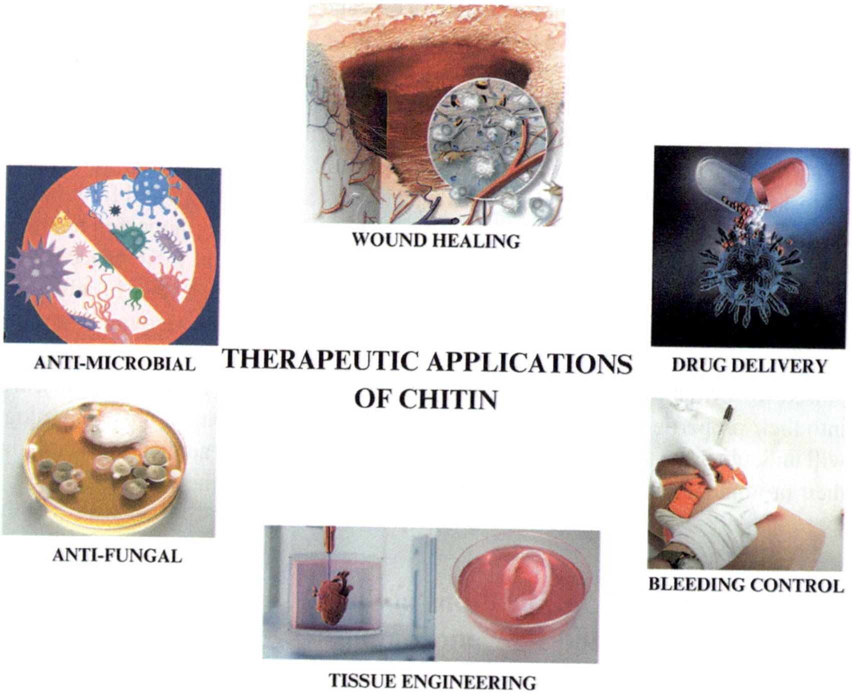

Fig. 10.2 Therapeutic applications of chitin

10.2.1 Drug Delivery Systems

Chitin derivatives have proven particularly beneficial in drug delivery systems, where their stability and ability to encapsulate bioactive compounds enable controlled release and target specificity (Parhi 2020). A major advancement in chitin involves its modification with functional groups to enhance its solubility and bioactivity

(Arjun et al. 2023). This allows its application in drug and gene delivery where drugs and nucleic acids can be transported effectively to target cells. Nanoparticles formed from chitin-based materials ensure drug protection from premature degradation while optimizing bioavailability, particularly in challenging environments such as acidic conditions. Studies also explore chitin pH-sensitive properties and enable site-specific drug release, such as in the gastrointestinal tract (Osak et al. 2024). Chitin nanofibers derived through mechanical or enzymatic processing exhibit high surface area and unique binding capacities, making them suitable carriers for drug molecules (Yang et al. 2020).

10.2.2 Gene Therapy and Tissue Engineering

Chitin potential extends into gene therapy, offering a biocompatible alternative for transporting nucleic acids efficiently and safely to target cells without eliciting significant immune responses (Gholap et al. 2024). These derivatives are employed in advanced cancer treatments, as they facilitate the delivery of chemotherapeutics, immunomodulators and small interfering RNAs (siRNA) to tumors, reducing systemic toxicity and enhancing therapeutic efficacy (Subhan et al. 2021). The wound healing and tissue engineering sectors have also seen substantial contributions from chitin nanocomposites. Wound dressings incorporating chitin derivatives promote rapid healing by providing an antimicrobial barrier and facilitating the skin natural regenerative processes (Jayakumar et al. 2011). In bone and cartilage repair, scaffolds designed from chitin composites have been shown to mimic the extracellular matrix, promoting cell adhesion, proliferation and differentiation crucial for tissue regeneration (Tao et al. 2020). These scaffolds, often fortified with other biocompatible materials like hydroxyapatite, are particularly effective in orthopedic applications where biomimetic properties are essential for long-term integration and stability (Bhat et al. 2021). Chitin nanocomposites have also shown significant promise in gastrointestinal therapies. Their ability to bind lipids and bile acids has made them effective in making formulations to lower cholesterol and support gut health (Wijesekara and Xu 2024). As prebiotics, they positively influence gut microbiota composition, further highlighting their relevance in nutrition-focused therapeutics. Their incorporation into delivery vehicles for vaccines and antimicrobial peptides provides a stable, efficient and cost-effective means to enhance immunization programs and address antibiotic-resistant infections.

10.2.3 Antimicrobial and Antifungal Agent

In the realm of antimicrobial treatments, chitin offers broad-spectrum activity against pathogenic bacteria, fungi and even viruses. Their antimicrobial mechanisms, combined with their ability to enhance the action of encapsulated bio-actives, make them indispensable in creating coatings for surgical implants, devices and hospital environments (Croitoru et al. 2024). Its natural origin suggests fewer side effects compared to chemical agents, making it a valuable candidate in pharmaceuticals and medicine. The specific compound 2-acetamido-2-deoxy-hexopyranose derived from chitin has demonstrated binding affinity to penicillin-binding proteins in *S. aureus*, indicating a novel mechanism of action against antibiotic-resistant strains (Vijayaraj et al. 2022). Chitin nanofibers when combined with copper nanoparticles showed enhanced antimicrobial activity against plant pathogens and improved crop protection and yield (Egusa et al. 2023). Its incorporation into nano-biocomposites enhances its antifungal properties, making it useful for early detection and treatment of fungal infections (Gómez-Gaviria and Mora-Montes 2024). Chitin has antibacterial and antiviral properties and can induce plant disease resistance and promote growth. Its application is limited due to insolubility, but when modified, it can effectively combat pathogens like the tobacco mosaic virus (Cao et al. 2023).

10.2.4 Immunomodulation and Vaccine Adjuvants

Chitin fragments depending on their size can act as immune modulators by interacting with pattern recognition receptors such as Toll-like receptors (TLRs) and nucleotide-binding oligomerization domain-containing receptors (NODs) (Jiang et al. 2004). These interactions initiate immune signaling cascades, resulting in the activation or suppression of specific immune pathways. Studies have shown that chitin microparticles enhance the recruitment and activation of macrophages, dendritic cells and neutrophils which are crucial for innate immunity. The nano-sized chitin derivatives have been investigated for their ability to reduce inflammatory responses, highlighting their anti-inflammatory properties in conditions such as asthma, allergies and autoimmune disorders. Chitin-based nanoparticles also serve as delivery systems for immunomodulatory agents, effectively enhancing the efficacy of vaccines by targeting antigen-presenting cells (Woodrow et al. 2012). As an adjuvant, chitin enhances the body immune response to antigens making vaccines more effective by promoting prolonged and robust immunity (Hung et al. 2018). Chitin and its derivatives and its nanoparticles have been widely explored for mucosal vaccine delivery, particularly for nasal and oral immunization. Their mucoadhesive properties facilitate prolonged retention at mucosal sites, enhancing antigen absorption and immune activation. Chitin-based adjuvants have shown promise in improving the efficacy of vaccines against infectious diseases, including influenza, tuberculosis and COVID-19 (Wike et al. 2023). Chitin nanoparticles enable controlled antigen release, ensuring sustained immune activation and reducing the need for booster doses. Studies have also explored the potential of chitin-based adjuvants in cancer immunotherapy by promoting anti-tumor immune responses. Chitin-derived polymers like C100 activate the stimulator of interferon genes (STING) pathway, leading to robust anti-tumor immunity through CD8+ T cell responses (Turley et al. 2024). Chitin can neutralize chitinase-like proteins (CLPs), reducing immunosuppression and enhancing the efficacy of immune checkpoint blockade therapies. The use of chitin in conjunction with other immunotherapies has shown promise in overcoming resistance mechanisms, particularly in triple-negative breast cancer models (Salembier et al. 2024). With continued research, chitin-based adjuvants could revolutionize vaccine formulations, providing safer, more effective and sustainable immunization strategies. Chitin promotes wound healing through its immunostimulatory effects, including the stimulation of pro-inflammatory cytokines such as interleukin-1β (IL-1β) and tumor necrosis factor-alpha (TNF-α) in early wound responses (Lee et al. 2008).

This immunomodulatory property is employed in tissue engineering and regenerative medicine to facilitate repair. Its role in vaccine adjuvant technology is particularly noteworthy, as it enhances the immune response to weak antigens by improving antigen delivery and presentation. Chitin multifunctionality offers new avenues for sustainable, safe and efficient immune regulation strategies (Mu et al. 2024).

10.2.5 Wound Healing and Bleeding Control

Chitin plays a crucial role in wound healing by promoting cell proliferation, collagen synthesis and angiogenesis, all of which accelerate tissue regeneration (Azuma et al. 2015). When applied as a wound dressing, chitin maintains a moist wound environment which aids in faster epithelialization and reduces scar formation. Chitin-based dressings in the form of hydrogels, sponges and membranes have been widely used for treating chronic wounds, burns and diabetic ulcers (Tachaboonyakiat 2021). The incorporation of nanomaterials into chitin structures has further optimized its therapeutic efficacy. Nanocomposites incorporating chitin and nanoparticles, such as silver (AgNPs), zinc oxide (ZnO) and titanium dioxide (TiO_2), enhance antimicrobial activity and promote rapid tissue regeneration (Singh 2019). Silver nanoparticles, for instance, exhibit potent bactericidal effects, preventing wound infections and facilitating healing. Chitin graphene oxide nanocomposites provide additional mechanical strength and bioactivity, while chitin-based nanofibers loaded with growth factors or anti-inflammatory agents can modulate immune responses for improved healing (Sivanesan et al. 2021). Chitin hydrogel nanocomposites ensure continuous therapeutic effects for chronic wound management. Chitin is an effective hemostatic agent due to its ability to induce rapid blood clotting (Whang et al. 2005). Upon application chitin interacts with platelets and erythrocytes, facilitating clot formation independently of the body coagulation pathways. This property makes chitin-based hemostatic materials particularly valuable for trauma care, military applications and surgical procedures, especially for patients with bleeding disorders. Chitin dressings not only absorb excess exudates but also create a physical barrier that minimizes further bleeding and infection risk.

10.3 Therapeutic Applications of Sustainable Chitosan Derivatives and Its Nanocomposites

The advancement in biopolymer technology has positioned chitosan derivatives and their nanocomposites as pivotal elements in therapeutic applications, leveraging their sustainability, biocompatibility and adaptability. Derived from the deacetylation of chitin, chitosan possesses unique properties such as biodegradability, mucoadhesiveness and antimicrobial activity, making it suitable for diverse medical and pharmaceutical innovations. Chitosan is structurally composed of glucosamine and *N*-acetylglucosamine units linked by β-(1 → 4) glycosidic bonds (Fig. 10.3). The ability to modify its structure through chemical and enzymatic modifications enhances its application potential (Gao et al. 2023). Chemical modifications of chitosan, such as quaternization, sulfation and carboxymethylation, enhance its solubility, mechanical strength and functional specificity allowing its use in advanced therapeutic systems (Mourya and Inamdar 2008). Chitosan and its derivatives play significant roles in drug delivery, wound healing, tissue engineering, antimicrobial applications, immunomodulation and cancer therapy (Xia et al. 2022) (Fig. 10.4). When incorporated into nanocomposites, these derivatives exhibit further improved efficacy targeted delivery capabilities making them invaluable for modern medicine.

10.3.1 Wound Healing and Antimicrobial Applications

Chitosan plays a pivotal role in wound healing and antimicrobial applications due to its biocompatibility, biodegradability and intrinsic bioactive properties. Its ability to enhance wound repair is linked to several key mechanisms which include its hemostatic property and promotes clot formation and its ability to stimulate fibroblast proliferation and extracellular matrix deposition. Chitosan has been widely utilized in wound dressings in the form of films, hydrogels and sponges, which provide a moist environment crucial for wound healing while also serving as a barrier against microbial infections (Elangwe et al. 2022). They support the proliferation and migration of fibroblasts, essential for tissue repair while providing an antimicrobial barrier that prevents infection. Moreover, chitosan activates macrophages and neutrophils,

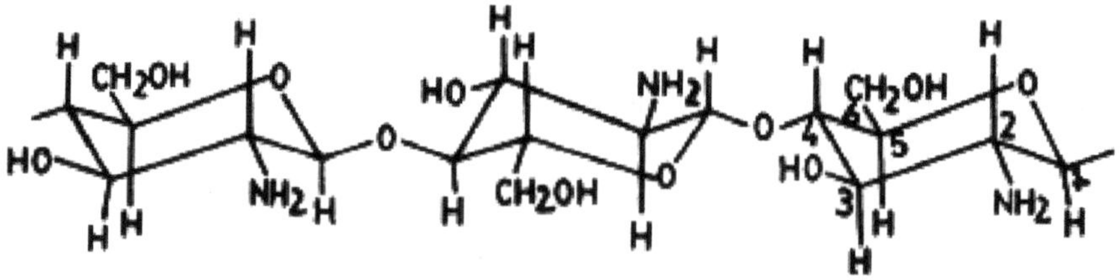

Fig. 10.3 Chemical structure of chitosan

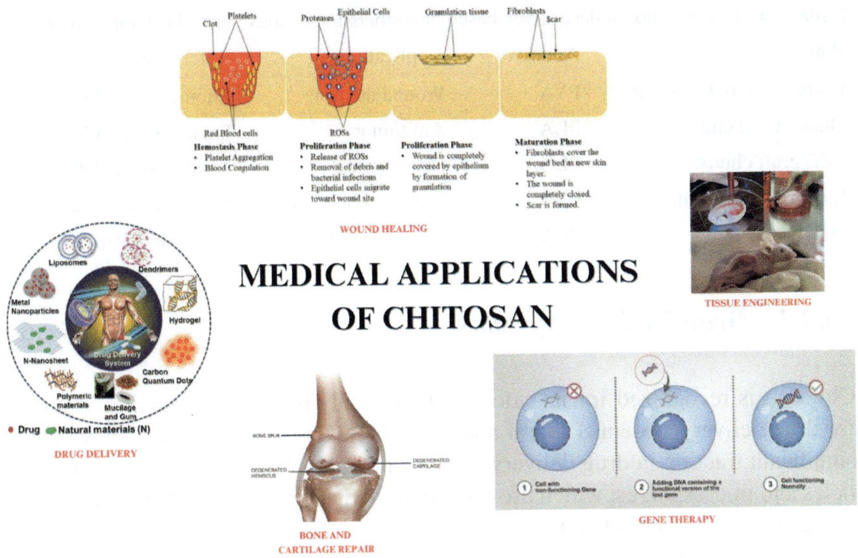

Fig. 10.4 Medical applications of chitosan

enhancing the release of growth factors that accelerate the re-epithelialization process (Saini et al. 2020). As an antimicrobial agent, chitosan demonstrates broad-spectrum activity against Gram-positive and Gram-negative bacteria, as well as fungi (Ke et al. 2021). Its antimicrobial effect arises from its positive charge, which interacts with negatively charged microbial membranes, leading to increased permeability and eventual cell lysis. This property makes chitosan particularly effective in preventing infections in chronic wounds or burns. Chitosan-based wound dressings are further enhanced by combining chitosan with nanoparticles of silver, copper or zinc oxide, which amplify its antimicrobial efficacy (Shekede 2018). Such nanocomposites address multidrug-resistant pathogens and provide sustained release of therapeutic agents at the wound site. Beyond antimicrobial action, chitosan and derivatives contribute to wound healing by reducing inflammation and oxidative stress, which are crucial for the resolution of chronic wounds. Chitosan derivatives like carboxymethyl chitosan or glycol chitosan are also explored for their enhanced solubility and bioactivity, improving their applicability in advanced therapeutic formulations (Edo et al. 2024) (Table 10.1). Chitosan has been incorporated into scaffolds for tissue engineering to promote both antibacterial protection and tissue regeneration (Sivashankari and Prabaharan 2016). Its dual ability to repair tissue and inhibit infections establishes chitosan as a vital material in modern wound care innovations. For tissue engineering, chitosan nanocomposites are incorporated with materials like collagen, hydroxyapatite, and bioactive molecules to form scaffolds that mimic natural extracellular matrices, facilitating regeneration in skin, bone and cartilage repair.

Table 10.1 Chitin/chitosan derivatives-based nanofibers for various biomedical applications

Polymer	Additives	Applications	Reference
Carboxymethyl chitosan	PVA	Wound dressing	Zhou et al. (2008)
Quaternized chitosan	PLA	Antitumor	Ignatova et al. (2010)
PEG graft chitosan	PLGA	Drug release	Jiang et al. (2019)
Carboxymethyl chitin	PVA	Tissue engineering	Shalumon et al. (2009)
Dibutyryl chitin	PLA	Wound healing	Jang et al. (2012)

10.3.2 Drug Delivery Systems

Chitosan is recognized for its safety in biological applications, making it suitable for drug delivery systems (Sharaf and Liu 2024). Chitosan derivatives have gained significant traction in drug delivery systems owing to their ability to form nanoparticles and microspheres, which provide controlled and sustained drug release. These properties are particularly advantageous in cancer therapy, where chitosan nanocarriers can encapsulate chemotherapeutic agents, ensuring their stability and targeted release to tumor cells while minimizing systemic toxicity and chitosan nanoparticles serve as effective drug delivery agents in brain delivery, improving bioavailability and minimizing systemic side effects. They facilitate drug transport across the blood–brain barrier, making them promising for treating neurodegenerative diseases and other medical conditions (Omidian et al. 2024). Chitosan is effective in prolonging the release of local anesthetics, addressing the challenge of short duration in pain management (Yu et al. 2024). Chitosan-based drug delivery systems are being explored in gene therapy, where they form complexes with nucleic acids such as DNA and siRNA, ensuring efficient cellular uptake and minimizing degradation risks. The non-toxic and non-immunogenic nature of chitosan-derived vectors makes them promising substitutes for traditional viral delivery systems.

10.3.3 Cardiovascular and Neural Therapy

Chitosan role in cardiovascular and neural therapies is steadily expanding. In cardiovascular medicine, chitosan-based hydrogels and nanoparticles are used to deliver growth factors and therapeutics for myocardial regeneration and vascular repair (Patel et al. 2021). The polymer ability to promote angiogenesis and its compatibility with endothelial cells make it a valuable material in repairing heart tissue (Deng et al. 2010). In neural applications, chitosan scaffolds and hydrogels facilitate neural growth and repair following injury or degenerative disorders, serving as a conduit for nerve regeneration while delivering neuroprotective drugs effectively (Pai et al. 2024).

10.3.4 Vaccination and Immunotherapy

Chitosan has emerged as a promising material in vaccination and immunotherapy due to its muco-adhesiveness and ability to enhance immune responses. Its unique properties allow it to act as an adjuvant and carrier for antigens improving antigen stability and delivery efficiency. Chitosan nanoparticles are especially effective in mucosal vaccination as they enhance the uptake of antigens by mucosal cells and stimulate both systemic and mucosal immunity (Renu and Renukaradhya 2020). This is particularly advantageous for oral, nasal and pulmonary vaccines, where chitosan ability to protect antigens from enzymatic degradation and facilitate cellular uptake plays a critical role. Chitosan-based delivery systems trigger a controlled release of antigens, prolonging immune activation and enhancing the immunogenicity of subunit vaccines (Gong et al. 2022). In immunotherapy, chitosan capacity to modulate immune responses has been leveraged to stimulate dendritic cells and macrophages, boosting the production of pro-inflammatory cytokines such as interleukin-6 (IL-6) and tumor necrosis factor-alpha (TNF-α) (Lin et al. 2024). This makes chitosan an excellent candidate for cancer immunotherapy, where targeted delivery of immune modulators can potentiate anti-tumor immunity. Chitosan derivatives, such as trimethylated chitosan, have shown superior efficacy in improving the uptake of RNA and DNA vaccines. With its ability to act both as a delivery system and an immunostimulant, chitosan holds great potential for advancing vaccination strategies and immunotherapeutic interventions (Kumar et al. 2024).

10.3.5 Dietary and Metabolic Health

Chitosan is significant in promoting dietary and metabolic health due to its ability to bind to lipids and prevent their absorption in the gastrointestinal tract. Acting as a natural dietary fiber, chitosan contributes to weight management by inhibiting fat absorption and reducing calorie intake (Preuss and Kaats 2006). This lipid-binding property has also been linked to its ability to lower serum cholesterol and triglyceride levels, thereby improving cardiovascular health. Studies have shown that chitosan can modulate lipid metabolism by promoting the excretion of bile acids, which triggers cholesterol breakdown to maintain balance in the body such as it enhances the excretion of bile acids, which is crucial for cholesterol metabolism. Increased bile acid levels in feces indicate enhanced cholesterol breakdown (Xu et al. 2007). Clinical trials have demonstrated that chitosan supplementation significantly lowers total cholesterol (TC) and LDL-C levels in participants, with reductions of approximately 0.20 mmol/L for TC and 0.20 mmol/L for LDL-C (Huang et al. 2018). Chitosan has been found to regulate blood glucose levels by slowing the absorption of glucose and enhancing insulin sensitivity, making it a promising adjunct in managing diabetes (Tzeng et al. 2022). Its prebiotic effects are another important aspect, as chitosan derivatives promote the growth of beneficial gut microbiota, which contributes to improved metabolic health. Chitosan supports digestive health by enhancing gut motility and reducing gastrointestinal inflammation (Zhang et al. 2022). Sustainable chitosan-based supplements and functional foods have gained increasing attention as safe and effective strategies for addressing obesity, metabolic syndrome and related chronic conditions (Galea and Pavli 2024). This makes it a valuable tool in advancing personalized and preventive healthcare approaches.

10.3.6 Cancer Therapy

Chitosan-based nanoparticles have shown great potential in cancer therapy by improving drug solubility, enhancing cellular uptake and reducing systemic toxicity. As a promising biomaterial, chitosan and its derivatives have been extensively investigated for their role in drug delivery, gene therapy, photothermal therapy and direct anticancer effects (Adhikari and Yadav 2018). One of the most prominent applications of chitosan in oncology is its use as a drug delivery carrier. Chitosan enhances the bioavailability of chemotherapeutic drugs, allowing for better absorption and prolonged circulation time. Its ability to form nanoparticles, hydrogels and microspheres provides a controlled and sustained drug release mechanism, reducing the systemic toxicity commonly associated with conventional chemotherapy (Tian et al. 2020). The functionalization of chitosan with ligands such as folic acid and peptides enables targeted drug delivery, specifically to cancer cells, minimizing damage to healthy tissues (Kesharwani et al. 2024). Chitosan-based nanoparticles have been utilized to encapsulate drugs like doxorubicin, paclitaxel and curcumin significantly improving their therapeutic efficacy (Kurczewska 2023). Chitosan is being explored in gene therapy for cancer treatment. It serves as a non-viral vector for the delivery of tumor-suppressing genes, RNA interference molecules and CRISPR/Cas9 components (Shivhare et al. 2023). Its cationic nature facilitates binding with negatively charged nucleic acids, promoting cellular uptake and efficient gene transfection (Mao et al. 2001). Studies have shown that chitosan-based gene delivery systems can silence oncogenes and enhance the expression of tumor suppressor genes, leading to cancer cell apoptosis. Delivery systems targeting genes like CASP8 and microRNAs (miRs) can restore the expression of tumor suppressor genes and promote apoptosis in resistant cancer cells (Chattopadhyay et al. 2023). It can induce apoptosis and inhibit the proliferation of cancer cells by disrupting mitochondrial membrane potential and activating caspase pathways. Chitosan oligosaccharides have also been reported to suppress tumor angiogenesis by downregulating vascular endothelial growth factor (VEGF) expression, thereby preventing tumor progression (Muanprasat and Chatsudthipong 2017). Chitosan-based hydrogels and scaffolds are being utilized in photothermal and photodynamic therapies, where they act as carriers for photosensitizers and nanoparticles, facilitating precise and effective cancer cell ablation with minimal side effects (Alsaikhan and Farhood 2024). The crucial application of chitosan in oncology is its role in immunotherapy. Chitosan-based adjuvants have been used to stimulate immune responses against cancer cells by enhancing antigen presentation and cytokine production (Lima et al. 2021). They have shown potential in vaccine development, particularly in dendritic cell-based immunotherapy. Moreover, chitosan-coated nanoparticles loaded with immunomodulators have been employed to modulate the tumor microenvironment, making it more susceptible to immune cell attack (Moghaddam et al. 2024). Chitosan has been utilized in postoperative cancer treatment, particularly in wound healing and tissue regeneration after tumor resection. Its hemostatic properties aid in controlling bleeding, while its antimicrobial nature reduces the risk of post-surgical infections (Cassano et al. 2024). The use of chitosan-based scaffolds in tissue engineering has further contributed to the development of biomimetic tumor models for drug screening and personalized cancer therapy.

10.3.7 Bone and Cartilage Regeneration

Chitosan role in bone and cartilage regeneration is multifaceted, encompassing its use as a scaffold, its ability to enhance cellular functions and its potential for promoting tissue repair and regeneration (Oryan and Sahvieh 2017). In bone regeneration, chitosan serves as an excellent scaffold material. The polymer is capable of forming a three-dimensional structure that mimics the extracellular matrix (ECM) of natural bone (Chahal et al. 2019). This structure provides mechanical support to the growing bone tissue, facilitating cell adhesion, migration and proliferation. Osteoblasts, the cells responsible for bone formation, are particularly attracted to chitosan scaffolds, where they are encouraged to differentiate and secrete bone matrix proteins such as collagen and osteocalcin (Muzzarelli 2011). Chitosan has been shown to enhance osteoblast activity, promote mineralization and facilitate the deposition of hydroxyapatite, a key mineral component of bone (Li et al. 2019). The ability of chitosan to deliver growth factors such as bone morphogenetic proteins (BMPs) or vascular endothelial growth factors (VEGFs) further accelerates bone healing and promotes angiogenesis, which is crucial for the successful regeneration of large bone defects (Zhang et al. 2019). In cartilage regeneration, chitosan plays a crucial role in supporting the repair of damaged cartilage by providing a conducive environment for chondrocyte survival and function (Eftekhari et al. 2020). Chitosan-based scaffolds have been demonstrated to enhance chondrocyte adhesion, proliferation and the synthesis of cartilage-specific extracellular matrix components, including collagen type II and proteoglycans. These ECM components are vital for the structural integrity and function of cartilage tissue. Chitosan ability to maintain an appropriate hydrated environment within the scaffold mimics the natural conditions of cartilage, which is essential for the proper functioning of chondrocytes (Chicatun et al. 2017). Chitosan also facilitates the formation of a cartilage-like matrix, thereby promoting tissue repair in areas of injury or degeneration. Chitosan antimicrobial properties proved beneficial in tissue regeneration, as it helps prevent infection during the healing process, a common complication in bone and cartilage repair surgeries. This property is especially important in orthopedic and reconstructive surgeries, where infection can delay healing and complicate recovery. The versatility of chitosan can be further enhanced by chemical modifications, such as grafting, crosslinking or blending with other materials, to improve its mechanical properties, degradation rate and biological activity. By adjusting the degree of deacetylation or introducing other bioactive molecules, the material functionality can be tailored to suit specific therapeutic needs, further improving its efficacy in bone and cartilage regeneration.

10.4 Conclusion

The therapeutic applications of sustainable chitin, chitosan derivatives and their nanocomposites represent a significant advancement in modern biomedical science, highlighting their immense potential in addressing diverse healthcare challenges. These materials, derived from renewable resources, offer remarkable biocompatibility, biodegradability and versatility making them ideal for developing innovative and eco-friendly medical solutions. Their ability to be chemically modified and engineered into various nanoscale formulations enhances their effectiveness in targeted drug delivery, tissue regeneration, antimicrobial interventions and immunotherapy.

Chitosan-based nanocomposites have revolutionized drug delivery by enabling controlled, sustained and site-specific drug release, thereby improving therapeutic outcomes while minimizing side effects. Their applications in wound healing, supported by antimicrobial properties and bio-adhesive capabilities, have led to breakthroughs in regenerative medicine and infection prevention. In cancer therapy, these materials provide targeted and less toxic alternatives by delivering chemotherapeutics and immunomodulatory agents specifically to tumor sites. Chitin and chitosan derivatives also show promise in managing metabolic disorders, reducing oxidative stress and enhancing immune system responses, expanding their applicability to chronic diseases. Beyond medical applications, the sustainable production of these materials aligns with global efforts toward environmentally responsible solutions in healthcare. Their multifunctionality, combined with cost-effective sourcing from natural byproducts, underscores their role in advancing green biotechnology. Emerging research into the synergistic integration of chitin and chitosan nanocomposites with other biomaterials further broadens their potential applications, particularly in vaccine delivery, diagnostics and personalized medicine.

Sustainable chitin, chitosan derivatives and their nanocomposites represent a promising and transformative class of materials. By bridging the gap between efficacy, environmental sustainability and cost efficiency, they hold the potential to shape the future of therapeutic and biomedical innovations, ultimately contributing to more accessible and effective healthcare solutions.

10.5 Future Perspectives

The future perspectives of sustainable chitin, chitosan derivatives and their nanocomposites are highly promising, with ongoing advancements opening new avenues in therapeutic and biomedical applications. A key area of focus is their integration into advanced drug delivery systems. Future research is likely to enhance their multifunctionality by incorporating stimuli-responsive mechanisms such as pH, temperature or enzyme-triggered drug release, enabling even more precise targeting for diseases like cancer, neurodegenerative disorders and infectious diseases. The development of hybrid nanocomposites by combining chitin and chitosan with other nanomaterials, such as metal nanoparticles, carbon nanostructures and polymers, could significantly expand their potential in diagnostics, imaging and theranostics.

In the realm of immunotherapy and vaccination, these materials have the potential to revolutionize antigen delivery, especially for emerging diseases, by enhancing stability, cellular uptake, and controlled release. The use of chitosan derivatives as adjuvants and carriers for DNA/RNA-based vaccines is particularly relevant in the post-pandemic era. Moreover, the rise of personalized medicine may be furthered by tailored chitin/chitosan-based formulations, leveraging their tunable properties to meet individual therapeutic needs.

Another promising direction involves their application in regenerative medicine and tissue engineering. Progress in 3D printing technology and the engineering of scaffolds with chitin and chitosan derivatives can facilitate the creation of biomimetic constructs for complex tissue repairs, such as cartilage, bone or nerve regeneration. Their use in sustainable and eco-friendly healthcare solutions, such as biodegradable medical implants and antimicrobial coatings, aligns with growing environmental and regulatory demands. Continued exploration of biofunctionalization strategies, green synthesis methods and cost-effective scaling of production will be critical in fully realizing the potential of these materials. As research continues to evolve, chitin and chitosan-based innovations will remain at the forefront of next-generation biomedicine and sustainable therapeutics.

References

Adhikari HS, Yadav PN (2018) Anticancer activity of chitosan, chitosan derivatives, and their mechanism of action. Int J Biomater 2018(1):2952085

Ahmad SI, Ahmad R, Khan MS, Kant R, Shahid S, Gautam L, Hasan GM, Hassan MI (2020) Chitin and its derivatives: structural properties and biomedical applications. Int J Biol Macromol 164:526–539

Alsaikhan F, Farhood B (2024) Recent advances on chitosan/hyaluronic acid-based stimuli-responsive hydrogels and composites for cancer treatment: a comprehensive review. Int J Biol Macromol 280:135893. https://doi.org/10.1016/j.ijbiomac.2024.135893

Andoh V, Ocansey DKW, Naveed H, Wang N, Chen L, Chen K, Mao F (2024) The advancing role of nanocomposites in cancer diagnosis and treatment. Int J Nanomed 19:6099–6126

Arjun B, Krishnendu PR, Anishma PN, Zachariah SM (2023) Chitin in drug delivery. Elsevier BV, pp 101–125. https://doi.org/10.1016/b978-0-323-98827-8.00022-9

Azuma K, Izumi R, Osaki T, Ifuku S, Morimoto M, Saimoto H, Okamoto Y (2015) Chitin, chitosan, and its derivatives for wound healing: old and new materials. J Func Biomaterials 6(1):104–142

Baharlouei P, Rahman A (2022) Chitin and chitosan: prospective biomedical applications in drug delivery, cancer treatment, and wound healing. Mar Drugs 20:460

Begum RF, Singh S, Prajapati B, Sumithra M, Patel RJ (2024) Advanced targeted drug delivery of bioactive agents fortified with graft chitosan in management of cancer: a review. Curr Med Chem 31.https://doi.org/10.2174/0109298673285334240112104709

Bhat S, Uthappa UT, Altalhi T, Jung HY, Kurkuri MD (2021) Functionalized porous hydroxyapatite scaffolds for tissue engineering applications: a focused review. ACS Biomater Sci Eng 8:4039–4076

Cao Z, Ma XZ, Lv D, Wang J, Shen Y, Peng S, Yang S, Huang J, Sun X (2023) Synthesis of chitin nanocrystals supported Zn^{2}+ with high activity against tobacco mosaic virus. Int J Biol Macromol 250:126168. https://doi.org/10.1016/j.ijbiomac.2023.126168

Cao Y, Fan R, Zhu K, Gao Y (2024) Advances in functionalized hydrogels in the treatment of myocardial infarction and drug-delivery strategies. ACS Appl Mater Interfaces 16:48880–48894

Cassano R, Perri P, Scarcello E, Piro P, Sole R, Curcio F, Trombino S (2024) Chitosan hemostatic dressings: properties and surgical applications. Polymers 16:1770. https://doi.org/10.3390/polym16131770

Chahal S, Kumar A, Hussian FSJ (2019) Development of biomimetic electrospun polymeric biomaterials for bone tissue engineering. A review. J Biomater Sci Polym Ed 30:1308–1355

Chattopadhyay S, Sarkar SS, Saproo S, Yadav S, Das B, Naidu S (2023) Apoptosis-targeted gene therapy for non-small cell lung cancer using chitosan-poly-lactic-co-glycolic acid -based nano-delivery system and CASP8 and miRs 29A–B1 and 34A. Front Bioeng Biotechnol 11:1188652. https://doi.org/10.3389/fbioe.2023.1188652

Chicatun F, Griffanti G, McKee MD, Nazhat SN (2017) Collagen/chitosan composite scaffolds for bone and cartilage tissue engineering. In: Biomedical composites. Woodhead Publishing, pp 163–198

Croitoru GA, Pîrvulescu DC, Niculescu AG, Rădulescu M, Grumezescu AM, Nicolae CL (2024) Advancements in aerogel technology for antimicrobial therapy: a review. Nanomaterials 14:1110. https://doi.org/10.3390/nano14131110

Deng C, Zhang P, Vulesevic B, Kuraitis D, Li F, Yang AF, Griffith M, Ruel M, Suuronen EJ (2010) A collagen–chitosan hydrogel for endothelial differentiation and angiogenesis. Tissue Eng Part A 16:3099–3109

Desai N, Rana D, Salave S, Gupta R, Patel P, Karunakaran B, Sharma A, Giri J, Benival D, Kommineni N (2023) Chitosan: a potential biopolymer in drug delivery and biomedical applications. Pharmaceutics 15:1313. https://doi.org/10.3390/pharmaceutics15041313

Edo GI, Yousif E, Al-Mashhadani MH (2024) Chitosan: an overview of biological activities, derivatives, properties, and current advancements in biomedical applications. Carbohyd Res 542:109199. https://doi.org/10.1016/j.carres.2024.109199

Eftekhari A, MalekiDizaj S, Sharifi S, Salatin S, Rahbar Saadat Y, Zununi Vahed S, Samiei M, Ardalan M, Rameshrad M, Ahmadian E, Cucchiarini M (2020) The use of nanomaterials in tissue engineering for cartilage regeneration; current approaches and future perspectives. Int J Mol Sci 21:536. https://doi.org/10.3390/ijms21020536

Egusa M, Watanabe S, Li HJ, Zewude DA, Ifuku S, Kaminaka H (2023) Production of copper nanoparticle-immobilized chitin nanofibers and their role in plant disease control. J Pestic Sci 48:86–92

Elangwe CN, Morozkina SN, Olekhnovich RO, Krasichkov A, Polyakova VO, Uspenskaya MV (2022) A review on chitosan and cellulose hydrogels for wound dressings. Polymers 14:5163. https://doi.org/10.3390/polym14235163

Fathi M, Majidi S, Zangabad PS, Barar J, Erfan-Niya H, Omidi Y (2018) Chitosan-based multi-functional nanomedicines and theranostics for targeted therapy of cancer. Med Res Rev 38:2110–2136

Galea S, Pavli F (2024) An overview of chitosan and chitosan oligosaccharides applications in the health sector and the food industry

Gao K, Qin Y, Liu S, Wang L, Xing R, Yu H, Chen X, Li P (2023) A review of the preparation, derivatization and functions of glucosamine and N-acetyl-glucosamine from chitin. Carbohydr Poly Technol Appl 5:100296. https://doi.org/10.1016/j.carpta.2023.100296

Gholap AD, Kapare HS, Pagar S, Kamandar P, Bhowmik D, Vishwakarma N, Raikwar S, Garkal A, Mehta TA, Rojekar S, Hatvate N, Mohanto S (2024) Exploring modified chitosan-based gene delivery technologies for therapeutic advancements. Int J Biol Macromol 260:129581. https://doi.org/10.1016/j.ijbiomac.2024.129581

Gómez-Gaviria M, Mora-Montes HM (2024) Exploring the potential of chitin and chitosan in nanobiocomposites for fungal immunological detection and antifungal action. Carbohyd Res 543:109220. https://doi.org/10.1016/j.carres.2024.109220

Gong X, Gao Y, Shu J, Zhang C, Zhao K (2022) Chitosan-based nanomaterial as immune adjuvant and delivery carrier for vaccines. Vaccines 10:1906. https://doi.org/10.3390/vaccines10111906

Haider A, Khan S, Iqbal DN, Khan SU, Haider S, Mohammad K, Mustfa G, Rizwan M, Haider A (2024) Chitosan as a tool for tissue engineering and rehabilitation: recent developments and future perspectives—a review. Int J Biol Macromol 278:134172. https://doi.org/10.1016/j.ijbiomac.2024.134172

Herdiana Y, Wathoni N, Shamsuddin S, Joni IM, Muchtaridi M (2021) Chitosan-based nanoparticles of targeted drug delivery system in breast cancer treatment. Polymers 13:1717. https://doi.org/10.3390/polym13111717

Herdiana Y, Wathoni N, Shamsuddin S, Muchtaridi M (2022) Drug release study of the chitosan-based nanoparticles. Heliyon 8:e08674. https://doi.org/10.1016/j.heliyon.2021.e08674

Huang H, Zou Y, Chi H, Liao D (2018) Lipid-Modifying effects of chitosan supplementation in humans: a pooled analysis with trial sequential analysis. Mol Nutr Food Res 62:1700842. https://doi.org/10.1002/MNFR.201700842

Hung CY, Zhang H, Castro-Lopez N, Ostroff GR, Khoshlenar P, Abraham A, Cole GT, Negron A, Forsthuber T, Peng T, Galgiani JN, Yu JJ (2018) Glucan-chitin particles enhance Th17 response and improve protective efficacy of a multivalent antigen (rCpa1) against pulmonary Coccidioides posadasii infection. Infect Immun 86:10–1128

Ignatova MG, Manolova NE, Toshkova RA, Rashkov IB, Gardeva EG, Yossifova LS, Alexandrov MT (2010) Electrospun nanofibrous mats containing quaternized chitosan and polylactide with in vitro antitumor activity against HeLa cells. Biomacromol 11(6):1633–1645

Islam S, Bhuiyan MR, Islam MN (2017) Chitin and chitosan: structure, properties and applications in biomedical engineering. J Polym Environ 25:854–866

Jang SI, Mok JY, Jeon IH, Park KH, Nguyen TTT, Park JS, Chai KY (2012) Effect of electrospun non-woven mats of dibutyryl chitin/poly (lactic acid) blends on wound healing in hairless mice. Molecules 17(3):2992–3007

Jayakumar R, Prabaharan M, Kumar PS, Nair SV, Tamura HJBA (2011) Biomaterials based on chitin and chitosan in wound dressing applications. Biotechnol Adv 29:322–337

Jiang H, Fang D, Hsiao B, Chu B, Chen W (2004) Preparation and characterization of ibuprofen-loaded poly (lactide-co-glycolide)/poly (ethylene glycol)-g-chitosan electrospun membranes. J Biomater Sci Polym Ed 15(3):279–296

Jiang X, Bao H, Merzendorfer H, Yang Q (2019) Immune responses of mammals and plants to chitin-containing pathogens. In: Targeting chitin-containing organisms, vol 1142, pp 61–81

Ke CL, Deng FS, Chuang CY, Lin CH (2021) Antimicrobial actions and applications of chitosan. Polymers 13:904. https://doi.org/10.3390/polym13060904

Kesharwani P, Halwai K, Jha SK, Al Mughram MH, Almujri SS, Almalki WH, Sahebkar A (2024) Folate-engineered chitosan nanoparticles: next-generation anticancer nanocarriers. Mol Cancer 23:244. https://doi.org/10.1186/s12943-024-02163-z

Khodaverdi K, Bakhshi A, Mozafari MR, Naghib SM (2024) A review of chitosan-based nanocarriers as drug delivery systems for brain diseases: critical challenges, outlooks and promises. Int J Biol Macromol 278:134962. https://doi.org/10.1016/j.ijbiomac.2024.134962

Kritchenkov AS, Andranovitš S, Skorik YA (2017) Chitosan and its derivatives: vectors in gene therapy. Russ Chem Rev 86:231

Kumar G, Virmani T, Misra SK, Sharma A, Pathak K (2024) Exploration of chitosan and its modified derivatives as vaccine adjuvant: a review. Carbohydr Polym Technol Appl 100537.https://doi.org/10.1016/j.carpta.2024.100537

Kurczewska J (2023) Chitosan-based nanoparticles with optimized parameters for targeted delivery of a specific anticancer drug—a comprehensive review. Pharmaceutics 15:503. https://doi.org/10.3390/pharmaceutics15020503

Lanka S (2020) Crustacean shell waste as a sustainable source of biodegradable biopolymers. Role Chem Sci Technol Dev Sustain

Lee CG, Da Silva CA, Lee JY, Hartl D, Elias JA (2008) Chitin regulation of immune responses: an old molecule with new roles. Curr Opin Immunol 20:684–689

Li N, Zhou L, Xie W, Zeng D, Cai D, Wang H, Zhou C, Wang J, Li L (2019) Alkaline phosphatase enzyme-induced biomineralization of chitosan scaffolds with enhanced osteogenesis for bone tissue engineering. Chem Eng J 371:618–630

Lima BV, Oliveira MJ, Barbosa MA, Goncalves RM, Castro F (2021) Immunomodulatory potential of chitosan-based materials for cancer therapy: a systematic review of in vitro, in vivo and clinical studies. Biomater Sci 9:3209–3227

Lin L, Liu H, Zhang D, Du L, Zhang H (2024) Nanolevel immunomodulators in sepsis: novel roles, current perspectives, and future directions. Int J Nanomed 19:12529–12556

Mao HQ, Roy K, Troung-Le VL, Janes KA, Lin KY, Wang Y, August JT, Leong KW (2001) Chitosan-DNA nanoparticles as gene carriers: synthesis, characterization and transfection efficiency. J Control Release 70(3):399–421

Mawazi SM, Kumar M, Ahmad N, Ge Y, Mahmood S (2024) Recent applications of chitosan and its derivatives in antibacterial, anticancer, wound healing, and tissue engineering fields. Polymers 16:1351. https://doi.org/10.3390/polym16101351

Moghaddam FD, Zare EN, Hassanpour M, Bertani FR, Serajian A, Ziaei SF, Paiva-Santos AC, Neisiany RE, Makvandi P, Iravani S, Xu Y (2024) Chitosan-based nanosystems for cancer diagnosis and therapy: stimuli-responsive, immune response, and clinical studies. Carbohyd Polym 330:121839. https://doi.org/10.1016/j.carbpol.2024.121839

Mourya VK, Inamdar NN (2008) Chitosan-modifications and applications: opportunities galore. React Funct Polym 68:1013–1051

Mu L, Wu L, Wu S, Ye Q, Zhong Z (2024) Progress in chitin/chitosan and their derivatives for biomedical applications: where we stand. Carbohyd Polym 343:122233. https://doi.org/10.1016/j.carbpol.2024.122233

Muanprasat C, Chatsudthipong V (2017) Chitosan oligosaccharide: biological activities and potential therapeutic applications. Pharmacol Ther 170:80–97

Mura P, Maestrelli F, Cirri M, Mennini N (2022) Multiple roles of chitosan in mucosal drug delivery: an updated review. Mar Drugs 20:335

Muzzarelli RA (2011) Chitosan composites with inorganics, morphogenetic proteins and stem cells, for bone regeneration. Carbohyd Polym 83:1433–1445

Narmani A, Jafari SM (2021) Chitosan-based nanodelivery systems for cancer therapy: recent advances. Carbohyd Polym 272:118464

Omidian H, Gill EJ, Chowdhury SR, Cubeddu LX (2024) Chitosan nanoparticles for intranasal drug delivery. Pharmaceuticals 16:746. https://doi.org/10.3390/pharmaceutics16060746

Oryan A, Sahvieh S (2017) Effectiveness of chitosan scaffold in skin, bone and cartilage healing. Int J Biol Macromol 104:1003–1011

Osak P, Łosiewicz B, Nowińska D (2024) Chitosan as a drug carrier. Prog Chem Appl Chitin Deriv 29:30–43

Pai V, Singh BN, Singh AK (2024) Insights into advances and applications of biomaterials for nerve tissue injuries and neurodegenerative disorders. Macromol Biosci 24:2400150. https://doi.org/10.1002/mabi.202400150
Pal D, Kahawas K (2024) Potential sources and uses of chitin and its polymers. J Discoveries
Parhi R (2020) Drug delivery applications of chitin and chitosan: a review. Environ Chem Lett 18:577–594
Patel B, Manne R, Patel DB, Gorityala S, Palaniappan A, Kurakula M (2021) Chitosan as functional biomaterial for designing delivery systems in cardiac therapies. Gels 7:253
Preuss HG, Kaats GR (2006) Chitosan as a dietary supplement for weight loss: a review. Curr Nutr Food Sci 2:297–311
Rai M, Dos Santos CA (eds) (2021) Biopolymer-based nano films: applications in food packaging and wound healing. Elsevier
Renu S, Renukaradhya GJ (2020) Chitosan nanoparticle based mucosal vaccines delivered against infectious diseases of poultry and pigs. Front Bioeng Biotechnol 8:558349. https://doi.org/10.3389/fbioe.2020.558349
Saini S, Dhiman A, Nanda S (2020) Immunomodulatory properties of chitosan: impact on wound healing and tissue repair. In: Endocrine, metabolic and immune disorders-drug targets (formerly current drug targets-immune, endocrine and metabolic disorders), vol 20, pp 1611–1623
Salembier R, De Haes C, Bellemans J, Demeyere K, Van Den Broeck W, Sanders NN, Van Laere S, Lyons TR, Meyer E, Steenbrugge J (2024) Chitin-mediated blockade of chitinase-like proteins reduces tumor immunosuppression, inhibits lymphatic metastasis and enhances anti-PD-1 efficacy in complementary TNBC models. Breast Cancer Res 26:63. https://doi.org/10.1186/s13058-024-01815-8
Shalumon KT, Binulal NS, Selvamurugan N, Nair SV, Menon D, Furuike T, Jayakumar R (2009) Electrospinning of carboxymethyl chitin/poly (vinyl alcohol) nanofibrous scaffolds for tissue engineering applications. Carbohyd Polym 77(4):863–869
Sharaf MH, Liu C (2024) Chitin and chitosan applications in medication delivery systems, vol 2025, pp 283–319
Sharma D, Arora S, dos Santos Rodrigues B, Lakkadwala S, Banerjee A, Singh J (2019) Chitosan-based systems for gene delivery. In: Functional chitosan: drug delivery and biomedical applications, pp 229–267
Shekede BT (2018) Bactericidal efficacy of wound gauze treated with chitosan nanomaterial hybrids of zinc, silver and copper on common wound bacteria. Doctoral dissertation, Cape Peninsula University of Technology
Shivhare R, Sabale V, Ingole A, Jain N (2023) Nanoparticles for tumor gene therapy: recent advances and perspective. Integr Biomaterials for Gene Ther 139–182
Silva SS, Mano JF, Reis RL (2017) Ionic liquids in the processing and chemical modification of chitin and chitosan for biomedical applications. Green Chem 19:1208–1220
Singh, R. P. (2019). Nanocomposites: recent trends, developments and applications. In: Advances in nanostructured composites, pp 16–47
Sivanesan I, Gopal J, Muthu M, Shin J, Oh JW (2021) Reviewing chitin/chitosan nanofibers and associated nanocomposites and their attained medical milestones. Polymers 13(14):2330
Sivashankari PR, Prabaharan M (2016) Prospects of chitosan-based scaffolds for growth factor release in tissue engineering. Int J Biol Macromol 93:1382–1389
Subhan MA, Attia SA, Torchilin VP (2021) Advances in siRNA delivery strategies for the treatment of MDR cancer. Life Sci 274:119337. https://doi.org/10.1016/j.lfs.2021.119337
Tachaboonyakiat W (2021) Physical and chemical modification of chitin/chitosan for functional wound dressings. Chitosan for biomaterials III: structure-property relationships. Springer International Publishing, Cham, pp 257–299
Tao F, Cheng Y, Shi X, Zheng H, Du Y, Xiang W, Deng H (2020) Applications of chitin and chitosan nanofibers in bone regenerative engineering. Carbohyd Polym 230:115658. https://doi.org/10.1016/j.carbpol.2019.115658

Tian B, Hua S, Tian Y, Liu J (2020) Chemical and physical chitosan hydrogels as prospective carriers for drug delivery: a review. J Mater Chem B 8(44):10050–10064

Turley JL, Ward RW, Huete-Carrasco J, Muñoz-Wolf N, Roche K, Bowie A, Andersson M, Lavelle EC (2024) Intratumoral delivery of the chitin-derived C100 adjuvant promotes robust STING, IFNAR, and CD8+ T cell-dependent anti-tumor immunity. Cell Rep Med 5:101560. https://doi.org/10.1016/j.xcrm.2024.101560

Tzeng HP, Liu SH, Chiang MT (2022) Antidiabetic properties of chitosan and its derivatives. Mar Drugs 20:784. https://doi.org/10.3390/md20120784

Venkatesan J, Revathi S, Sasirekha R, Sudha PN, Pazhanisamy P, Gomathi T (2023) Chitosan based nanobioceramics in bone tissue engineering application. In: Nanobiomaterials. CRC Press, pp. 161–181

Vijayaraj R, Altaff K, Lakshmanan G, Jayaprakashvel M, Mickymaray S, Gunapriya R, Palanisamy M, Alothaim AS (2022) Inhibitory activity of chitin, (2-acetamido-2-deoxy-hexopyranose) against penicillin-binding proteins of staphylococcus aureus. Coatings 12:1854. https://doi.org/10.3390/coatings12121854

Wang YW, Liu CC, Cherng JH, Lin CS, Chang SJ, Hong ZJ, Liu CC, Chiu YK, Hsu SD, Chang H (2019) Biological effects of chitosan-based dressing on hemostasis mechanism. Polymers 11:1906. https://doi.org/10.3390/polym11111906

Whang HS, Kirsch W, Zhu YH, Yang CZ, Hudson SM (2005) Hemostatic agents derived from chitin and chitosan. J Macromol Sci Part C: Polym Rev 45(4):309–323

Wijesekara T, Xu B (2024) New insights into sources, bioavailability, health-promoting effects, and applications of chitin and chitosan. J Agric Food Chem 72:17138–17152

Wike NY, Akinbo O, Olaniyan OT, Adetunji CO, Adetunji JB (2023) The role of nano chitosan for effective delivery of nutrients and drugs including hormones and vaccines in cattle. In: Next generation nanochitosan. Academic Press, pp 171–202

Woodrow KA, Bennett KM, Lo DD (2012) Mucosal vaccine design and delivery. Annu Rev Biomed Eng 14:17–46

Xia Y, Wang D, Liu D, Su J, Jin Y, Wang D, Han B, Jiang Z, Liu B (2022) Applications of chitosan and its derivatives in skin and soft tissue diseases. Front Bioeng Biotechnol 10:894667

Xu G, Huang X, Qiu L, Wu J, Hu Y (2007) Mechanism study of chitosan on lipid metabolism in hyperlipidemic rats. Asia Pac J Clin Nutr 16:313–317. https://doi.org/10.6133/APJCN.2007.16.S1.60

Yang X, Liu J, Pei Y, Zheng X, Tang K (2020) Recent progress in preparation and application of nano-chitin materials. Energy Environ Mater 3:492–515

Yu YM, Long Y, Zhu Z (2024) Chitosan, a natural polymer, is an excellent sustained-release carrier for amide local anesthetics. J Pain Res 17:3539–3551

Zargar V, Asghari M, Dashti A (2015) A review on chitin and chitosan polymers: structure, chemistry, solubility, derivatives, and applications. Chem Biochem Eng Rev 2:204–226

Zhang S, Chen J, Yu Y, Dai K, Wang J, Liu C (2019) Accelerated bone regenerative efficiency by regulating sequential release of BMP-2 and VEGF and synergism with sulfated chitosan. ACS Biomater Sci Eng 5:944–1955

Zhang HY, Zhang Y, Zhang Y, Jiang ZP, Cui YL, Wang QS (2022) ROS-responsive thioketal-linked alginate/chitosan carriers for irritable bowel syndrome with diarrhea therapy. Int J Biol Macromol 209:70–82

Zhou Y, Yang D, Chen X, Xu Q, Lu F, Nie J (2008) Electrospun water-soluble carboxyethyl chitosan/poly (vinyl alcohol) nanofibrous membrane as potential wound dressing for skin regeneration. Biomacromol 9(1):349–354

Chapter 11
Thermoplastic Starch-Based LLDPE Films for Active Packaging

Seifollah Jamalpour, Sahar Yavarimehr, Aida Fooladi, Fatemeh Sharafi, Mohammad Ali Sattari Tabrizi, and Mohammad Mehdi Seraji

Abstract Active packaging refers to a type of food packaging that interacts with the food or the surrounding environment to extend shelf life, preserve food quality, and improve food safety. Unlike traditional passive packaging, which provides only a barrier to external factors, active packaging actively modifies the conditions inside the package. Active packaging performance can include: oxygen scavengers, moisture absorbers, antimicrobial packaging, ethylene absorbers, and carbon dioxide emitters. Thermoplastic starch (TPS)-based low-linear density polyethylene (LLDPE) films have been studied for their potential use in active packaging. TPS-LLDPE films can be prepared by blending TPS with LLDPE and incorporating active agents. These active agents help to scavenge oxygen, absorb moisture, inhibit microbial growth, or absorb ethylene gas, which can help to maintain the quality of food products. TPS-LLDPE films can provide a biodegradable and environmentally friendly alternative to conventional petroleum-based packaging materials, while also offering active packaging functionalities. By optimizing the composition of these films and incorporating appropriate active agents, TPS-LLDPE films have the potential to be used as an effective and sustainable solution for food packaging applications. In this study, firstly, thermoplastic starch and packaging technologies are reviewed. Then, TPS-LLDPE films for active packaging are described.

Keywords Active packaging · Thermoplastic starch · Low-linear density polyethylene · Shelf life · Food quality · Food safety

S. Jamalpour (✉) · A. Fooladi · F. Sharafi · M. M. Seraji
Department of Chemical Engineering, Faculty of Engineering, Shahid Chamran University of Ahvaz, Ahvaz, Iran
e-mail: jamalpour@scu.ac.ir

S. Yavarimehr
Department of Polymer Chemistry, Faculty of Science, Shahid Chamran University of Ahvaz, Ahvaz, Iran

M. A. S. Tabrizi
Environmental Science and Technology Research Center, Department of Environmental Health Engineering, Shahid Sadoughi University of Medical Sciences, Yazd, Iran

P. M. Visakh (ed.), *Engineering Applications of Polymer based Nano Blends*, Advanced Structured Materials 235, https://doi.org/10.1007/978-3-031-90979-5_11

11.1 Introduction

In recent years, regarding the growing global demand for healthy and fresh food and consumer desire for high-quality food products, attention to packaging issue has been increasing. In this context, one of the important issues is that the packaging materials should not be detrimental to humans or the environment. That is why the consumer attention toward having effective packaging, that also minimizes environmental damage, has increased significantly. Packaging plays an essential role in the protection of foods products (Mazerolles et al. 2020). The primary purpose of packaging is to hold and safeguard the product, ensuring convenience for consumer form the moment of production until consumption (Deshmukh et al. 2023). In simple words, the packaging is designed to protect, preserve, manage, transport, and store of products (Loucanova et al. 2017). Nowadays, packaging also play a significant role in promoting the product and provides a more accessible way to purchase it. The most important advantages of using packaging can include the following:

- Protection: such as chemical protection, biological protection, and physical protection
- Food waste reduction: Packaging minimizes total waste by expanding the life span of products (Marsh and Bugusu 2007).

Traditional packaging keeps the product's form, color, and flavor and protects it against mechanical, microbiological, physical, and chemical contamination and hinders the loss of product ingredients and prohibit the entry of unwanted substances from outward. Also, they promote the product by choosing marketing elements of the packaging effectively, including form, color, and typography. The traditional packaging is designed to shield food and maintain its shelf life for the maximum duration, as reducing the interaction between the packaging and the product (Wyrwa and Barska 2017). Traditional objectives of packaging have been to maintain the food from biochemical contamination originating from external source and to supply consumers with convenience during each stage of storage, transport, and delivery. This packaging is recognized for its passive activities, and its characteristics are defined based on this (Siciliano et al. 2024). The common packaging consists of: glass, metal such as aluminum, papers, and plastics such as polyolefins, polyesters, and polyvinyl chloride. Although each of them has advantages, such as accessibility like the ease of accessing aluminum-, non-reactivity with food like glass as well as its recyclability (Marsh and Bugusu 2007), they also have disadvantages like pollution caused by plastic and its harm to human health. A recent study revealed that humans utilize tens of thousands of microplastic particles yearly. Also, plastic can pollute soil as plastic fragments decompose. They disintegrate into microplastics, and consequently they taint the soil and impressing its richness and soil health. Soil pollution is merely one of the many challenging resulting from using of plastic packaging, emphasize that the essential requirement for more green or renewable substitute and more effective waste control technique (Siciliano et al. 2024). In recent years, contemporary consumers tend to products made from natural ingredients and materials obtained from renewable resource (Promsorn and Harnkarnsujarit 2022a).

Regard to the environmental needs and consumer concerns, new and alternative methods for packaging have emerged as substitutes for usual packaging. One of these methods involves using active agent in the packaging materials. Based on explanation active packaging materials, they can be categorized based on how each of them influences the characteristics of the product as follows: absorbers—active packaging systems based on absorption, and emitters—active packaging systems based on the release of substances (Nosáľová et al. 2018). There are two types of systems typically incorporate active packaging: ingredients that are directly incorporated into packaging materials and sachets and pads put into packages. Active packaging is regarded as an innovative packaging system because it purposely engages changing and controlling the atmosphere within the package (Deshmukh et al. 2023). In addition, the trend is moving toward packaging materials that is based on natural and eco-friendly alternative, as well as indicated in active agents utilized for active packaging (Vilela et al. 2018). The advantages of active packaging include the control of internal conditions. This control condition is archived through reacting to them accordingly by emitting advantageous agents or absorbing those that damage the product. The potential of longer shelf life and preservation of the sensory properties of products, the environmental protection due to use of biodegradable and biocompatible components, and also increased microbiological safety of foods is the advantages of active packaging. The use of active packaging also has several disadvantages. Active packaging is more expensive than traditional packaging, and the active substances released in the packaging can affect the food composition. Additionally, the stronger characteristic smell of these active material can be unpleasant for consumer (Wyrwa and Barska 2017).

Overall, in this study, active agents in packaging, thermoplastic starch, packaging technologies, and TPS-LLDPE films as active packaging are reviewed.

11.2 Active Agents in Packaging

In Fig. 11.1, the active agents are mentioned. In the following, the performance of each of them will be explained (Marsh and Bugusu 2007).

Oxygen Scavengers

Scavengers are created for eliminate unwanted components from internal the atmosphere of packaging (Wang et al. 1994). Advanced level of oxygen diminishes the nutritional quality of food and shorten its shelf life. The oxygen in packing gases interacts with easy spoiled foods in the package and speeding up the deterioration of various food products (Prasad and Kochhar 2014). The goal of food industry remove oxygen from food packaging, this is largely done through gas flushing or modified atmosphere packaging (MAP) technique, or oxygen scavengers (OS) can be used with enhanced performance to manage residual oxygen levels (Vilela et al. 2018).

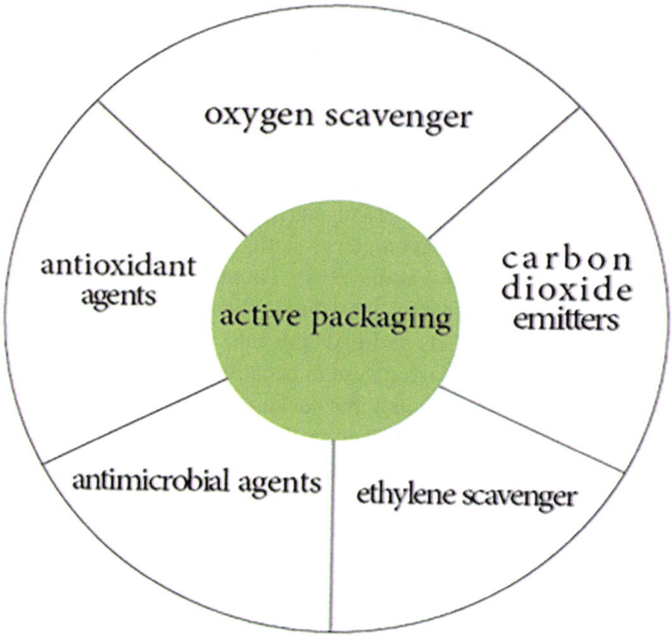

Fig. 11.1 Active agents

Most oxygen scavengers (OS) used in the food industry are sachets made of iron, generally comprising a permeable sachet that filled with iron powder. The mechanism of action is trigged by moisture, according the theory of iron oxidation. For numerous of food packaging, iron-based OS-sachets can be effective; however, there are some disadvantages including the possibility of tearing. For this reason, efforts are being undertaken to create alternative solution utilizing polymeric films for example: enzymes as active agents in polymeric matrices [e.g., silicon, polyethylene terephthalate (PET), low-density polyethylene (LDPE), and polylactic acid (PLA)] applied to food (Vilela et al. 2018).

Carbon Dioxide Emitter

The advantageous impact and function of CO_2 include preventing of microbial growth, oxidation–reduction, sensory characteristics, and increasing the duration that products remain usable. The oxygen amount to be observed, emitted, or effused by the packaging device must be aligned with kinetics and equilibrium properties of CO_2 created within the package by food. Expand a multifunctional biopolymer based on agar as packaging for fresh produce, which includes mechanism for CO_2 absorption and absorption of water vapor. This film was designed with sodium carbonate/sodium glycinate to function as a CO_2 absorbent (Deshmukh et al. 2023).

Antimicrobial Agents

Antimicrobial packaging refers to packaging system that is designed to kill or prevent spoilage and pathogenic or harmful microorganisms that contaminate food products. The antimicrobial agents used in active packaging can be classified into two main categories material based on; organic or inorganic and chemical agents or natural agents or probiotics (Hosseinnejad 2014).

Ethylene Scavengers

Ethylene is a reactive molecule specious that is emitted by fruits and leads to the ripening of fruits. Controlling of ethylene levels in environment around fresh food products after harvesting is significant. Substances based on potassium permanganate ($KMnO_4$) that are most popular ethylene scavenger, are supported on inert substrates, in oxidizing ethylene, and in changing its color from purple to brown. This ethylene scavenger is placed within a matrix made of silica (SiO_2) and alumina (Al_2O_3) nanoparticles (Vilela et al. 2018).

11.3 Thermoplastic Starch (TPS)

Polysaccharides such as starch, cellulose, and glycogen are agricultural products that have recently been researched and used for food and non-food uses. Starch is considered a renewable resource due to its benefits, including ease of production, environmental friendliness, availability, and cost-effectiveness. The source of starch production is various cereals and roots such as manioc, potatoes, corn, green peas, and rice. In addition, aquatic ecosystems are other sources of starch and can be used to obtain this material. In the granular state, starch is a substance made up of linear amylose and branched amylopectin and has a crystalline structure. Amylose and amylopectin are present in significant amounts in starch, and their structure leads to many differences in the physical and chemical properties of these two compounds. According to the natural starch source, the morphology and crystal structures A, B, and C can be identified (Martinez Villadiego et al. 2022).

In Fig. 11.2, the structure of starch is shown. Amylose is one of the components of starch and constitutes approximately 5–35% of it. The amount of amylose varies in different types of starch. It is composed of a linear chain consisting of 500–2000 glucose units. Starches rich in amylose exhibit enhanced thermal stability and can form more durable amylose–lipid complexes. These thermal characteristics are influenced by physical interactions. Around 70–80% of starch consists of amylopectin. This molecule, as a result of its formation from numerous glucose units with extensive branching, is one of the greatest natural molecules, and its molecular structure is more complex than that of amylose. The chains of amylopectin molecules are classified into three categories: A, B, and C, each possessing specific characteristics (Falua et al. 2022).

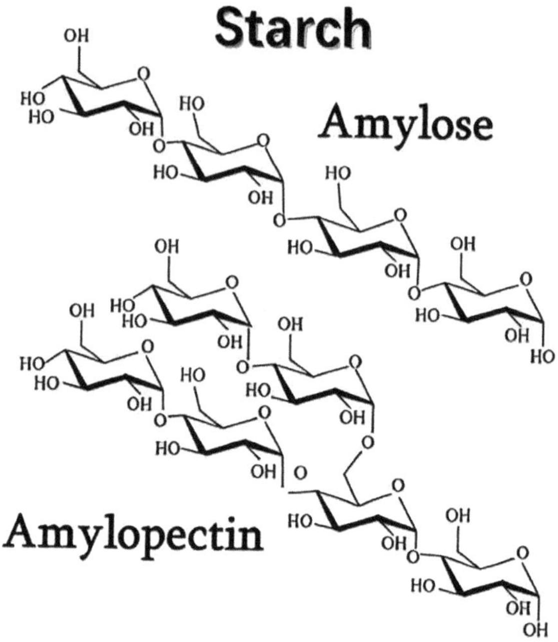

Fig. 11.2 Starch structure (Pan et al. 2022) with permission, copyright memberanes, 2025

The presence of hydroxyl groups, along with the generation of intramolecular and intermolecular hydrogen links, contributes to the hydrophilic properties of starch. This characteristic causes starch to degrade when exposed to water, resulting in weak mechanical properties. Consequently, starch, in its natural form, is not suitable for direct use as a plastic material (Martinez Villadiego et al. 2022).

When starch is mixed with a certain amount of water or plasticizers and subjected to heat and shear, its structure breaks down, resulting in a homogeneous material known as thermoplastic starch (TPS) (as shown in Fig. 11.2). The creation of thermoplastic starch occurs when starch granules undergo structural decomposition in the presence of water, heat, plasticizers, and mechanical forces that resist evaporation. In the production of thermoplastic starch, a variety of plasticizers such as formamides, urea, acetamide, sugars, polyols, amino acids, lipids, and phosphate sorbates are used. Water and glycerol are plasticizers that are primarily used in the preparation of TPS. Water and glycerol are needed to increase the flexibility of thermoplastic starch (TPS). Factors such as the amount of plasticizer and shear force, as well as process conditions, affect the physical properties of the processed starch. In addition, it can be processed by thermoplastic equipment, such as extrusion and molding (injection, compression, and pneumatic (Mohammadi Nafchi et al. 2013) (Fig. 11.3).

The plasticizer greatly affects the physical properties of processed starch. This effect appears in two ways:

1. Regulation of structural degradation and depolymerization of starch

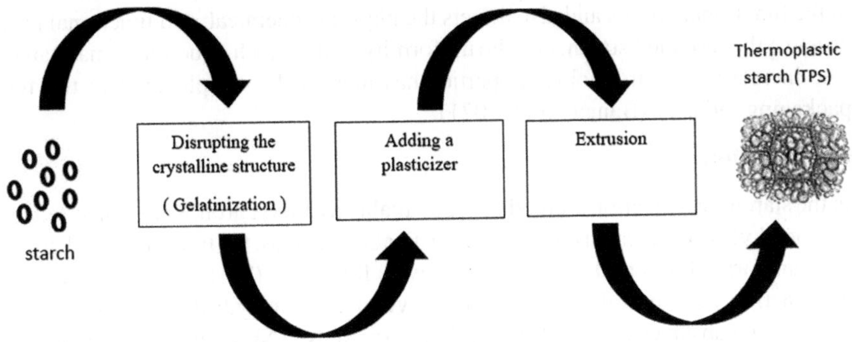

Fig. 11.3 Conversion of starch to thermoplastic starch (Martinez Villadiego et al. 2022) with permission, copyright memberanes, 2025

2. Creation of differences in the characteristics and final properties of the material.

Shear forces and plasticizers are factors that impact the physical properties of processed starch (Róz et al. 2006). Thermoplastic materials derived from starch are utilized in plastic applications due to their characteristics such as biodegradability and renewability. Additionally, they are cheaper and more cost-effective compared to synthetic thermoplastics. However, issues such as low moisture resistance, unsuitable physical properties, and others have limited their applications. Choosing the right starch sources (with a higher amylose percentage) or modifying them, selecting the right plasticizers, combining TPS and polymers, developing nanofillers, and more can solve the existing challenges. Another challenge in the development of TPS is its high amylose content, which is less flexible compared to TPS made from materials with high amylopectin content (Bangar et al. 2021). In the following, we will further explore these methods to improve the properties of thermoplastic starch.

11.4 Modification of TPS for Using in the Food Packaging

Common methods for starch modification are physical, chemical, enzymatic, or hybrid methods. Chemical modification is usually carried out through etherification, esterification, and oxidation. However, physical modification also has significant advantages. As mentioned, the hydrophilic properties of starch have limited industrial applications, reducing its mechanical properties and stability. Replacing hydroxyl groups with bulky hydrophobic groups, such as esterified groups of acetyl, succinyl, carboxymethyl, and hydroxypropyl, increases the hydrophobic properties, and thus, chemical modification of starch takes place. Also, things such as cross-linking, grafting, and condensation reactions increase the hydrophobic properties. Large hydrophobic groups hinder the reorganization of starch chains, thus preventing or minimizing the process of recrystallization. Chemical treatment leads to a change in the functional groups and also affects the physical, chemical, and functional properties of the modified starch, and the uniformity of the starch structure is maintained. The increase in hydrophobic properties has increased its application in the food packaging industry (Bangar et al. 2021).

Esterification

In the starch esterification method, three available hydroxyl groups are converted into aryl or alkyl derivatives. In the starch esterification process, the three hydroxyl groups are transformed into aryl or alkyl derivatives. Ren et al. (2016) employed water as the reaction medium, along with dodecenyl succinic anhydride (DDSA) or other esterifying substances, to modify starch nanocrystals (SNCs) and impart hydrophobicity to starch nanocrystals. Based on the results, it was found that esterification improved the dispersion stability of SNCs in water and increased their dispersion in non-polar solvents. Also, they found that starch modification with cross-linking before esterification improved hydrophobicity more than single-stage esterification. In the esterification and acetylation method, the starch is in the form of suspended particles and cannot come into contact with the reactant completely, which reduces the efficiency of the starch esterification reaction. But this problem is solved by organic solvents (Wang et al. 2020).

Acetylation

In the starch acetylation process, hydroxyl groups are removed and acetate groups are added. The acetylation reaction of starch was used to improve the hydrophobic properties. In a report, it is mentioned that a polymer material with high hydrophobic properties was produced using cassava starch and water as solvents. Also, research has shown that amending starch with solvent-free acetylation and esterification (using maleic anhydride) reduces water absorption compared to natural starch. When natural starch is combined with stearyl chloride and ethyl acetate solvent, acetylated starch is obtained. It was confirmed by the findings that stearyl chloride is an efficient substance for producing starch with a high degree of esterification. The esterified starch imparted hydrophobic properties to PCL/esterified starch composites. By increasing the acetylation (DS) degree of the modified starch, the equilibrium water absorption in the PCL/esterified starch composites is significantly reduced. This indicates improved water resistance and increased hydrophobicity in composites by increasing the acetylation rate (Wang et al. 2020).

When natural starch is combined with stearyl chloride and ethyl acetate solvent, acetylated starch is obtained. It was confirmed by the findings that stearyl chloride is an efficient substance for producing starch with a high degree of esterification. The esterified starch imparted hydrophobic properties to PCL/esterified starch composites. By increasing the acetylation (DS) degree of the modified starch, the equilibrium water absorption in the PCL/esterified starch composites is significantly reduced. This indicates improved water resistance and increased hydrophobicity in composites by increasing the acetylation rate.

Octenyl Succinic Anhydride (OSA)

Esterification with dicarboxylic acids causes the starch to become hydrophobic. Modified starch with succinic actinyl anhydride (OSA), one of the most common types is called OSA starch. The hydroxyl groups of glucose units are modified with OSA. In an aqueous solution, the presence of OSA groups in starch chains causes the formation of esters that give rise to hydrophilic and hydrophobic groups, creating amphiphilic properties. OSA starch is produced by the esterification process between the OH groups of starch and the OSA reactant. The reaction medium is aqueous and takes place under mild alkaline conditions (Agama-Acevedo and Bello-Perez 2017).

OSA starch can be sourced from a wide variety of manufacturers. Existing patents related to OSA starch products are often owned by the National Starch Corporation (now Corn Products International). Since receiving the initial patents in 1952–1953, it has expanded these products. In the first study of OSA starch synthesis in a non-aqueous medium, pyridine was substituted for water, which acts as a catalyst by increasing the initial reactivity of starch grains, as it forms a succinyl-pyridinium intermediate that is more reactive with starch hydroxyl groups than inactive OSA. Another proposed method involves corn starch reacting through reflux in pyridine at 115 °C. To prevent starch burns, a weight ratio of 2:1 (pyridine/starch) was important. Also, in this process, the starch chains are partially degraded. Despite the increased efficiency of the reaction, industrial use of this method is not possible due to the toxicity of pyridine (Sweedman et al. 2013).

Cross-linking

Cross-linking is used to stabilize the granular structure of the modified starch. One of its pre-considered applications is to improve thermal stability. Also, starches with cross-binding are more hydrophobic than primary starches. Cross-linking in starch film chemistry enhances the mechanical properties of polymeric materials. For instance, the cross-linking of wheat starch with two chemicals (STMP and STPP) increases tensile strength (TS) and water vapor permeability (WVP) while reducing the stretchability of starch. Cross-linking starch with other materials such as citric acid, boric acid, and sodium hexametaphosphate (SHMP) is common. Citric acid increases the hardness and reduces the hydrophilicity of starch, but it does not affect strength. Boric acid enhances the thermoplasticity of starch but decreases water resistance. Due to the toxicity of glutaraldehyde and epichlorohydrin, these materials are not commonly used; however, SHMP is a better choice as it is safe and non-toxic. One of the main challenges of chemical modification is environmental issues and economic concerns. For this reason, efforts have been made to naturally cross-link polymer chains. Ultraviolet (UV) light is a natural and environmentally friendly method for this purpose. In this method, photo-initiators are used to create cross-links. Research results have shown that UV radiation at a wavelength of 380 nm has the greatest impact on improving the mechanical properties of thermoplastic starch films (Bangar et al. 2021).

Cross-linking is one of the methods of natural starch modification, and the use of this method is common. In this method, various substances such as sodium trimetaphosphate, sodium tripolyphosphate, epichlorohydrin, and phosphoryl chloride are used, which lead to intermolecular interactions of ether or ester within the hydroxyl groups of starch. The cross-linking of starch depends on many factors, such as the type of starch source, the concentration and composition of the cross-linking agent, the rate of substitution, pH, reaction time, and temperature. In the food industry, this method is used to create a stable grain structure of starch and also has beneficial nutritional effects (Koo et al. 2010).

Based on the conducted research, the extent of cross-linking in the modified starch was evaluated using viscosity. The peak viscosity of modified starch-suspended particles was measured using a stress control rheometer and starch adhesion cell (AR1000 model, TA Instruments, New Castle, DE, USA). For the measurement, a planned heating and cooling cycle was used. The starch-suspended particles were heated at a certain speed. After that, the mixture was cooled at the same initial speed and finally stored at a certain temperature. The degree of transverse connection was calculated using the following equation:

$$\text{Degree of cross–linking}(\%) = \frac{(A - B)}{A} \times 100 \tag{11.1}$$

In this equation, A is the peak viscosity of the control sample (no STMP/STPP) and B is the peak viscosity of the starch crossed (Koo et al. 2010).

Hydroxypropylation (HPT)

Hydroxypropylation (HPT) is a chemical treatment method in which starch reacts with propylene oxide to produce starch ether. This process helps to alter the chemical properties of starch and improve its characteristics. Hydroxypropylated starches display a reduced gelatinization temperature in comparison with native starch, making it easier to create starch films with high amylose content. High amylose corn starch films that have been hydroxypropylated show adequate flexibility, tensile strength, and moisture vapor permeability (Bangar et al. 2021). Hydroxypropyl starch (HPS) is a modified starch product called stabilized or substituted starches and is the only type of food stabilized starch whose substitute group is an ether group. Ether groups are quite stable in acidic and alkaline environments, while ester groups are unstable (Fu et al. 2019). For the design of biodegradable food packaging, the selection of the appropriate type of starch is of great importance. High amylose starch (HAS) has greater tensile strength and water resistance, but it causes brittleness in films. Hydroxypropylation (HPT) increases flexibility and water vapor permeability but reduces tensile strength. The combination of HAS and HPT can produce improved films (Bangar et al. 2021).

Oxidation

Oxidation of starch is one of the processes of modifying physical and chemical properties. Hydrogen peroxide, sodium hypochlorite, permanganate, and UV radiation are oxidizers that can be used in this process. By carrying out this reaction, two phenomena occur, at first hydroxy groups are oxidized and converted to carbonyl groups and then to carboxyl. Also, starch undergoes depolymerization due to the hydrolysis of glycosidic. Oxidation concentration, reaction time, pH, and presence of catalyst metal ions affect chemical changes. Hydrogen peroxide is used to produce starches with different degrees of oxidation. According to a report, potato starch was oxidized using hydrogen peroxide at 30 °C. In this method, starch and distilled water were stirred in a glass reactor equipped with a mechanical stirrer and thermometer. The reaction temperature was regulated by the circulation of water from a temperature-controlled water bath in the reactor jack. After adjusting the pH to 4 with acetic acid, hydrogen peroxide was added to the reaction mixture in different amounts per starch content, and this reaction was mixed for 8 h under continuous stirring. At the end of the reaction, the product was washed with distilled water until the pH reached 6. Finally, the mixture was filtered under vacuum to separate the starch from it and then dried in the oven at 45 °C for 2 days. Based on the obtained results, the content of carbonyl and carboxyl groups is dependent on the starch, and potato starch has the highest carboxyl content compared to the other starches (Barbosa et al. 2019).

In the oxidation method with sodium hypochlorite, the hydroxyl groups in amylose and amylopectin are converted to carbonyl and carboxyl. In oxidation, it is possible for part of the amylose and amylopectin to be depolymerized. The amount of carbonyl and carboxyl in oxidized starches affects their physicochemical properties and oxidized starches can be used for various industrial applications. In recent years, extensive research has been conducted on the oxidation of starch using sodium hypochlorite, leading to significant successes. Research has shown that pH influences the oxidation reaction of starch with sodium hypochlorite. The reaction rate is faster at pH close to 7 and slow at pH 10. Oxidation of starch under mild to moderate alkaline conditions produces more carboxyl groups that prevent starch retrograde. Recent studies have shown that at higher pH, there are more carboxyl groups than carbonyl groups, which improves the application of oxidized starch. The effects of oxidation with sodium hypochlorite on the expansion properties and texture of fermented cassava starch were investigated. It was found that the expansion characteristics of cassava starch are controlled by oxidation with sodium hypochlorite, as this method is not dependent on climatic conditions and produces a more uniform product. However, the oxidation process is not conducted using the sunlight drying method. At different concentrations of activated chlorine (0, 1, 2, 3, and 4%). Increasing the concentration of activated chlorine decreased the zeta potential in starch nanocrystalline suspensions. Also, a relatively more uniform size distribution was created, which improved the redegradability of starch nanocrystals (Vanier et al. 2017).

Blending of TPS with Other Polymer

Thermoplastic starch (TPS) becomes brittle in the presence of moisture; however, its combination with hydrophobic polymers can mitigate this issue. According to conducted research, the combination of TPS with biodegradable and non-biodegradable polymers can enhance mechanical properties and processability, thereby increasing the commercial applications of these compounds. Next, we will examine the combination of TPS with various polymers (Bangar et al. 2021).

TPS-Polyethylene

Polyethylene (PE) is one of the most important thermoplastics that have numerous applications today, and the reasons for this can be attributed to its low cost, lightweight, durability, chemical inertness, and so on (Khanoonkon et al. 2016). Polyolefin plastics are stable against oxidation and optical degradation and are not easily degraded. To accelerate this process, their combination with carbohydrate polymers such as thermoplastic starch (TPS) has been proposed. One of these polyolefin polymers is polyethylene. Based on research conducted, the combination of TPS with various types of polyethylene, such as LDPE, LLDPE, and HDPE, enhances the properties of food packaging. These compounds, particularly when combined with plasticizers and specific processes such as twin-screw extrusion, maintain mechanical properties at high concentrations of TPS. Therefore, the combination of TPS with polyethylene affects the physical properties and processability of plastics and contributes to their improvement (Bangar et al. 2021).

One of the significant challenges of TPS is its hydrophilic properties. This property makes it highly sensitive to moisture. To address this issue, TPS can be combined with hydrophobic materials such as polyolefins. Blends of thermoplastic starch and polyethylene were studied, demonstrating the properties of a fully immiscible polymer blend with a dispersed phase morphology in the matrix. A one-step process has been developed that effectively controls morphology, demonstrating that high loadings of thermoplastic starch can still yield polyethylene/TPS blends with excellent elongation at break. To address the compatibility issues between TPS and polyethylene (PE), various interfacial modifiers have been employed in the literature. Among them, PE-grafted maleic anhydride is one of the most widely used materials, showing reasonable improvements in both morphology and mechanical properties. The modification of other polymers has also been reported to affect the morphology in polymer blends. Similar to all classical compatibilized systems, these compatibilizers are used to reduce interfacial tension in thermoplastic starch (TPS) and polyethylene (PE) blends, which in turn decreases the phase size (Taghizadeh et al. 2013).

Exposure of agrarian polyethylene (PE) films to heat and UV radiation in an oxygen-rich environment often leads to a decline in their mechanical properties. Alterations in mechanical behavior are associated with adjustments in chemical and molecular composition, which manifest as differences in crystallinity and physical properties. The effect of corn starch and a pro-oxidant on the photo-oxidation of LDPE films was investigated. The results showed that simply adding starch did not impact the material's sensitivity to photolysis. In a study, combining LDPE and corn starch reduced its degradation rate compared to pure LDPE, indicating that corn starch enhances the stability of LDPE against degradation. Based on the studies by Ratanakamnuan and Aht-Ong, combining corn starch with LDPE reduces its degradation rate compared to pure LDPE. However, adding banana starch to LDPE with banana starch films increases the photodegradation rate and decreases tensile properties. Combining synthetic polymers with thermoplastic starch such as polyethylene is good in terms of stability and cost (Méndez-Hernández et al. 2011).

TPS-Polybutylene Adipate-Co-Terephthalate (PBAT)

PBAT is employed in the manufacturing of biodegradable films because of its excellent tensile strength and elasticity. The combination of PP/PBAT/TPS in the recycling process alters their mechanical properties, resulting in the degradation of PBAT with TPS occurring more rapidly than that of PP. Furthermore, the use of fungi such as Aspergillus and Penicillium for biodegradability research and the application of sorbate in packaging have led to improved mechanical properties, delayed fungal growth, and enhanced packaging quality. The properties of this polymer are mentioned in Table 11.1 (Bangar et al. 2021).

One clear advantage of PBAT is that it has been approved for food contact and compostability based on EN13432. Its primary application is in blown film production, while another common use is in extrusion coatings. Generally, monolayer films are more prevalent in the realm of biodegradable plastics compared to complex multilayer barrier films. This is because the biodegradability of a single layer mixed with other layers of non-biodegradable bioplastics offers little added value. PBAT is an appealing alternative to LDPE for blown film, particularly for producing thin foils used in bags. PBAT/TPS blends are an effective way to incorporate bio-based carbon content into PBAT formulations. Manufacturers such as BASF and Jinhui Zhaolong offer their pure PBAT products under the trade names Ecovio™ and Ecoworld™, respectively. Additionally, their PBAT/TPS compounds are marketed under the names Ecoflex™ and Ecowill™ (Lackner et al. 2021).

TPS-Propylene (PP)

Polypropylene is one of the most commonly used synthetic polymers, applied in the manufacturing of a wide range of products. However, at high amounts of TPS, these properties are weak. The addition of clays such as MMT and Cloisite 30B along with various modifiers like malleated polypropylene (mPP) has a significant impact on enhancing and improving the tensile strength and modulus of TPS/PP composites. Additionally, the use of advanced and modern methods such as nuclear magnetic resonance (NMR) spectroscopy and dielectric relaxation spectroscopy (DRS) can be beneficial. These techniques assist in understanding the

Table 11.1 Properties of PBAT (Lackner et al. 2021) with permission, copyright memberanes, 2025

Property	Unit	Standard	Value
Melt flow rate	g/10 min	ISO 1133	≤ 5 T = 190 °C,w = 2.16 kg
Density	g/cm^3	ISO 1183	1.26
Tensile strength	MPa	ISO 527	> 17
Melting temperature	°C	DSC	110–120
	%	ISO 527	≥ 600
Elongation at break			
Vicat softening point	°C	ISO 306	89

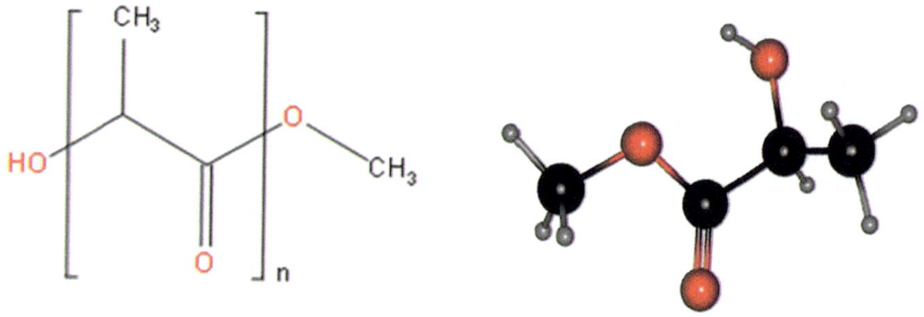

Fig. 11.4 Chemical structure of polylactic acid (Martinez Villadiego et al. 2022) with permission, copyright memberanes, 2025

molecular dynamics and dynamic processes of TPS/PP compounds. These results can contribute to improving the faster biodegradation process and the environmental applications of these compounds in the future. In a process, starch was modified by adding glycerol and polypropylene in conjunction with acetyl tributyl citrate using a co-rotating extruder. Materials like organoclay and maleated polypropylene were used as compatibilizers in the experiment. By conducting analyses such as XRD and SEM, the results showed that the organoclay nanofiller improves compatibility between the immiscible components (Bangar et al. 2021).

TPS-Polylactic Acid (PLA)

PLA is degraded by hydrolysis is biodegradable and is used in medicine and pharmaceuticals. As shown in Fig. 11.4, its molecular structure consists of oxygen, hydrogen, and carbon. The starch is combined with polylactic acid to reduce cost and improve biodegradability properties. However, the two polymers have differences in polarity, creating challenges in compatibility and mechanical properties. However, with the presence of additives such as glycidyl methacrylate, compatibility can be improved. It should be noted that emulsifiers are toxic and limit the use of combinations as food substances (Bangar et al. 2021).

The thermal, mechanical, shape, structure, and appearance properties of the thermoplastic starch and polylactic acid blend were investigated. Based on the findings, the ratio of different components in the blend influences the glass transition temperature of polylactic acid. Particularly, with a higher TPS content in the blend, the Tg progressively declined, suggesting a decrease in the immiscibility of the two components. It was determined that phase separation, which led to a lack of interaction between TPS and PLA, presented a significant limitation. The need for a strategy to improve compatibility was emphasized. The process of blending thermoplastic starch with polylactic acid is shown in Fig. 11.4. The influence of enhancing glycerol on starch and its mixtures with polylactic acid was studied. It was observed that voids formed due to the separation of starch particles in the combination of PLA and pure starch. However, these voids were not present in the PLA and gelatinized starch blends, which indicated improved attachment intermediate to the starch and polylactic acid. Furthermore, the structural characteristics of the PLA and gelatinized

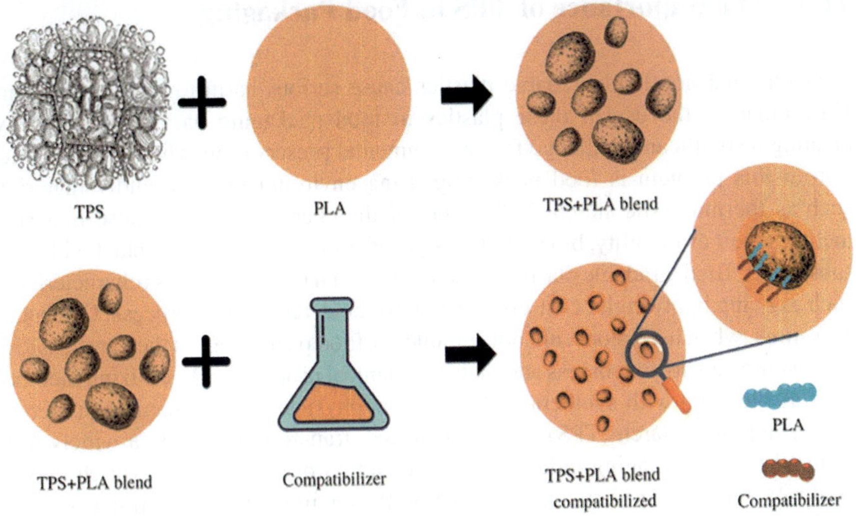

Fig. 11.5 Compatibilization process of TPS with PLA (Martinez Villadiego et al. 2022) with permission, copyright memberanes, 2025

starch combinations were found to be superior (Martinez Villadiego et al. 2022) (Fig. 11.5).

TPS-Polyvinyl Alcohol (PVOH)

Polyvinyl alcohol (PVOH) is a polymer that has significant applications and is used in food packaging due to its inhibition of oxygen passage. This polymer combines with low-cost starch and reduces the total costs and improves the properties of starch. For example, the combination of PVOH with corn starch increases thermal stability and prevents surface cracks. Also, combining thermoplastic starch with this polymer can be used instead of aluminum foil and metallized films. A report indicates that a 30 wt% starch blend exhibited the highest tensile strength among PVOH/starch composites. When corn starch/PVOH blends are plasticized with glycerol, they can be extruded using a single-screw extruder. The resulting samples demonstrated a tensile strength of 1.8 MPa and an elongation at break of 113% at 50% relative humidity. The addition of PVOH to the starch/glycerol blends helped prevent surface cracks, indicating good compatibility. The tensile strength of commercial-grade PVOH/TPS films varied based on whether they were tested in dry or wet conditions and was influenced by the degree of hydrolysis and molecular weight (Bangar et al. 2021).

11.5 The Importance of TPS in Food Packaging

Synthetic and non-biodegradable plastics cause serious harm to the environment. Unfortunately, the use of these plastics in food packaging has increased today, creating a significant challenge for environmental preservation. The proposed solution to this problem is food packaging using environmentally friendly materials such as thermoplastic starch (TPS). One of the characteristics of these materials includes their availability, biodegradability, and good properties in contact with food substances. Biodegradable plastics, unlike traditional plastics, possess characteristics such as being lightweight, cost-effective, recyclable, and easy to process. Of course, TPS films, when used alone, are not suitable for food packaging, as they are sensitive to moisture and do not possess adequate mechanical properties. By utilizing methods such as blending with synthetic polymers, the physical and mechanical properties of thermoplastic starch (TPS) can be enhanced, transforming it into a widely used material in the packaging industry. The production of thermoplastic starch plays a vital role in reducing waste and minimizing the environmental pollution caused by it (Tedeschi et al. 2022).

One of the most important factors in the preservation and maintenance of food quality is its packaging. By incorporating various active compounds into the packaging of food products, additional functionalities can be provided. These active compounds create antioxidant and antimicrobial capacities, contributing to an extended shelf life (Panrong et al. 2019).

11.6 Linear Low-Density Polyethylene (LLDPE)

The previous section introduced starch polymers, and this section described linear low-density polyethylene as a base for polymer films used in active packaging. Linear low-density polyethylene is a versatile polymer synthesized by the polymerization of ethylene combined with small proportions of alpha olefins, including short chain comonomers such as octene or hexane. This polymer possesses a linear structure, which contributes to mechanical and chemical properties. As a member of the polyethylene family, LLDPE enjoys widespread application across various industries, owing to its distinctive characteristics (Khumkomgool et al. 2020). The structure of LLDPE is shown in Fig. 11.6.

It is noteworthy that LLDPE exhibits remarkable flexibility and adaptability, enabling it to function effectively under a variety of conditions. Its excellent impact resistance is essential for protecting sensitive materials during packaging processes. In addition, LLDPE's strong chemical resistance enhances its application in the chemical sector. Thermal stability is another key strength (Khumkomgool et al. 2020). LLDPE is widely used in the packaging industry to produce base films, plastic sheets and also required bags. In addition, it is used in agricultural environments for traditional and innovative coatings as well as in the manufacture of household goods, toys,

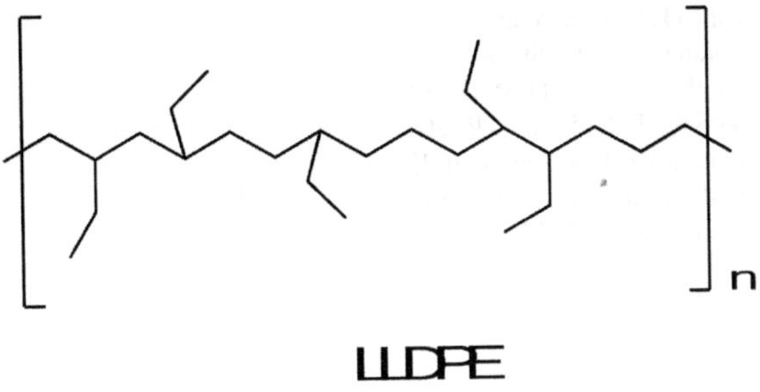

Fig. 11.6 Structure of LLDPE (Muzakkar 2013) with permission, copyright journal of physics conference series, 2013

and industrial products. Overall, LLDPE stands out as a highly efficient and high-performance polymer material, making it an excellent choice for many applications. LLDPE is a widely utilized, cost-effective polymer in the food-packaging industry, renowned for its excellent mechanical properties and high production efficiency. Its ability to serve as a high-quality water barrier makes it an ideal choice for a variety of food packaging applications (Khumkomgool et al. 2020). Despite all the unique characteristics of LLDPE, this polymer exhibits weak barrier properties against oxygen and ultraviolet radiation, which exacerbates concerns about plastic pollution and its environmental impact. Given these challenges, various studies have sought to alternative petroleum-based plastics with biopolymers derived from rechargeable resources such as chitosan, starch, and cellulose. Incorporating biopolymers to conventional plastics such as polyethylene (PE), polyethylene terephthalate (PET), and polypropylene (PP) is a promising and efficient way to reduce the environmental impacts associated with plastics while adding value to agricultural productions and by-products. In the following section, we will examine various research initiatives focused on LLDPE/starch-based packaging. Combining starch with plant extracts with antimicrobial properties has the potential to create active packaging that enhances food quality preservation. In a study conducted by Aruchida Khumkomgool et al., active packaging for meat was fabricated from TPS infused with plant extracts, specifically saponin and cinnamon powders, using LLDPE-blown film extrusion techniques. Increasing the concentration of TPS and plant extracts resulted in the dispersion of aggregates in the film matrices, which in turn resulted in a decrease in mechanical properties—specifically, a 60–80% decrease regarding tensile strength and a decrease in elongation of 30–90%. When at rest, this was ascribed to the development of heterogeneous microstructures. Surface topography analysis revealed an increase in roughness by factors of 2.0–4.0 following the addition of TPS and plant extracts, along with the presence of nanovoids in films with high levels of saponin. Infrared spectroscopy revealed the creation of hydrogen bonds between starch and plant extracts, which reduced the disruption of starch granules during the extrusion process (Khumkomgool et al. 2020). A recent study on TPS produced using potato starch extracted from surplus potato chips snack industry showed that TPS was incorporated into LLDPE in various forms to produce processed films and that incorporation of TPS into LLDPE films resulted in visible changes, notably a change in the color of the films from a bright yellow to a pale yellow appearance. These

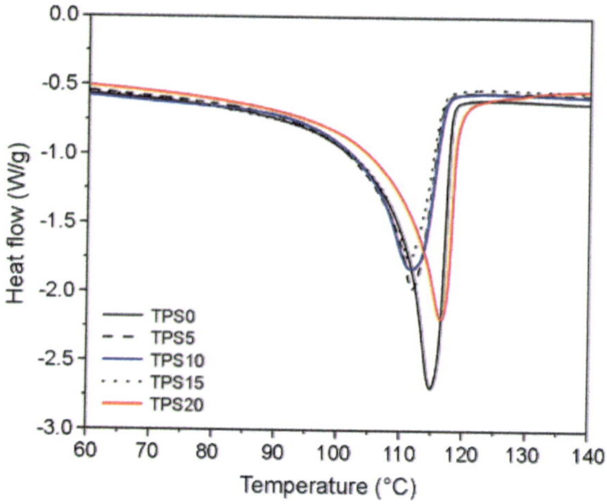

Fig. 11.7 Thermograms obtained from DSC analysis for the samples TPS0, TPS5, TPS10, TPS15, and TPS20 (Montoya-Anaya et al. 2023) with permission, copyright memberanes, 2025

results indicate that the presence of TPS creates pores in the films, which leads to an increase in water vapor transmission rates (WVTR), water vapor permeability (WVP), water absorption, and solubility parameters. Scanning electron microscopy (SEM) coupled with Fourier-transform infrared spectroscopy using attenuated total reflectance analysis (FTIR-ATR) images showed that the interaction between the polymers results in a reduction in the mechanical properties. As shown in Fig. 11.7, the thermal analysis results also showed that the maximum degradation temperature (T_{max}) of the formulated films increased with the addition of TPS, and differential scanning calorimetry (DSC) findings indicated that TPS acted as a nucleating agent in the LLDPE matrix. Among the different formulations tested, the blend with 15 wt% TPS showed mechanical property values comparable to the LLDPE matrix, indicating that this concentration is optimal for producing films suitable for flexible packaging applications.

11.7 Scavengers Systems for Active Packaging

An approach utilized in active packaging to safeguard food products involves the use of oxygen scavengers. In the following sections, we will examine this approach in detail. Food spoilage can be attributed to several factors, including microbial growth, chemical interactions between food components, as well as the activity of enzymes present in the products, One of the major factors in spoilage is the existence of oxygen in the atmosphere around the food. Oxygen can cause fats to spoil, food to discolor, flavor to deteriorate, the proliferation of mold and aerobic bacteria, and ultimately reduce the food's nutritional value. The use of oxygen scavengers or absorbers has several benefits, such as preventing the growth of microorganisms, preserving the quality of foods high in fat, thus averting rancidity, avoiding color

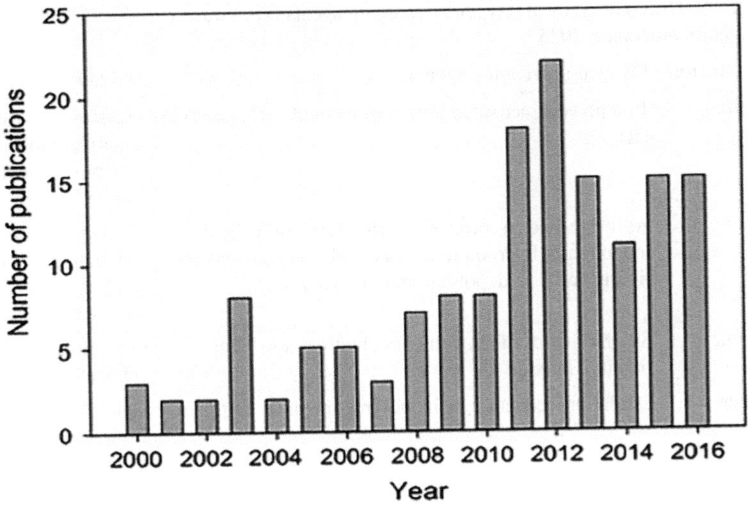

Fig. 11.8 Trends in publications regarding oxygen-scavenging films utilized in food packaging from 2000 to 2016. Information gathered from Scopus using the terms "oxygen-scavenging packaging" and "oxygen scavenging polymer" (Gaikwad et al. 2018) with permission, copyright memberanes, 2025

change, and inhibiting oxidation. These oxygen scavengers have become popular in the food industry, where the shelf life of products can be prolonged from 3–4 days to 14 days or beyond. Currently, commercial oxygen scavengers come in various formats, such as pouches, films integrated into the packaging, and labels, all crafted to effectively enhance the shelf life of food products (Gaikwad et al. 2018). As can be seen in Fig. 11.8, it shows the diffusion trend related to oxygen barrier films used in food packaging.

Oxygen-Scavenging Agents

Oxygen-scavenging agents may consist of any oxidizing substrate, whether it be organic or inorganic. The choice of appropriate agents for food packaging is affected by various factors, such as cost, safety, and their reaction rate and capacity with oxygen. Recently, there has been increasing interest in novel nonmetallic agents, especially those that are organic or polymer-based. Different potential oxygen-scavenging agents suitable for food packaging are listed in Table 11.2 (Gaikwad et al. 2018).

Metal Scavengers

Studies show that metal scavengers represent the dominant segment in the oxygen-scavenging market and have been used for commercial purposes for decades (Irkin and Esmer 2015). In another study, Matche et al. investigated modified LDPE films including zinc, iron, or ascorbic acid and evaluated their physical and mechanical characteristics, as well as oxygen absorption capabilities through shelf-life analysis of

Table 11.2 Different types of oxygen-scavenging agents (Gaikwad et al. 2018) with permission, copyright memberanes, 2025

Classification	Oxygen-scavenging agents	Oxidation mechanism
Metallic	Iron powder, activated iron, ferrous oxide, iron salt, Co (II), Zn	Oxidation of iron with supply of moisture and action of optional catalyst
Organic	Ascorbic acid, ascorbic acid salts, isoascorbic acid, tocoph-erol, hydroquinone, catechol, rongalit, sorbose, lignin, gallic acid, polyunsaturated fatty acids	Oxidation of organic substrate with metallic catalyst or alkaline substance
Inorganic	Sulphite, thiosulfate, dithionite, hydrogen sulphite, titanium dioxide	Oxidation of inorganic substrate by UV light
Polymer based	Oxidation–reduction resin, polymer metallic complex	Oxidation of polymer components with metallic catalyst (mostly cobalt)
Enzyme based	Glucose oxidase, laccase, ethanol oxidase	Immobilization

breads and loaves. Although the presence of iron-based coatings affected the mechanical and barrier properties, packaged bread showed minimal microbial growth within acceptable limits (Irkin and Esmer 2015). However, films with metal-based oxygen scavengers are not without drawbacks, including potential metal contamination, unintended activation of in-line metal detectors, and complications upon microwave heating. Furthermore, the temperature-dependent behavior of metal oxidation poses challenges, leading to premature oxygen inhibition at high temperatures and reduced collection efficiency at lower temperatures, which can complicate food packaging applications (Gaikwad et al. 2018).

Organic Scavengers

Recent investigations in both academic and industrial fields have increasingly emphasized the development and utilization of organic oxygen scavengers intended for oxygen-scavenging films. Given the challenges that metallic agents face in terms of detection in packaging production lines, there has been a significant shift toward organic alternatives. Typically, these organic scavengers are composed of low molecular weight oligomers that can be incorporated into polymer layers or added to the structure as oxygen-scavenging polymers. In the latter case, the polymer backbone can be cleaved in response to oxygen interaction, or the side chains may directly interact with oxygen (Gaikwad et al. 2018). Compared to inorganic powders, organic oxygen scavengers demonstrate superior dispersion and compatibility within polymer matrices, aiding in the production of optically active films. A notable example is provided by Damaj et al. who investigated the incorporation of a Co(II) threonine complex, functioning as an amino acid-based adsorbent, incorporated into a polyvinyl alcohol (PVA) matrix. The resulting films showed activation dependent on relative humidity levels (Damaj et al. 2015). Additionally, Janjarasskul et al. created an edible film made from whey protein isolate and ascorbic acid that efficiently harnesses their oxygen-scavenging capabilities while alleviating worries related to accidental ingestion and recycling difficulties typically faced in conventional systems (Janjarasskul et al. 2011). Another important approach to biodegradable oxygen-scavenging involves the use of 40 wt% α-tocopherol, a natural antioxidant, encapsulated in PLA microparticles via the solvent-emulsion evaporation method, which reportedly preserves the biodegradability of the food ingredient (Scarfato et al. 2015). Consequently, transparent films based on organic oxygen scavengers offer significant potential for creating innovative packaging solutions that meet consumer preferences and sustainability challenges (Byun et al. 2012).

Polymer-based Scavengers

A significant amount of research has been directed toward polymer-based oxygen scavengers, which are composed of a suitable starting polymer required for packaging applications, modified with an unsaturated side chain attached to the main chain (column-column) (Gaikwad et al. 2018). Active oxygen scavengers, such as polyamides and polyolefins, can be incorporated into the base polymer structure to form oxygen-scavenging polymers suitable for packaging walls. These scavenging polymers may be blended with other polymers or utilized as oxygen-scavenging layers in multilayer containers. It is common to add photo-initiators to these formulations to inhibit premature oxidation during processing and storage. However, a major concern with this technology is that reactions between polyunsaturated compounds and oxygen can produce by-products like organic acids, aldehydes, or ketones, which may compromise food quality and lead to off-flavors and changes in sensory characteristics (Li et al. 2012). A number of these by-products can actually play a role in monitoring the integrity and shelf stability of lipid-containing products due to their association with rancidity (Jo et al. 2002). Apart from that, volatile compounds can be absorbed by zeolites and activated within a short period to reduce off-flavors, while functional barrier technologies developed for this purpose are much superior because they actually inhibit the migration of oxidation products. This can be achieved by modifying the polymer structure to contain unsaturated functional groups, such as those in LDPE and PET, which also serve as oxygen scavengers. During this process, the oxidation of unsaturated hydrocarbons within the polymer matrix serves in the role of efficient method for removing oxygen (Ferrari et al. 2009). The auto-oxidation process can be initiated in polymers through UV light in the vicinity of metallic catalysts. High-performance polymers like LDPE and PET are infused with unsaturated hydrocarbons to create an oxygen barrier (Galdi et al. 2008).

Enzymatic Scavengers

Enzyme-based oxidation represents a new way to manage and control oxygen levels in food packaging. The approach depends on certain enzymes that react with substrates to absorb the incoming oxygen. One of the enzymes most used in this respect is glucose oxidase. In addition to this, Vermeiren et al. (1999) suggested that the combination of glucose oxidase and catalase is capable of scavenging oxygen.

Glucose oxidase catalyzes the oxidation of glucose into gluconate and hydrogen peroxide when moisture is present (Mabeck and Malliaras 2006). Very few patents exist specifying polymeric materials containing enzyme-based oxygen absorbers for film applications (Labuza and Breene 1989), and glucose oxidase and catalase are among the enzymes that have been widely studied in this regard. These enzymes can be immobilized on a wide variety of polymers including polystyrene, polyethylene, and polypropylene. But in order for the immobilized enzymes to catalyze the desired redox reactions, the active enzyme must contact food directly, which implies that the benefits the use of enzyme-based films in food packaging is restricted.

Natural Oxygen-Scavenging Systems

Natural products have gained much greater preference over synthetic products in recent years due to their proven safety benefits. Therefore, oxygen-scavenging systems derived from natural materials are increasingly gaining attention (Irkin and Esmer 2015). An attempt was made to take inspiration from oxidation-sensitive natural substrates and use them to enhance the efficiency of scavengers that are metal based. These natural foundations may also be explored for their development as oxygen scavengers by immobilization or mixing with plastic films (Janjarasskul and Suppakul 2018). An et al. recently introduced an innovative oxygen absorption system that merges gallic acid with potassium carbonate and is triggered by moisture. For this reason, this system was incorporated into LDPE film for packaging moist food. The reaction mechanism of potassium carbonate and gallic acid utilized as an oxygen scavenger is shown in Fig. 11.9. Their observations showed that when the gallic acid-based system was incorporated, the films exhibited mechanical and physical characteristics that were lower than compared to pure LDPE. Nevertheless, the film incorporating 20% (w/w) gallic acid as the oxygen absorption agent demonstrated an oxygen-scavenging capability of 0.709 mL/g over a duration of 7 days (Janjarasskul and Suppakul 2018).

The oxygen-scavenging film developed by Bion et al., which contains natural α-tocopherol and a transition metal, was applied using a film applicator (Fig. 11.10) (Byun et al. 2012).

Their system produced a capacity for oxygen scavenging of 6.72 mL/g, along with an average absorption rate of 0.11 O_2/g daily, which showed particular efficiency in the presence of moisture. Antioxidant compounds that have the ability to function as oxygen absorbers include erythorbic acid, quinones, and catechols. Other potential natural scavengers are hydroxyamines and ketoximes, whose use in polymer matrices and unit weight reactive power has been limited so far (Cardona et al. 2011). Natural oxygen scavengers are a promising avenue for new forms of packaging that offer desirable properties and many sustainability benefits, in line with consumer expectations. The active food packaging sector is anticipated to experience significant demand for these natural absorbers (Gaikwad et al. 2018).

Oxygen-Scavenging Film Manufacturing Techniques

Detergents can be incorporated into packaging films in a variety of ways, as listed below. It is important to note that while creating a packaging film that absorbs

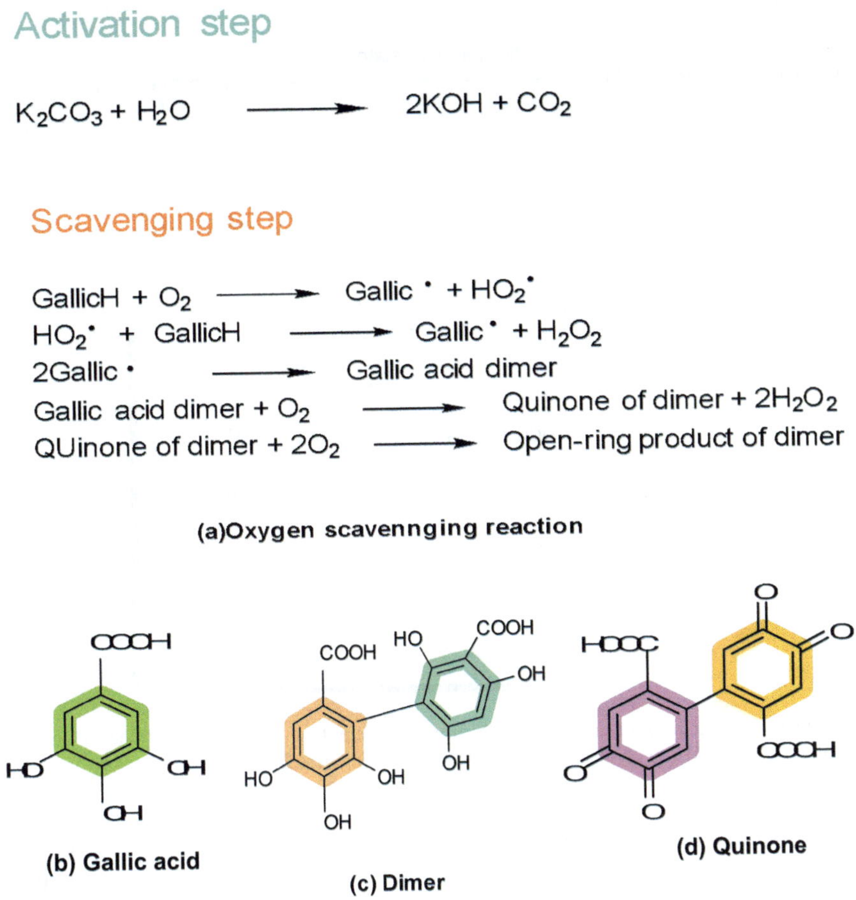

Fig. 11.9 Interactions of an oxygen-scavenging system that includes gallic acid and potassium carbonate (Gaikwad et al. 2018) with permission, copyright memberanes, 2025

oxygen, the activation mechanism of the scavenging compounds should be developed/considered in relation to the specific product application (Gaikwad et al. 2018).

Coating

New oxygen scavengers are incorporated into packaging materials through physical adsorption or coatings, allowing the scavengers to migrate from the packaging into the food. This migration therefore prevents oxidative reactions in food products, as shown in Fig. 11.11a. These active substances can engage directly with the food or generate a vapor phase that can convey antioxidant properties to the vapor space (Gaikwad et al. 2018). In recent studies, Gaikwad et al. introduced moisture-responsive oxygen-scavenging films by applying a pyrogallol coating on LLDPE, in

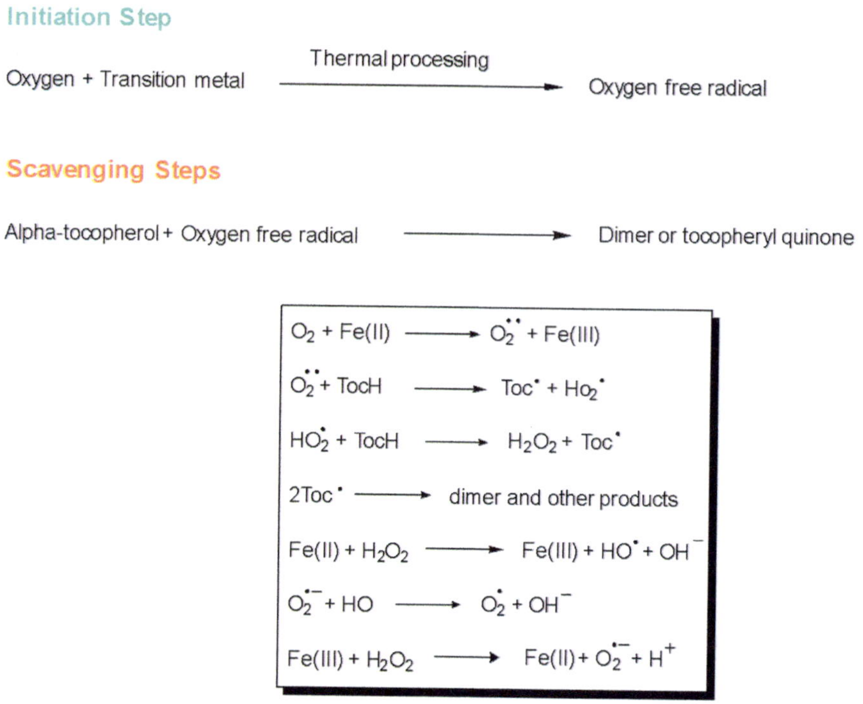

Fig. 11.10 Interactions of an oxygen-absorbing system that includes α-tocopherol and iron (Gaikwad et al. 2018) with permission, copyright memberanes, 2025

combination with sodium carbonate. The roughness on the film increased the water and oxygen permeability and decreased the tensile strength due to the incorporation and clustering of sodium carbonate and pyrogallol. A film with a coating of 20% pyrogallol demonstrated an oxygen-scavenging efficiency of 0.443 mL/g at a temperature of 23 °C (Gaikwad et al. 2017a).

Integration into Packaging

Oxygen absorption systems has the ability to uniformly spread across the packaging as shown in Fig. 11.11b. Employing an oxygen-scavenging agent integrated within the polymer matrix facilitates the capture of oxygen present in both the package's, the vapor space, and the food item. This limits or stops the processes that cause oxidation, especially oxygen, to enter the package from the environment (Gaikwad et al. 2018). This technique is suitable for preparing active films using a diverse range of inhibitor compounds, either by casting or by extrusion (Janjarasskul et al. 2011; Ahn et al. 2016). In addition to the positive effects of these films, these may introduce an unpleasant flavor in food products, particularly when iron-based oxygen scavengers are utilized (Renato Souza et al. 2012). Furthermore, iron can migrate into foods and stimulate lipid oxidative reactions. Also, significant quantities of antioxidants are

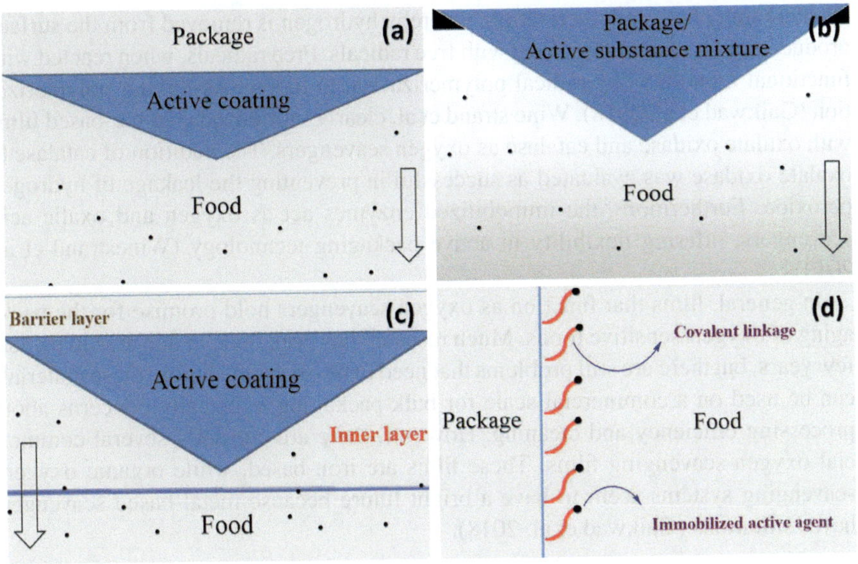

Fig. 11.11 Different techniques for applying oxygen-scavenging films and the movement of oxygen-scavenging agents within each system include: **a** applying an active layer coating to the surface of the package, **b** integrating the active compound into the polymeric matrix of the package, **c** using a multilayer active film as well as a non-immobilized active film, and **d** utilizing an immobilized active film (Gaikwad et al. 2018) with permission, copyright memberanes, 2025

necessary to guarantee their distribution across the packaging, which is associated with large losses in mass and activity during production (Johnson and Decker 2015).

Immobilization

In general, surface stabilization or stabilization involves the use including widely used polymers like low-density polyethylene (LDPE), high-density polyethylene (HDPE), polypropylene (PP), polyethylene terephthalate (PET), polyvinyl alcohol (PVA), and polystyrene. These polymers are available in the market and, since they are inactive materials, require some kind of pretreatment to create reactive sites on their surfaces so that many functional compounds can be attached to them. (Fig. 11.11d) (Gaikwad et al. 2018). Many techniques for surface modification have been created to achieve surface functionalization, one of the most efficient of which is UV irradiation (Goddard and Hotchkiss 2007). Ultraviolet (UV) radiation can effectively generate functional groups or free radicals serve to trigger the cross-polymerization of functional monomers. Ultraviolet light in the short-wave range (UVC 185 nm) interacts with atmospheric oxygen, resulting in the production of atomic oxygen and ozone. Upon exposure to 254 nm UV radiation, ozone is further decomposed transforming into molecular oxygen and atomic oxygen. It has been documented that UV irradiation is a very effective method for surface functionalization (Barish and Goddard 2011). Long-wave ultraviolet (UVA) light assists in polymerization by achieving photo-polymerization, whereby hydrogen is removed from the surface produces a polymer matrix along with free radicals. Free radicals, when reacted with functional monomers by radical polymerization, initiate cross-linking polymerization (Gaikwad et al. 2018). Wine strand et al. clearly introduced enzyme-based films with oxalate oxidase and catalase as oxygen scavengers. The addition of catalase to oxalate oxidase was evaluated as successful in preventing the leakage of hydrogen peroxide. Furthermore, the immobilized enzymes act as oxygen and oxalic acid scavengers, offering flexibility in active packaging technology (Winestrand et al. 2013).

In general, films that function as oxygen scavengers hold promise for the packaging of oxygen-sensitive foods. Much research has been done in this area in the last few years, but there are still problems that need to be overcome before these materials can be used on a commercial scale for bulk packaging, especially concerns about processing efficiency and cleaning. However, there are currently several commercial oxygen-scavenging films. These films are iron-based, while organic oxygen-scavenging systems seem to have a bright future because metal-based scavengers have limitations (Gaikwad et al. 2018).

Moisture Absorption

Although moisture is required to enhance the shelf life of packaged items like food, it is important to control excessive moisture in the packaging, as it can negatively affect product quality and destroy the structural integrity of the package. This is most noticeable in foods that have high water activity which is associated with items like fresh fruits and raw meat. In these instances, moisture-absorbing agents can play a very useful role in packaging (Gaikwad et al. 2019). As explained above, the presence of moisture is of great importance. This will be discussed in more detail below. Packaging holds significant importance in maintaining the quality and value of food from the point of production to its final consumption. This is achieved primarily through the prevention of undesirable chemical and biological alterations (Gaikwad et al. 2017a). In today's world, food safety and the types and amounts of preservatives used in food have become important consumer concerns. Consumers prefer to consume natural, healthy, and high-quality foods that are minimally processed or unprocessed and free of preservatives, while having a good shelf life (Gaikwad et al. 2018). As mentioned earlier, food packaging has many functions: it provides a barrier against contamination, informs consumers about the product, and assists in marketing efforts. The most important function of food packaging is to isolate food from the external environment to reduce contact with spoilage factors such as microorganisms, moisture in the air, oxygen, and undesirable flavors is significantly reduced. At the same time, it prevents the loss of nutritious and desirable compounds including aromatic volatiles in order to increase adequate life span of the food (Gaikwad et al. 2017b). Moisture constitutes to the primary food spoilage. Hence, controlling moisture content in food packaging has been of great importance for many years in preventing microbial growth and improving sensory perception. Relative humidity during storage is another major challenge for many dry food products. Even the presence of minor moisture potentially compromises quality. High relative humidity is likely to increase susceptibility to microbial growth and subsequent undesirable alterations in texture and visual aspects, which in turn can lead to reduced shelf life (Murmu and Mishra 2018). On the other hand, maintaining high levels of controlled relative humidity (RH) in packaging is very beneficial to prevent dehydration of products with foods with high moisture content, uncooked meat, including fresh seafood and fruits and vegetables (Rux et al. 2016). Desiccants are an effective and excellent method for reducing excess moisture in packaging and have a significant impact in controlling the water activity of food products, thereby preventing microbial growth (Choi and Lee 2013). Desiccants are produced in various designs and forms, including bags, films, trays, and pads, which remove liquid secretions from food or maintain a suitable relative humidity inside the package. In most cases, desiccants or moisture collectors are enclosed within bags and pads that are incorporated into food packaging systems. These components are added to the package to effectively balance the excess moisture that accumulates due to the saturation of the food during storage. Desiccants work through a physical adsorption process in which they interact favorably with moisture.

In general, humidity control techniques can be divided into three main approaches:

1. Humidity reduction:

This technique can be accomplished through the use of modified atmosphere packaging (MAP) which replaces the moist air present in the package's headspace and can be replaced with dry modified atmosphere (MA) gas or through vacuum sealing, which eliminates the damp air.

2. Moisture prevention:

This technique is achieved by using packaging materials with a high barrier to moisture penetration.

3. Moisture removal:

This method involves the use utilizing adsorbents for eliminate moisture.

Furthermore, the control of relative humidity levels inside the packaged material can be managed very effectively by selecting materials that have a high barrier to water vapor (Bovi et al. 2018).

Classification of Moisture Absorbers

Desiccants are materials that are composed of suitable hygroscopic substrates or materials and are designed to absorb and retain moisture from the surrounding environment. These materials have hygroscopic properties that allow them to effectively control moisture levels. Active desiccant agents can be further classified into two main types:

1. Relative humidity (RH) controllers: Absorbents that primarily address the levels of relative moisture content within the headspace of the package. Such agents include desiccants.

2. Moisture eliminators: Absorbents that are intended to absorb free liquids that can escape from food products. They are commonly used as pads and sheets placed under fresh produce and meat in various packaging configurations (Yildirim et al. 2018).

Acceptable desiccants for packaging applications are influenced by selection criteria such as cost, safety, and moisture absorption capacity. Also, there has been recent interest in new superabsorbent composite materials, consisting of organic and polymer-based composites.

Advances in Organic-Based Desiccants for Packaging

Recently, there has been significant progress and utilization of organic-based desiccants in desiccant packaging has been increasingly studied by academic and industrial researchers (Gaikwad et al. 2018). These organic-based adsorbents can be promising in the development of new packaging systems, offering desirable properties and benefits in terms of consumer perception and sustainability. In recent years, various types of moisture-absorbing composite materials have been evaluated, the most important of which are super absorbents in the form of films and trays. Future research will focus on natural moisture absorbers encapsulated in polymer matrices.

For example, the use of cellulose and its derivatives as desiccants has also been investigated conducted by numerous researchers for applications in food packaging (Pinming et al. 2016). Carboxy methyl cellulose (CMC) is one such desiccant, capable of taking in nearly eight times the amount of moisture compared to regular cotton. Nonetheless, CMC has significant disadvantages, including limited resistance to degradation, insect damage, and light exposure (Bovi and Mahajan 2017).

Despite the advantages offered by organic-based desiccants, there are several significant drawbacks. The high-cost and lower absorption capacity compared to their inorganic counterparts necessitate the use of larger sachets than traditional silica gel desiccants. Future research efforts are needed to increase the technical and commercial viability of this type of organic-based desiccant, focusing on increasing their stability during processing and optimizing their effectiveness regarding moisture absorption (Gaikwad et al. 2019).

Inorganic-Based Moisture Absorbers

Inorganic-based desiccants constitute the largest portion of the market and have historically been employed to retain water across various applications. For edible items that need exceptionally low relative humidity levels, bags including calcium oxide and silica gel are frequently utilized (Yildirim et al. 2018). Other widely used mineral desiccants include calcium chloride ($CaCl_2$), potassium chloride (KCl), potassium carbonate, and bentonite, all being noteworthy because of their non-toxic and non-corrosive characteristics. The overall reactivity of calcium oxide when interacting with water enhances its effectiveness in this function.

$$CaO + H_2O \rightarrow Ca(OH)_2$$

These minerals have very high thermal resistance, which allows them to be incorporated into polymer matrices through the process of melt extrusion processes to create plastic-based desiccant/desiccant structures (Gaikwad et al. 2019).

Polymer-Based Desiccants

Polymer-based desiccants have emerged as a novel field of focus for manufacturers of packaging materials. These desiccants typically consist of a foundational polymer appropriate for packaging purposes and modified with a desiccant polymer (Gaikwad et al. 2019). As mentioned earlier, starch is a bio-based material obtained from renewable sources like cassava, potatoes, and maize that can be converted into TPS through a plasticization process. TPS is beneficial for the thermal treatment of packaging materials. Nevertheless, it is very susceptible to humidity, which can lead to processing challenges and instability during storage. Blending TPS with LLDPE enhances processing capabilities and improves the stability of biomaterials for potential commercialization (Khanoonkon et al. 2016). Also, pyrogallol (PG) is a natural phenolic compound found in hardwoods such as oak and eucalyptus (Mohammed-Ziegler and Billes 2002). Its molecular structure consists of a benzene ring with three hydroxyl groups, which allows the formation of non-covalent hydrogen bonds and hydrophobic interactions. The auto-oxidation of hydroxyl groups can lead to the consumption of dissolved oxygen under alkaline conditions, thereby removing free oxygen from the surrounding environment (Shin et al. 2019) Research has shown that the auto-oxidation of PG in a liquid medium leads to the formation of various by-products, including hydrogen peroxide and quinone derivatives (Ramasarma et al. 2015).

In this section, we will review the performance and results of the effect of different PG concentrations on the morphology, properties, and oxygen-scavenging efficacy of films made from TPS and LLDPE via twin-spiral extrusion in oxygen scavenging.

As mentioned, films produced from TPS and LLDPE containing pyrogallol (PG) have shown effective oxygen-scavenging capabilities. In these films, PG hydrogen bonds with starch through its hydroxyl groups. In addition, the aromatic benzene rings of PG facilitate electron transfer to the LLDPE chains, leading to a modified C–H stretch of the terminal methyl group. This interaction allows PG to act as a compatibilizer and increase compatibility between polymers, especially when the PG concentration is maintained at 3% or less. However, at higher PG concentrations, self-aggregation tendencies can lead to the formation of heterogeneous matrices that negatively affect tensile strength by creating voids in the film structure. Furthermore, the presence of PG reduces the crystallinity of the films, as it causes the starch granules to melt due to the plasticizing effects exerted by the TPS phase. The oxygen absorption capacity, as assessed through the inhibition rates and the measurement of residual oxygen, is affected by the PG concentration and the storage temperature. High storage temperatures correspond to an increase in the oxygen inhibition rate. The results show that several factors influence the oxygen inhibition rate of PG-containing films, such as temperature, which increases molecular mobility, the activation energy required for the reactions, and the microstructural characteristics of the film matrices, all of which play an important role in the release of volatile molecules (Promsorn and Harnkarnsujarit 2022b).

Mechanisms of Moisture Absorption

The mechanisms of moisture uptake in organic and inorganic materials have been extensively studied. Key factors in the development of effective moisture absorption technologies include not only technical aspects, such as adsorption capacity, kinetics, activation mechanisms, and potential interactions with other functional packaging technologies and food products, but also regulatory and marketing considerations, including food hygiene, expenses, and acceptance by consumers.

Desiccants are defined as materials that physically absorb and retain water molecules from their environment, while desiccants are adsorbed through chemical reactions between two functional components. Desiccants can reduce the relative humidity of protected packages using both physical and chemical adsorption methods, where silica gel represents physical adsorption and calcium chloride represents a chemical interaction. The operation of silica gel can be seen schematically in the figure below (Gaikwad et al. 2019).

Among recent developments, synthesized attapulgite has been identified as a promising moisture-absorbing material, characterized by a hydrated magnesium aluminum silicate structure that has exchangeable cations and reactive –OH groups. Incorporation of attapulgite, such as LLDPE/starch-based films, into a polymer matrix has been shown to enhance moisture absorption and swelling capacity. In the realm of organic materials, cellulose, a natural polymer, has many sites for the adsorption of water molecules at the hydroxyl (–OH) and carboxyl (–COOH) groups due to its hygroscopic properties. Moisture absorption occurs in both the surface and amorphous regions of cellulose, with studies showing that moisture penetrates these disordered regions more readily than crystalline regions (Bovi and Mahajan 2017).

Forms of Desiccant

Desiccant materials are designed to absorb moisture within the package and free water, effectively reducing the relative humidity to a level that prevents condensation. Several factors can affect the amount of water absorbed by the desiccant, including: the type and amount of desiccant used, the type of bond between the water molecules and their source, the amount of water previously absorbed by the desiccant, and the ambient temperature (Bovi and Mahajan 2017).

When developing desiccant packaging systems for a variety of products with their own unique characteristics, it is essential to consider the moisture absorption capacity of the material in relation to the intended application and anticipated shelf life. In food packaging, the most common forms of desiccant systems include sachets, pads, humidity control trays, and polymeric films, as shown in Fig. 11.12. The integrity of the outer packaging materials of these formats allows for efficient moisture management without compromising the overall structure or performance of the packaging system (Gaikwad et al. 2019).

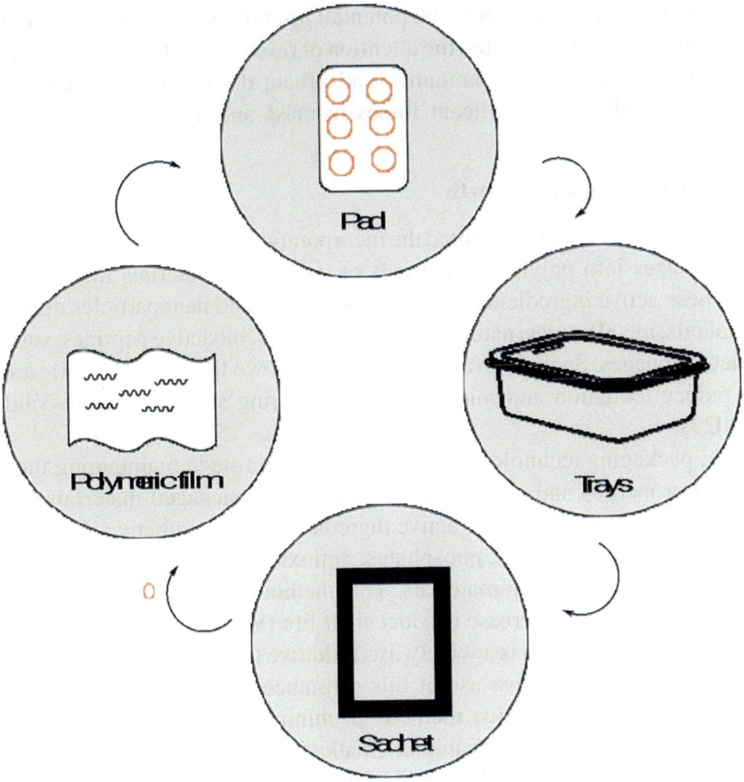

Fig. 11.12 Various types of moisture-absorbing agents for use in food packaging (Gaikwad et al. 2019) with permission, copyright memberanes, 2025

Polymeric Films

The wide range of plastic materials and their diverse compositions enables companies to design packaging solutions to meet the specific needs of individual products. Common polymer films for fresh produce packaging, such as polyethylene, polypropylene, polyvinyl chloride, and polyethylene terephthalate, typically exhibit low water vapor transmission rates compared to the moisture loss rates of fresh produce. Similarly, in meat packaging, the volume of moisture removed from the product often exceeds the ability of the films to transmit water vapor. As a solution, moisture-absorbing materials can be integrated throughout the packaging. By dispersing these sorbent compounds in the polymer matrix or placing humectants between two layers of plastic film, humidity can be managed in the headspace and effectively prevent the ingress of external moisture into the package.

Despite the commercial success of desiccant films, these types of adsorbents are not without limitations. For example, their moisture absorption capacity and rate are generally lower than those found in desiccant-based bags (Choi and Lee 2013).

Furthermore, concerns regarding the potential migration of desiccants from the film into food products have attracted the attention of researchers. Furthermore, achieving uniform dispersion of a high amount of adsorbent throughout the package can be challenging, leading to significant losses in mass and activity during production (Gaikwad et al. 2018).

Inhibit Microbiological Growth

Recent discussions have highlighted the incorporation of active ingredients of various nanosized sizes into polymeric matrices of packaging materials including muscle foods. These active ingredients include nanocarriers and nanoparticles derived from essential oils, metal oxides, natural extracts, enzymes, bioactive peptides, surfactants, and bacteriophages. Such innovations have been shown to extend shelf life and effectively reduce oxidation and microbial growth during storage (Jacinto-Valderrama et al. 2023).

Active packaging technologies play an important role in maintaining the physicochemical properties and extending the shelf life of packaged materials, including packaged muscle foods. Various active ingredients, both synthetic and natural, such as sodium chloride, vitamins, phosphates, antioxidants, and antimicrobials, can be incorporated into packaging materials. This method is effectively used in the meat industry as a method to increase product shelf life (Wrona et al. 2021).

Sodium nitrite ($NaNO_2$) is a widely used additive to preserve the red color in meat products. However, excessive use of this substance raises safety concerns. Recent studies have sought alternative methods to minimize or directly eliminate nitrite incorporation into meat. For example, Chatkitanan and Harnkarnsujarit created a thermoplastic film based on TPS and LLDPE that incorporated 5% sodium nitrite. The film effectively stabilized oxidation and microbial growth in vacuum-packed pork stored at 4 °C for six days, resulting in a reduction in thiobarbituric acid reactive substances (TBARS), methemoglobin content, total viable count (TVC), and other microbial indices (Chatkitanan and Harnkarnsujarit 2020). Notably, nitrite levels in encapsulated pork were measured at approximately 1.05 ppm, significantly below the FDA limit of 200 ppm and Chinese regulations of 150 ppm for meat products (Jacinto-Valderrama et al. 2023). The FTIR results of this study are as follows.

Infrared (FTIR) absorption spectra are useful in revealing the molecular vibrational properties of materials in the film as shown in Fig. 11.11b. The Fourier-transform infrared (FTIR) spectrum of LLDPE shows specific absorption bands at 718, 1470, 2850, and 2916 cm^{-1}, which correspond to different molecular vibrations.

It is noteworthy that blending LLDPE with TPS in a ratio of 50:50 (designated as LL50/TPS50) did not change the positions of these four bands, indicating that the LLDPE chains remained unaffected and confirming the phase separation between the starch and LLDPE matrices. The IR fingerprint region, which is 500–1700 cm^{-1}, shows peaks that represent individual film components. In particular, TPS showed a prominent absorption peak at 1024 cm^{-1}, associated with the C–O stretching vibrations of the glucose units, and a sharp peak at 1462 cm^{-1}, indicating the C–H bending of the methyl group. In the hydrogen bonding region (3000–3700 cm^{-1}), significant changes in peak positions were observed in the films containing nitrite. For the LL50/

TPS50 blend, an O–H stretching vibration peak was identified at 3337 cm^{-1}, while the blends containing 1 and 5% nitrite (LL50/TPS50-1%N and LL50/TPS50-5%N) showed peaks at 3327 cm^{-1}. This shift in the position of the band maximum indicates changes in the intermolecular hydrogen bonding associated with the hydroxyl groups in the starch (Wang et al. 1994).

Furthermore, the presence of nitrite in the TPS films was reflected in broader stretching bands in the range of 3000–3700 cm, indicating that the nitrogen derived from nitrite changes the strength of the hydrogen bond. Nitrogen atoms may also form hydrogen bonds with hydrogen atoms in the starch structure. Importantly, the FTIR spectrum showed no change in the absorption bands of LLDPE, meaning that while the interaction between nitrite and starch significantly affects the TPS matrices, there is no detectable chemical change in the LLDPE network itself (Chatkitanan and Harnkarnsujarit 2020) (Fig. 11.13).

Ethylene Gas Absorbing

Ethylene is a growth hormone that affects the growth of plants and the development of fruits. Ethylene helps speed up the ripening in fruit, and as a result, it can cause the aging and spoilage of fruits. Additionally, the production of ethylene from fruits raised during bacterial and fungal infections in fruits or vegetables, so it is important to protect them from physical and chemical damage (Brody et al. 2001). Active packaging that focuses on ethylene removal, can reduce the ethylene level around fresh produce during transportation and their sale marketing. To create a method for eliminating ethylene, it is crucial to consider several key factors related to the produce: (1) rate of ethylene production, (2) sensitivity of the product to ethylene. (3) how concentration of CO_2 and O_2 effect ethylene production (Awalgaonkar et al. 2020).

Ethylene Scavenger Based on ($KMnO_4$)

Ethylene scavenger based on potassium permanganate ($KMnO_4$) are the most popular type of scavengers in industries. $KMnO_4$ reacted with C_2H_4 and the products are H_2O and CO_2. During this process, manganese dioxide (MnO_2) and potassium hydroxide (KOH) are released as by-products (Álvarez-Hernández et al. 2018).

$$3C_2H_4 + 12KMnO_4 \rightarrow 6CO_2 + 12MnO_2 + 12KOH$$

It should be noted that the adsorption of $KMnO_4$ into a porous inert material with a large surface area, including clays, silica gel (SiO_2), zeolite, and alumina (Al_2O_3), has been carried out. Some of these materials have ability to absorb C_2H_4 and creating an adsorption–oxidation system, and whenever C_2H_4 and $KMnO_4$ are adsorbent onto these materials, an oxidation process happens (Álvarez-Hernández et al. 2018).

Considering that density of ethylene is lower than density of air, for example, at 22 °C and 1 atm, the densities of C_2H_4 and air are 1.16 and 1.20 g/mol, the best place for put the ethylene scavenger sachet is in the top of packaging, common materials for sachets such as polyethylene and cellulose (López-Gómez et al. 2023).

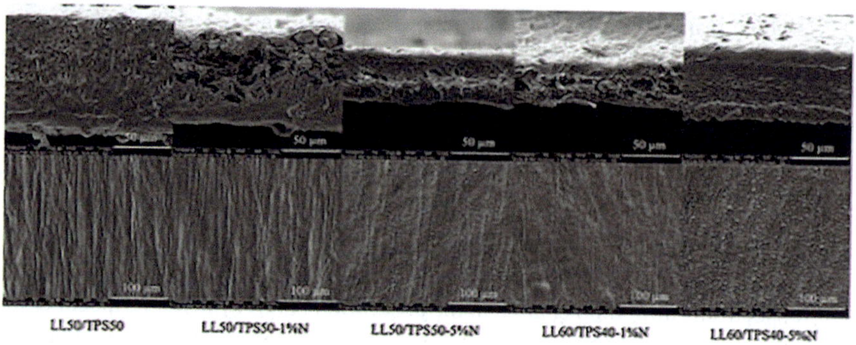

(a)

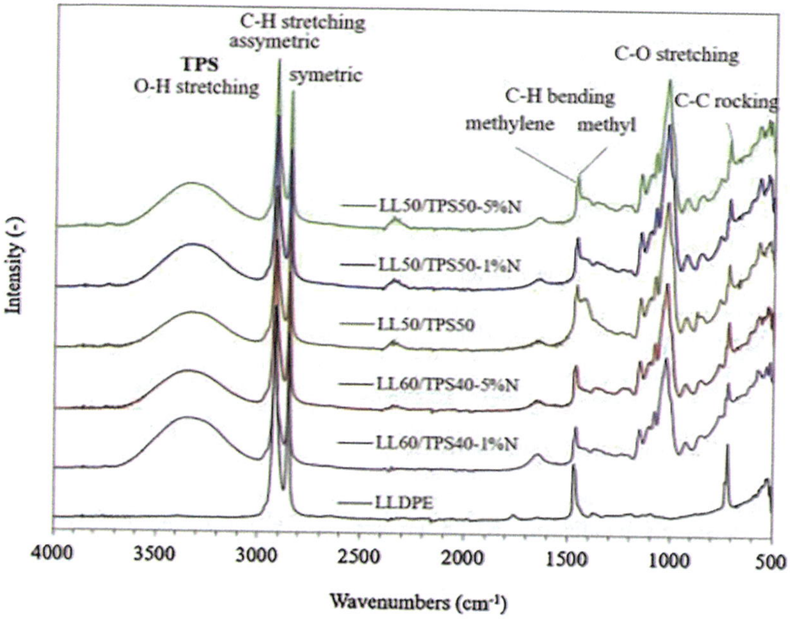

(b)

Fig. 11.13 Analysis of the morphology and spectroscopy of blended LLDPE films (Chatkitanan and Harnkarnsujarit 2020) with permission, copyright memberanes, 2025. **a** Scanning electron microscopy (SEM) images displaying the cross-sectional (top) and surface (bottom) microstructures of the compounded LLDPE films. **b** Fourier-transform infrared (FTIR) spectra of LLDPE and TPS-compounded LLDPE films include nitrite concentrations of 1 and 5%. The films examined feature LLDPE/TPS ratios of 50/50 (designated as LL50/TPS50) and 60/40 (denoted as LL60/TPS40), highlighting the influence of hydroxyl (O–H) interactions (Chatkitanan and Harnkarnsujarit 2020)

A study on the ability of potassium permanganate absorbent to remove low levels of ethylene has been conducted. The experiments show that at 20 °C temperature and 90% RH, use of potassium permanganate absorbent is suitable. However, it might not be effective for larger packages and atmosphere with higher levels of ethylene. Other researchers developed an active packaging by incorporating active carbon, either individually or impregnated with palladium as a catalyst. They studied the effect of this packaging on the shelf life of tomatoes and found that the shape, color, taste, and its aroma received higher ratings. The palladium used on activated carbon increases the ethylene removal rate both at high and low levels of ethylene in atmosphere. However, the downside is that a considerable amount of this catalyst and activated carbon is necessary, and palladium is high-cost metal (Dobrucka et al. 2017) (Fig. 11.14).

$KMnO_4$-based ethylene scavengers are generally used in the form of sachets in food packaging. Additionally, they exist in various other form such as filter tubes, blankets, or filter machines that mainly used during transportation. Sometimes, they

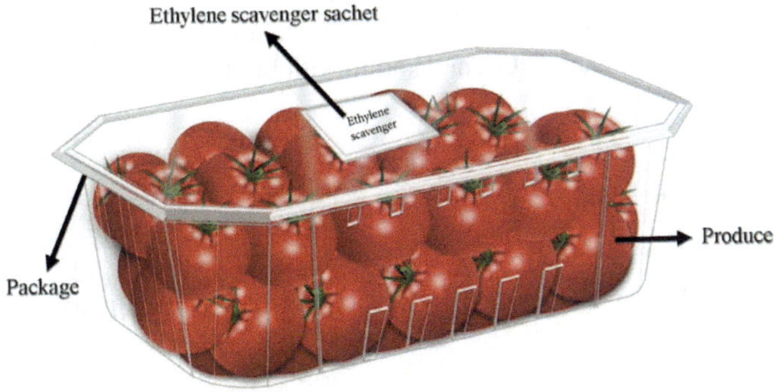

Fig. 11.14 Model an ethylene scavenger sachet used in tomato package

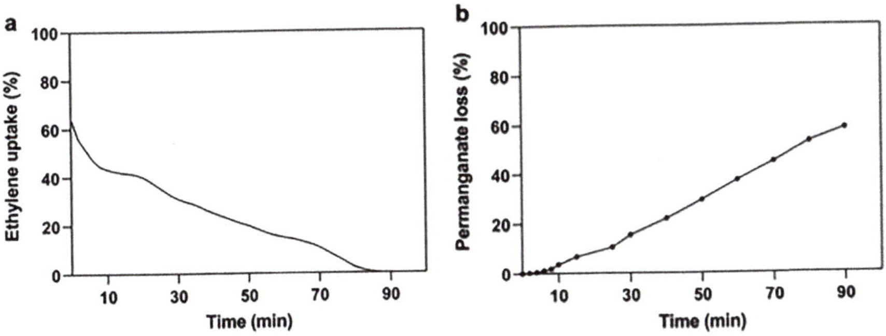

Fig. 11.15 **a** Ethylene uptake for food package based on $KMnO_4$, **b** permanganate loss for food package based on $KMnO_4$ (Álvarez-Hernández et al. 2018) with permission, copyright memberanes, 2025

also directly incorporate into packaging agents. Also, $KMnO_4$ is generally not placed in direct contact with food, because it is a toxic substance (Álvarez-Hernández et al. 2018).

In Fig. 11.15, graphs a and b show the rate of ethylene absorption over time and the rate of permanganate loss over time for $KMnO_4$-based packaging. As can be seen, over time, ethylene is absorbed by permanganate, which leads to oxidation of permanganate and its subsequent reduction, while ethylene continues to be absorbed (Álvarez-Hernández et al. 2018).

In Fig. 11.16, the color changes of the ethylene scavenger are visible. The color changes slowly from purple to dark brown as the permanganate ion (MnO_4^-) is transformed into manganese dioxide (MnO_2). This color change indicate the absorption of ethylene as well as the oxidation of permanganate (Álvarez-Hernández et al. 2018).

Ethylene Scavenger Films

Zeolite molecular sieve (ZMS) has been identified as suitable additive for incorporation into packaging films to absorb ethylene. Research shows that the incorporation of this additive to packaging films boost ethylene permeability. The reason is its crystalline porous-three-dimensional framework structure. Incorporation of nanopowders such as nano-Ag, nano-TiO_2, and kaolin with PE decreases the ethylene amount

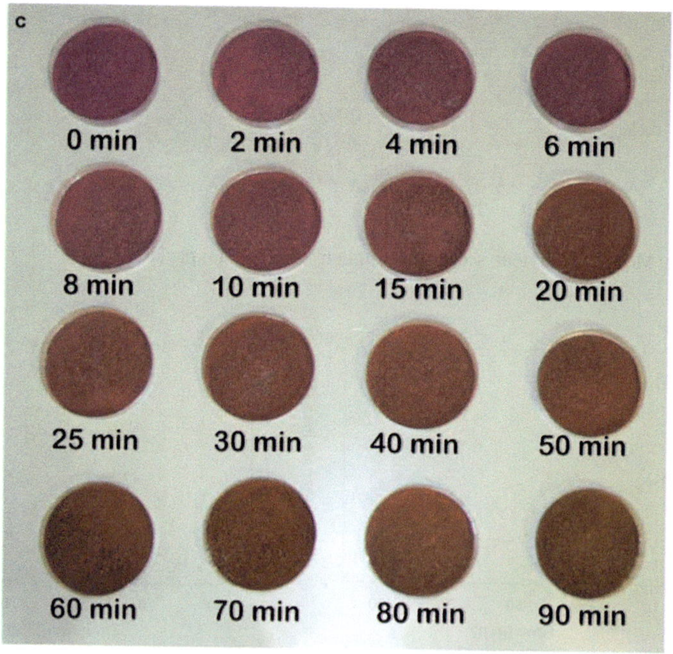

Fig. 11.16 Visual color change image (Álvarez-Hernández et al. 2018) with permission, copyright memberanes, 2025

in packaging atmosphere. Reassert indicate that the addition of sodium nanoparticles to the polyethylene layer significantly prevents softening and discoloration even 12 days after storage. Food companies are looking for replace petroleum-based packaging with source-based and biodegradable packages to avoid their drawbacks, like environmental hazard. Chitosan (CH) is one of the most abundant natural polymers and is used alongside cellulose for packaging because of its antimicrobial characteristics. Polylactic acid (PLA) is also used due to its biocompatibility. Another example for starch-based film, 4% of $Bi_2WO6–TiO_2$ (BT) was used to design starch nanocomposite films for the ethylene scavenger agent which increases the ethylene degradation rate of up to 12% (Wei et al. 2021).

LDPE Films

Active packaging film was created by incorporating $KMnO_4$ into low-density polyethylene (LDPE), which enhanced oxygen barrier performance and also extended shelf life of some fruits. LDPE is one of the popular polymers in packaging films, having flexibility, mechanical properties, non-toxicity, affordable, plentiful resources. It also has good adsorption characteristics and environmentally friendly attributes (Wang and Ajji 2022). The permeability of ethylene at 19 °C has been determined for three type of polyethylene films (LLDPE, LDPE, and HDPE) in the temperature range from 4 to 30 °C. At constant temperature, the permeability of ethylene and oxygen reduction as the density increases. Based on experiments on polyethene films and data acquired from these experiments, the gas permeability for the polyethylene film for different densities has been determined. The permeability of ethylene obtained for LDPE is 30.8, LLDPE is 20.6 and for HPDE is 5.75 (permeability constant unit: ml(STP) cm/m^2 24 h/atm). These data indicates that permeability decreases by an average factor of 5–6 for an increase in density (approximately 5% increase in density) (Wang et al. 1998).

The permeability of ethylene for continuous flow is calculated by Eq. 11.2 (Sonawane et al. 2023).

$$p = \frac{QL}{A \Delta P t} \tag{11.2}$$

Q is the quantity of ethylene (mol), which pass through the film over time t (s). L indicates the thickness of polymer film (m). A refers to the surface area (m^2), and ΔP shows the pressure difference of ethylene between inside and outside of film packaging (pa) (Sonawane et al. 2023).

Starch Film

Concerns regarding the negative impact of using plastic materials on the environment and human health led to efforts to utilize biodegradable polymer plastic. Various polymers based on biodegradable agent sourced from bioresources and synthetic materials based on fossil resources, such as polylactic acid (PLA), polybutylene adipate-co-terephthalate (PBAT), polybutylene succinate (PBS), starch, and cellulose, have been created in different kinds for use in packaging applications. Thermoplastic starch (TPS) is produced by plasticizing starch obtained different sources, e.g., corn, potato, and cassava. The advancement of cassava starch-based (TPS) supports the bioplastic packaging. Mixing TPS with other biodegradable polymers such as PBAT and PLA boost efficiency of using TPS. PBAT is very flexible and can be combined with TPS to enhance its stability. Titanium dioxide (TiO_2) is a photocatalyst that can break down ethylene and speed down the ripening fruits. UV-activated TiO_2 molecules interact with water and oxygen and produce hydroxyl radicals and reactive oxygen component, which successfully prevent the growth of microorganism and break down ethylene compounds. Research indicated that adding TiO_2 to the environment of biodegradable films, such as PBAT and TPS, increase shelf life of food products. When TiO_2 nanocrystals are exposed to UV light ($\leq$ 387.5 nm), they absorb photons (hν) on their surface, which excites electrons from the valence band (h_{vb}^+) to the conduction band (e_{cb}^-). This interaction produces highly reactive oxygen species (ROS), such as hydroperoxyl ($HO_2^{\cdot}$) and hydroxyl ($HO^{\cdot}$) radicals. These radicals then interact with unsaturated carbon bonds (C=C) in ethylene molecules, causing their breakdown and degradation (Phothisarattana et al. 2021). Using cellulose-based packaging along with modified atmosphere packaging (MAP) also enhances the freshness duration of food products. The utilization of cellulose fibers like paper products is more economical and provides greater effectiveness compared to polyolefin-based plastics. Starch nanocomposite films were developed with 4% Bi_2 WO_6-TiO_2 (BT) designed to ethylene oxidation, achieving a 12.4% enhancement in the rate of ethylene breakdown. Considering the significant of preventing microbial permeation into food products which tend to the increased adsorption of ethylene, it is essential to use antimicrobial films. ZnO nanoparticles are commonly used as an antimicrobial factor in nanocomposites made from chitosan and carboxymethyl cellulose. For example, a group of researchers coated filter paper with a film made of zinc oxide nanoparticles, carboxymethyl cellulose, and guanidine-based starch to produce an antimicrobial paper that prevents growth of *Escherichia coli* (Mariah et al. 2022).

11.8 Conclusions

Active packaging is regarded as an innovative packaging system because it can control the atmosphere within the package. Furthermore, the researchers and industrials have focused on natural and eco-friendly materials that act as active agents for packaging applications. The advantages of active packaging include the control of internal conditions. This control condition is archived through reacting to them accordingly by emitting advantageous agents or absorbing those that damage the product. There are various active agents for using in packaging including: oxygen scavengers, carbon dioxide emitter, antimicrobial agents, and ethylene scavengers. Among various polymers, thermoplastic starch (TPS)-based low-linear density polyethylene (LLDPE) films have been studied for their potential use in active packaging. TPS-LLDPE films can be prepared by blending TPS with LLDPE and incorporating active agents. These active agents help to scavenge oxygen, absorb moisture, inhibit microbial growth, or absorb ethylene gas, which can help to maintain the quality of food products. TPS-LLDPE films can provide a biodegradable and environmentally friendly alternative to conventional petroleum-based packaging materials, while also offering active packaging functionalities.

References

Agama-Acevedo E, Bello-Perez LA (2017) Starch as an emulsions stability: the case of octenyl succinic anhydride (OSA) starch. Curr Opin Food Sci 13:78–83

Ahn BJ, Gaikwad KK, Lee YS (2016) Characterization and properties of LDPE film with gallic-acid-based oxygen scavenging system useful as a functional packaging material. J Appl Polym Sci 133(43)

Álvarez-Hernández MH et al (2018) Current scenario of adsorbent materials used in ethylene scavenging systems to extend fruit and vegetable postharvest life. Food Bioprocess Technol 11(3):511–525

Awalgaonkar G, Beaudry R, Almenar E (2020) Ethylene-removing packaging: basis for development and latest advances. Comp Rev Food Sci Food Safe 19(6):3980–4007

Bangar SP et al (2021) Recent advances in thermoplastic starches for food packaging: a review. Food Packag Shelf Life 30:100743

Barbosa JV et al (2019) Effect of peroxide oxidation on the expansion of potato starch foam. Ind Crops Prod 137:428–435

Barish JA, Goddard JM (2011) Topographical and chemical characterization of polymer surfaces modified by physical and chemical processes. J Appl Polym Sci 120(5):2863–2871

Bovi GG et al (2018) Moisture absorption kinetics of FruitPad for packaging of fresh strawberry. J Food Eng 223:248–254

Bovi GG, Mahajan PV (2017) Regulation of humidity in fresh produce packaging. In: Reference module in food science. Elsevier

Brody AL, Strupinsky E, Kline LR (2001) Active packaging for food applications. CRC Press

Byun Y, Bae HJ, Whiteside S (2012) Active warm-water fish gelatin film containing oxygen scavenging system. Food Hydrocolloids 27(1):250–255

Cardona ED, del Pilar Noriega M, Sierra JD (2011) Oxygen scavengers impregnated in porous activated carbon matrix for food and beverage packaging applications. J Plast Film Sheeting 28(1):63–78

Chatkitanan T, Harnkarnsujarit N (2020) Development of nitrite compounded starch-based films to improve color and quality of vacuum-packaged pork. Food Packag Shelf Life 25:100521

Choi HY, Lee YS (2013) Characteristics of moisture-absorbing film impregnated with synthesized attapulgite with acrylamide and its effect on the quality of seasoned laver during storage. J Food Eng 116(4):829–839

Damaj Z, Joly C, Guillon E (2015) Toward new polymeric oxygen scavenging systems: formation of poly(vinyl alcohol) oxygen scavenger film. Packag Technol Sci 28(4):293–302

Deshmukh RK, Hakim L, Gaikwad KK (2023) Active packaging materials. Curr Food Sci Technol Rep 1(2):123–132

Dobrucka R, Leonowicz A, Cierpiszewski R (2017) Preparation of ethylene scavenger based on $KMnO_4$ to the extension of the storage time of tomatoes

Falua KJ et al (2022) Valorization of starch to biobased materials: a review. Polymers 14. https://doi.org/10.3390/polym14112215

Ferrari MC et al (2009) Modeling of oxygen scavenging for improved barrier behavior: blend films. J Membr Sci 329(1):183–192

Fu Z et al (2019) Developments in hydroxypropylation of starch: a review. Starch Stärke 71(1–2):1800167

Gaikwad KK, Singh S, Lee YS (2017a) A pyrogallol-coated modified LDPE film as an oxygen scavenging film for active packaging materials. Prog Org Coat 111:186–195

Gaikwad KK, Singh S, Lee YS (2017b) A new pyrogallol coated oxygen scavenging film and their effect on oxidative stability of soybean oil under different storage conditions. Food Sci Biotechnol 26(6):1535–1543

Gaikwad KK, Singh S, Lee YS (2018) Oxygen scavenging films in food packaging. Environ Chem Lett 16(2):523–538

Gaikwad KK, Singh S, Ajji A (2019) Moisture absorbers for food packaging applications. Environ Chem Lett 17(2):609–628

Galdi MR et al (2008) Production of active PET films: evaluation of scavenging activity. Packag Technol Sci 21(5):257–268

Goddard JM, Hotchkiss JH (2007) Polymer surface modification for the attachment of bioactive compounds. Prog Polym Sci 32(7):698–725

Hosseinnejad M (2014) Active packaging for food applications—a review. Int J Adv Biol Biomed Res 2(4):1174–1180

Irkin R, Esmer OK (2015) Novel food packaging systems with natural antimicrobial agents. J Food Sci Technol 52(10):6095–6111

Jacinto-Valderrama RA et al (2023) Recent trends in active packaging using nanotechnology to inhibit oxidation and microbiological growth in muscle foods. Foods 12(19):3662

Janjarasskul T, Suppakul P (2018) Active and intelligent packaging: the indication of quality and safety. Crit Rev Food Sci Nutr 58(5):808–831

Janjarasskul T, Tananuwong K, Krochta JM (2011) Whey protein film with oxygen scavenging function by incorporation of ascorbic acid. J Food Sci 76(9):E561–E568

Jo C, Ahn DU, Byun MW (2002) Irradiation-induced oxidative changes and production of volatile compounds in sausages prepared with vitamin E-enriched commercial soybean oil. Food Chem 76(3):299–305

Johnson DR, Decker EA (2015) The role of oxygen in lipid oxidation reactions: a review. Annu Rev Food Sci Technol 6:171–190

Khanoonkon N, Yoksan R, Ogale AA (2016) Morphological characteristics of stearic acid-grafted starch-compatibilized linear low density polyethylene/thermoplastic starch blown film. Eur Polymer J 76:266–277

Khumkomgool A, Saneluksana T, Harnkarnsujarit N (2020) Active meat packaging from thermoplastic cassava starch containing sappan and cinnamon herbal extracts via LLDPE blown-film extrusion. Food Packag Shelf Life 26:100557

Koo SH, Lee KY, Lee HG (2010) Effect of cross-linking on the physicochemical and physiological properties of corn starch. Food Hydrocolloids 24(6):619–625

Labuza TP, Breene WM (1989) Applications of "active packaging" for improvement of shelf-life and nutritional quality of fresh and extended shelf-life foods. J Food Process Preserv 13(1):1–69

Lackner M et al (2021) Mechanical properties and structure of mixtures of poly(butylene-adipate-co-terephthalate) (PBAT) with thermoplastic starch (TPS). Int J Biobased Plast 3(1):126–138

Li H et al (2012) Characterization of oxygen scavenging films based on 1,4-polybutadiene. Ind Eng Chem Res 51(21):7138–7145

López-Gómez A et al (2023) The potential of essential oils from active packaging to reduce ethylene biosynthesis in plant products. Part 2: fruits (blueberries and blackberries). Plants 12. https://doi.org/10.3390/plants12193418

Loucanova E, Kalamarova M, Parobek J (2017) The innovative approaches to packaging—comparison analysis of intelligent and active packaging perceptions in Slovakia. Studia Univ „Vasile Goldis" Arad—Econ Ser 27(2):33–44

Mabeck JT, Malliaras GG (2006) Chemical and biological sensors based on organic thin-film transistors. Anal Bioanal Chem 384(2):343–353

Mariah MA et al (2022) The emergence and impact of ethylene scavengers techniques in delaying the ripening of fruits and vegetables. Membranes 12. https://doi.org/10.3390/membranes12020117

Marsh K, Bugusu B (2007) Food packaging—roles, materials, and environmental issues. J Food Sci 72(3):R39–R55

Martinez Villadiego K et al (2022) Thermoplastic starch (TPS)/polylactic acid (PLA) blending methodologies: a review. J Polym Environ 30(1):75–91

Mazerolles T et al (2020) Development of multilayer barrier films of thermoplastic starch and low-density polyethylene. J Polym Res 27(2):44

Méndez-Hernández ML et al (2011) The effect of thermoplastic starch on the properties of HDPE/TPS blends during UV-accelerated aging. Polym Bull 67(5):903–914

Mohammadi Nafchi A et al (2013) Thermoplastic starches: properties, challenges, and prospects. Starch Stärke 65(1–2):61–72

Mohammed-Ziegler I, Billes F (2002) Vibrational spectroscopic calculations on pyrogallol and gallic acid. J Mol Struct (Thoechem) 618(3):259–265

Montoya-Anaya DG et al (2023) Films made of potato starch from industrial potato stragglers and polyethylene: physicochemical, thermal, and mechanical properties. Iran Polym J 32(10):1321–1333

Murmu SB, Mishra HN (2018) Selection of the best active modified atmosphere packaging with ethylene and moisture scavengers to maintain quality of guava during low-temperature storage. Food Chem 253:55–62

Muzakkar M et al (2013) Shear strength of single lap joint aluminium-thermoplastic natural rubber (Al-TPNR) laminated composite. J Phys Conf Ser 423

Nosáľová M, Loučanová E, Parobek J (2018) Perception of packaging functions and the interest in intelligent and active packaging. Probl Agric Econ 357(4):141–152

Pan Y et al (2022) Fine-tuned regulation of photosynthetic performance via γ-aminobutyric acid (GABA) supply coupled with high initial cell density culture for economic starch production in microalgae. Bioresour Bioprocess 9(1):52

Panrong T, Karbowiak T, Harnkarnsujarit N (2019) Thermoplastic starch and green tea blends with LLDPE films for active packaging of meat and oil-based products. Food Packag Shelf Life 21:100331

Phothisarattana D et al (2021) Biodegradable poly(butylene adipate-co-terephthalate) and thermoplastic starch-blended TiO_2 nanocomposite blown films as functional active packaging of fresh fruit. Polymers 13. https://doi.org/10.3390/polym13234192

Pinming C et al (2016) Humidity sensor using carboxymethyl cellulose hydrogel membrane. In: 13th international conference on electrical engineering/electronics, computer, telecommunications and information technology (ECTI-CON)

Prasad P, Kochhar A (2014) Active packaging in food industry: a review. IOSR J Environ Sci Toxicol Food Technol 8:01–07

Promsorn J, Harnkarnsujarit N (2022a) Oxygen absorbing food packaging made by extrusion compounding of thermoplastic cassava starch with gallic acid. Food Control 142:109273

Promsorn J, Harnkarnsujarit N (2022b) Pyrogallol loaded thermoplastic cassava starch based films as bio-based oxygen scavengers. Ind Crops Prod 186:115226

Ramasarma T et al (2015) New insights of superoxide dismutase inhibition of pyrogallol autoxidation. Mol Cell Biochem 400(1):277–285

Renato Souza C, Geany Peruch C, Ana Clarissa dos Santos P (2012) Oxygen scavengers: an approach on food preservation. In: Ayman Amer E (ed) Structure and function of food engineering. IntechOpen, Rijeka

Róz ALD et al (2006) The effect of plasticizers on thermoplastic starch compositions obtained by melt processing. Carbohyd Polym 63(3):417–424

Rux G et al (2016) Humidity-regulating trays: moisture absorption kinetics and applications for fresh produce packaging. Food Bioprocess Technol 9(4):709–716
Scarfato P, Di Maio L, Incarnato L (2015) Recent advances and migration issues in biodegradable polymers from renewable sources for food packaging. J Appl Polym Sci 132(48)
Shin M, Park E, Lee H (2019) Plant-inspired pyrogallol-containing functional materials. Adv Func Mater 29(43):1903022
Siciliano S, Lopresto CG, Lamonaca F (2024) From traditional packaging to smart bio-packaging for food safety: a review. Euro-Mediterr J Environ Integr 9(4):1971–1986
Sonawane AD et al (2023) Ethylene modelling in package headspace of fresh produce: a review. Packag Technol Sci 36(9):731–743
Sweedman MC et al (2013) Structure and physicochemical properties of octenyl succinic anhydride modified starches: a review. Carbohyd Polym 92(1):905–920
Taghizadeh A, Sarazin P, Favis BD (2013) High molecular weight plasticizers in thermoplastic starch/polyethylene blends. J Mater Sci 48(4):1799–1811
Tedeschi AM et al (2022) Sustainable bioactive packaging based on thermoplastic starch and microalgae. Int J Mol Sci 23. https://doi.org/10.3390/ijms23010178
Vanier NL et al (2017) Molecular structure, functionality and applications of oxidized starches: a review. Food Chem 221:1546–1559
Vermeiren L et al (1999) Developments in the active packaging of foods. Trends Food Sci Technol 10(3):77–86
Vilela C et al (2018) A concise guide to active agents for active food packaging. Trends Food Sci Technol 80:212–222
Wang C, Ajji A (2022) Ethylene scavenging film based on low-density polyethylene incorporating pumice and potassium permanganate and its application to preserve avocados. LWT 172:114200
Wang FC et al (1994) FTIR analysis of hydrogen bonding in amorphous linear aromatic polyurethanes. I. Influence of temperature. J Polym Sci B Polym Phys 32(8):1305–1313
Wang Y, Easteal AJ, Chen XD (1998) Ethylene and oxygen permeability through polyethylene packaging films. Packag Technol Sci 11(4):169–178
Wang X et al (2020) Research advances in chemical modifications of starch for hydrophobicity and its applications: a review. Carbohyd Polym 240:116292
Wei H et al (2021) Ethylene scavengers for the preservation of fruits and vegetables: a review. Food Chem 337:127750
Winestrand S et al (2013) Co-immobilization of oxalate oxidase and catalase in films for scavenging of oxygen or oxalic acid. Biochem Eng J 72:96–101
Wrona M et al (2021) Design of new natural antioxidant active packaging: screening flowsheet from pure essential oils and vegetable oils to ex vivo testing in meat samples. Food Control 120:107536
Wyrwa J, Barska A (2017) Innovations in the food packaging market: active packaging. Eur Food Res Technol 243(10):1681–1692
Yildirim S et al (2018) Active packaging applications for food. Comp Rev Food Sci Food Safe 17(1):165–199

9783031909788